Practical GIS Analysis

About the Author

David L. Verbyla is an Associate Professor of GIS/Remote Sensing with the Department of Forest Sciences, University of Alaska. He has a B.S. from Cook College, Rutgers University in Wildlife Management, an M.S. from Michigan State University in Park and Recreation Resources, and a Ph.D. from Utah State University in Forest Resources with minors in Applied Statistics and Soil Science. Dr. Verbyla has taught GIS workshops and university courses at Utah State University, University of New Hampshire, University of Idaho, University of Montana, and the University of Alaska.

Practical GIS Analysis

David L. Verbyla

London and New York

First published 2002
by Taylor & Francis
11 New Fetter Lane, London EC4P 4EE

Simultaneously published in the USA and Canada
by Taylor & Francis
29 West 35th Street, New York, NY 10001

Taylor & Francis is an imprint of the Taylor & Francis Group

Typeset in Sabon by Sans Serif Inc., Michigan, USA
Printed and bound in Great Britain by
TJ International Ltd, Padstow, Cornwall

British Library Cataloguing in Publishing Data
A catalogue record for this book is available from the
British Library

Library of Congress Cataloging-in-Publication Data
A catalog record for this title has been requested.

ISBN 0-415-28609-3

Contents

Preface

Geographic Information System (GIS) analysis is basically spatial problem-solving. It can be quite challenging at first, especially if you get bogged down in syntax and details. The hardest part is to conceptually understand the problem, and the GIS tools available to solve your problem. This book will help you understand conceptually how various GIS tools work. So for now, forget about workspaces, libraries, spatial database engines, dangling nodes, fuzzy tolerances, requests and instances. The syntax and details do vary by GIS . . . yet conceptually the analysis tools used by GISs are pretty much the same. Start by building a foundation of understanding about GIS tools at a conceptual level. Test your understanding by trying the exercises at the end of each chapter.

This book is written so you can learn important concepts away from your GIS . . . at the laundromat, airport, on a bus, or sitting under a tree in a park. My main GIS experience is with ESRI's ARC/INFO and ArcView. Therefore, some of the terms used in the text are from these products. However, you do not have to be a user of these products to learn conceptual GIS analysis from this book.

I have been teaching GIS at universities and workshops since 1989. Since then, my students have endured earlier drafts of this book and have taught me many GIS pitfalls and GIS applications that I would have never imagined. I thank these students for helping me. I also thank the faculty and staff at the University of Alaska, where I started this book, and also the folks at the Numerical Terradynamics Simulation Group (NTSG) Lab at the University of Montana, where I finished this book while on sabbatical there. I am grateful for their support, friendship, and kindness.

David Verbyla
Becker Ridge, Alaska

Chapter 1

GIS Data Models

INTRODUCTION

This is a book about GIS analysis tools. The hard part of GIS analysis is figuring out which tools to use to solve your GIS problem. Once you have learned conceptually how these tools work, then you can simply look up the correct syntax for using the tools in your specific GIS. Let's take an example. Imagine that you have a line theme of streams, and a polygon theme of ownership. You want to determine the ownership for each stream segment. Conceptually, this is a simple GIS problem to solve . . .

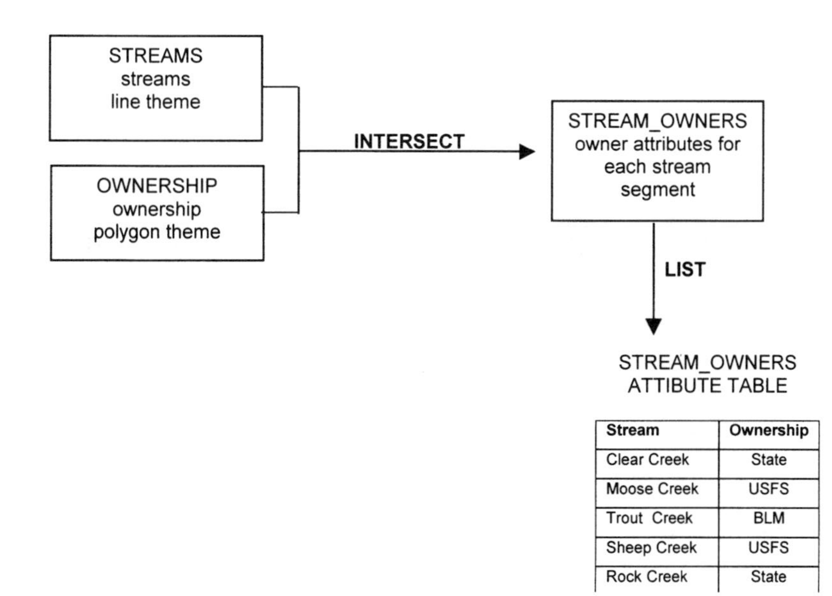

Stream	Ownership
Clear Creek	State
Moose Creek	USFS
Trout Creek	BLM
Sheep Creek	USFS
Rock Creek	State

Once you have figured out that **INTERSECT** is the tool to use to solve your problem, you can look up the specific usage of that tool in your GIS. By first conceptually solving the problem, you can avoid becoming confused with thousands of GIS commands and detailed options.

Once you know which tools to use, then you can concentrate on how to use them in your GIS by reading the help manuals. For example, the conceptual solution is applied in the ARC/INFO and Arcview GISs as follows:

Application of conceptual solution in ARC/INFO GIS:

Arc: **intersect** streams ownership stream_owners line

Intersecting streams with ownership to create stream_owners
Sorting...
Intersecting...
Assembling lines...
Creating stream_owners.AAT...

Arc: **list** stream_owners.aat stream_name,ownership_class,ownership_name

Record	Stream_name	Ownership_class	Ownership_name
1	Sage Flats River	2	Private
2	Sage Flats River	2	Private
3	Sage Flats River	2	Private
4	Sage Flats River	2	Private
5	Sage Flats River	2	Private
6	Sage Flats River	2	Private
7	Rock Creek	4	US Forest Service
8	Rock Creek	2	Private
9	Clear Creek	3	State Forest
10	Clear Creek	2	Private

Application of conceptual solution in ArcView GIS:

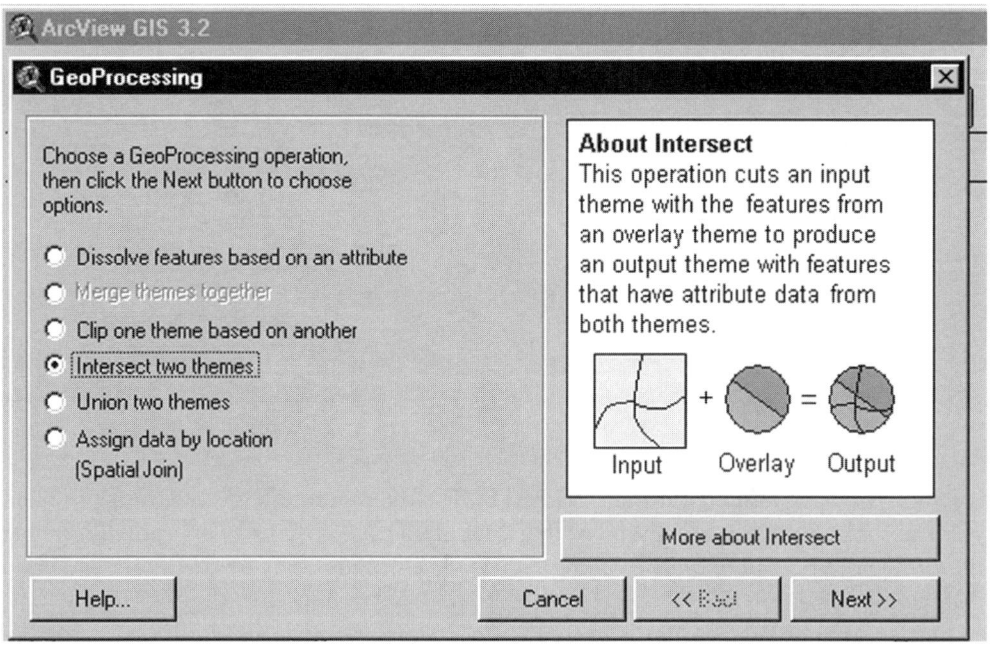

ArcView GIS 3.2
GeoProcessing

Choose a GeoProcessing operation, then click the Next button to choose options.

- Dissolve features based on an attribute
- Merge themes together
- Clip one theme based on another
- Intersect two themes
- Union two themes
- Assign data by location (Spatial Join)

About Intersect
This operation cuts an input theme with the features from an overlay theme to produce an output theme with features that have attribute data from both themes.

Input + Overlay = Output

More about Intersect

Help... Cancel << Back Next >>

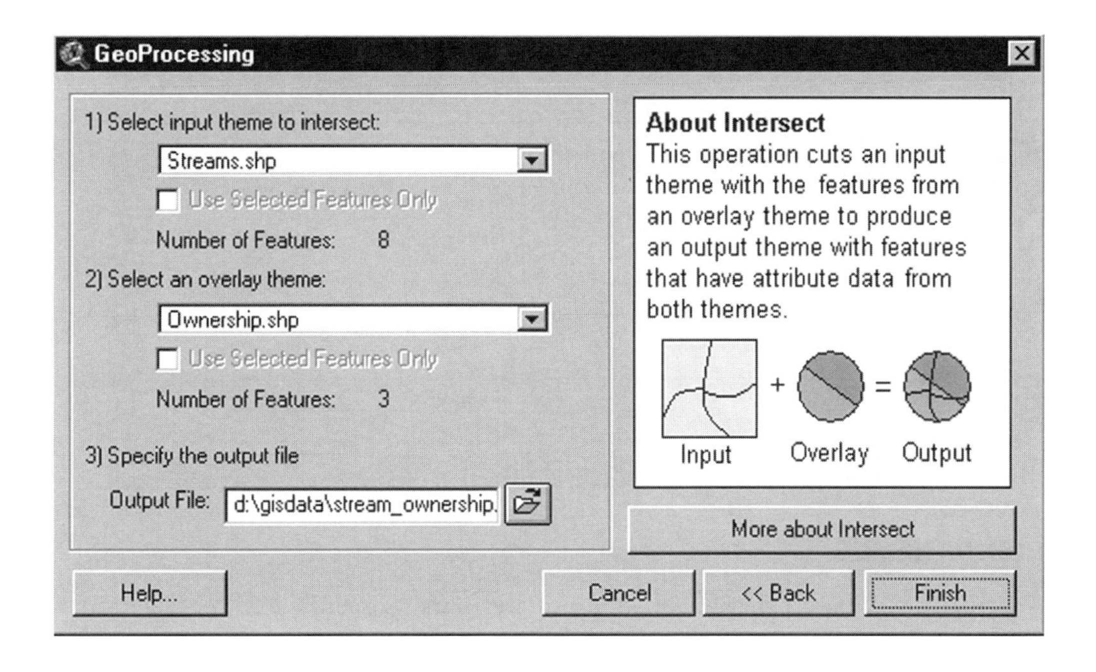
GeoProcessing
1) Select input theme to intersect:
Streams.shp
Use Selected Features Only
Number of Features: 8
2) Select an overlay theme:
Ownership.shp
Use Selected Features Only
Number of Features: 3
3) Specify the output file
Output File: d:\gisdata\stream_ownership.
About Intersect
This operation cuts an input theme with the features from an overlay theme to produce an output theme with features that have attribute data from both themes.
Input
Overlay
Output
More about Intersect
Help...
Cancel
<< Back
Finish

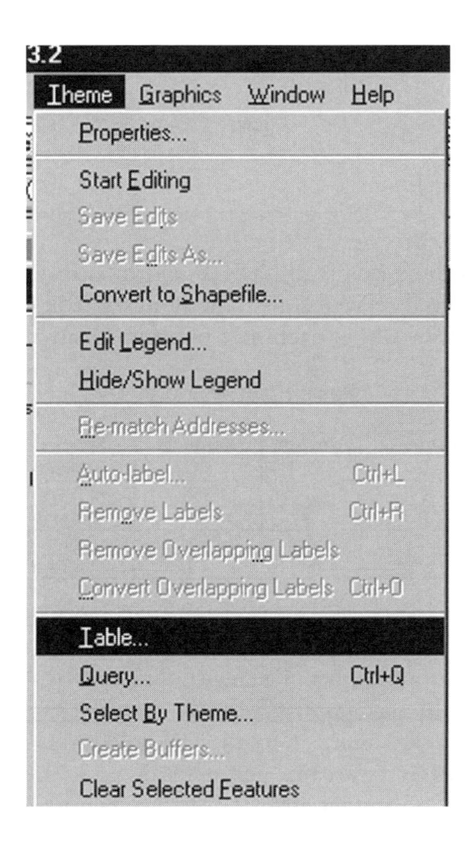
3.2
Theme
Graphics
Window
Help
Properties...
Start Editing
Save Edits
Save Edits As...
Convert to Shapefile...
Edit Legend...
Hide/Show Legend
Re-match Addresses...
Auto-label... Ctrl+L
Remove Labels Ctrl+R
Remove Overlapping Labels
Convert Overlapping Labels Ctrl+O
Table...
Query... Ctrl+Q
Select By Theme...
Create Buffers...
Clear Selected Features

Attributes of Stream_ownership.shp

Stream Name	Ownership Class	Ownership Name
Sage Flats River	2	Private
Sage Flats River	2	Private
Sage Flats River	2	Private
Sage Flats River	2	Private
Sage Flats River	2	Private
Sage Flats River	2	Private
Rock Creek	4	US Forest Service
Rock Creek	2	Private
Clear Creek	3	State Forest
Clear Creek	2	Private

POINT THEMES

A point is a GIS feature that has no length or area. It has a specific X,Y coordinate and attribute information associated with that location point.

For example, imagine that you have a point theme of waterfowl nests as follows:

+124 +76
+156
+34
+26

Since the GIS stores these points in an X,Y coordinate system, it knows the location of each nest and the spatial relationships among nests. You can assign an identification number and all information about each nest point in an attribute table.

Nests Point Attribute Table

Area	Perimeter	Nests#	Nests-ID	Species	N_Eggs
0	0	1	124	Mallard	10
0	0	2	156	Widgeon	11
0	0	3	34	Widgeon	13
0	0	4	26	Pintail	8
0	0	5	76	GW Teal	10

To avoid confusion when determining spatial relationships, each point feature must have a unique identification number. To ensure this, the GIS assigns a unique Nests# to each point. The user could make a mistake and assign the same Nests-ID to two or more nests. The internal ID Nests# assigned by the GIS solves this potential problem.

By definition, a point has no area or perimeter. Keep in mind that X,Y coordinates of points always have some positional error. Therefore, many GIS tools that analyze points use a fixed area, or a distance within a search radius. For example, you might create a buffer area of 1 km around an owl nest to determine the number of snags within 1 km of the nest. Or, in another application, you might ask for the distance from a well site to the nearest fuel tank located within a search radius of 2 km.

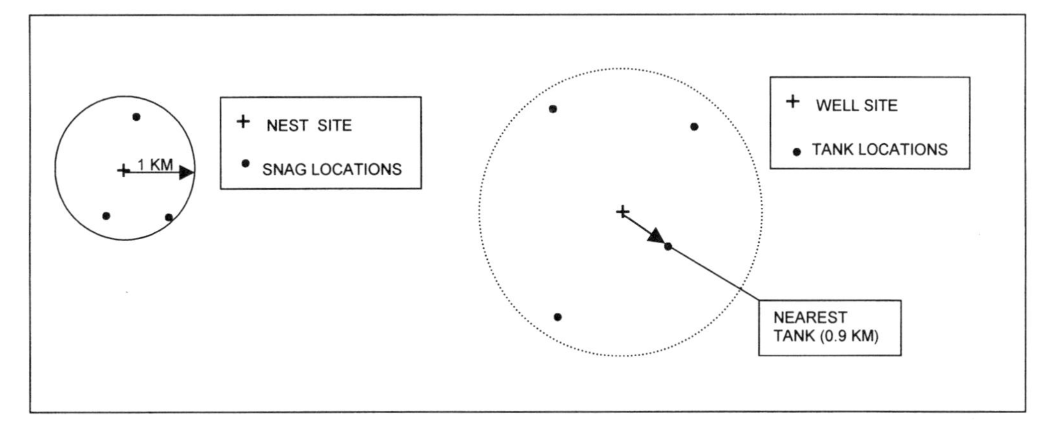

You will learn about GIS tools for point analysis in Chapter 3.

LINE THEMES

A line or arc is a GIS feature that has length but no width. The following is an example line theme of streams.

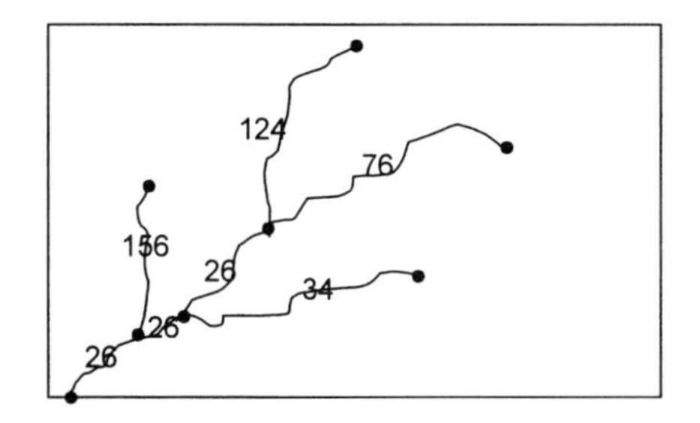

Streams Arc Attribute Table

FNODE#	TNODE#	Length	Streams#	Streams-ID	Name	Kings
1	2	1500	1	124	Clear Creek	Yes
7	5	800	2	156	Moose Creek	No
8	4	1400	3	34	Mud Creek	No
3	2	1800	4	76	Flat Creek	Yes
2	4	700	5	26	Milk River	No
4	5	300	6	26	Milk River	No
5	6	500	7	26	Milk River	No

Since the GIS stores each arc as a series of X,Y vertices, it can easily estimate the length of each stream arc. The GIS computes the length of each stream arc in the same units as your GIS coordinate system. And since each stream arc has a unique stream#, the GIS can determine spatial relationships among arcs. As a user, you can store information about each stream in the arc attribute table. Information could be quantities (stream pH), categories (stream class), character strings (stream name), and dates (month/day/year).

Each arc is composed of a series of X, Y coordinates called *vertices*. The beginning and ending vertex of each arc is called a *node*. The beginning node of an arc is called the from-node and is typically identified in an attribute table as the **FNODE#**. The ending node of an arc is the to-node and is typically identified in an attribute table as the **TNODE#**. The **FNODE#** and **TNODE#** describe the direction of an arc. Once the GIS knows the direction of an arc, it knows which is the "left side" and which is the "right side" of the arc.

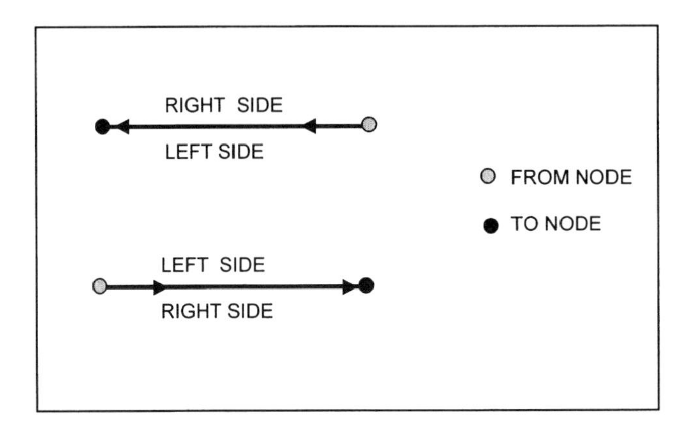

You will learn about GIS tools for analysis of line themes in Chapter 4.

NETWORK THEMES

A network is a special type of line theme consisting of connected arcs such as streets, utility lines, or stream networks. With network analysis tools you can find the fastest, cheapest, shortest, or "best" path to get from one location to another along a network. For example, the optimal path to get from Point A to Point B in the following network costs 75 + 10 + 10 +75.

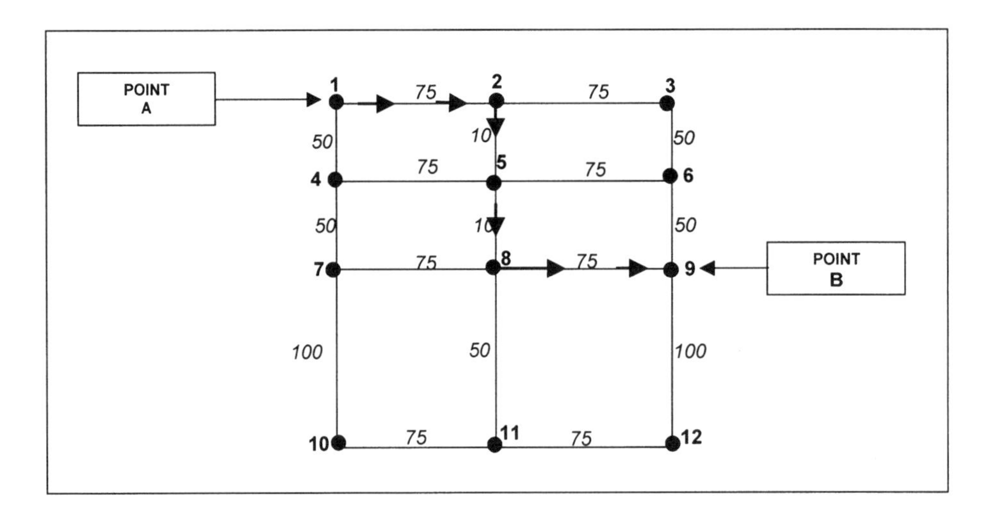

Other network applications include:

- Address geocoding to estimate the location of street addresses.
- Finding the closest resource facilities to any event location.
- Allocation of resources from supply centers to customers within the network.

You will learn about network analysis in Chapter 5.

DYNAMIC SEGMENTATION

Sometimes important line information is not available in X, Y coordinates, but instead is recorded as measurements along lines such as mileage along a road, meters along a transect, etc. This type of information can be translated into a GIS by using a technique called Dynamic Segmentation. The technique allows for segmentation of arcs into sections without changing the arc-node structure of a line theme.

Original Street Arc:

Begin	End	F-ArcPercent	T-ArcPercent	Section#	Surface
0	1.5	0	33..33	1	Dirt
1.5	3.5	33.33	77.77	2	Gravel
3.5	4.5	77.77	100	3	Paved

Dynamically Segmented Street Arc:
Dirt Gravel Paved

You will learn about Dynamic Segmentation Tools in Chapter 6.

POLYGON THEMES

A polygon is a GIS feature that has an area and a perimeter. The following example polygon theme represents four lake polygons.

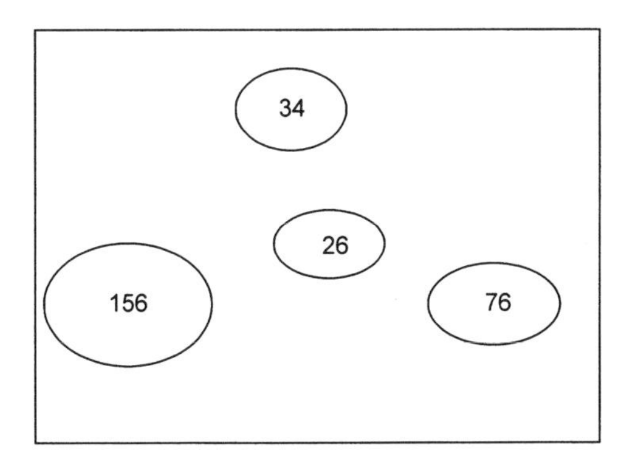

Lakes Polygon Attribute Table

Area	Perimeter	Lakes#	Lakes-ID	Lake_Type	PH
-25000		1	0	0	0
10000		2	156	1	6.5
5000		3	34	1	6.2
4000		4	26	2	7.2
6000		5	76	1	6.0

A polygon attribute table has a special record for an artificial polygon called the *universe polygon*. The universe polygon has an area that is the sum of the area of all polygons in the theme. It is always assigned a negative sign because it is an artificial polygon that is used by the GIS in computations. You will learn about tools for analysis of polygons in Chapter 7.

GRID THEMES

Grids are grid cells with a fixed number of rows and columns that have several tables associated with them. If the grid contains integer values, typically it is assumed that these values represent categories and a value attribute table is associated with the grid as follows:

TEXTURE GRID

1	1	2	3	1
1	1	2	3	1
2	2	2	3	1
2	2	2	2	3
2	2	2	2	3

TEXTURE.VAT (VAT stands for Value Attribute Table

Value	Count	Name
1	7	Loamy Sand
2	13	Sandy Loam
3	5	Loam

If the grid contains floating-point values, it is assumed that these values represent quantities rather than categories and a statistics table is associated with the grid. For example, imagine that we had a grid of elevation values as follows:

ELEVATION GRID

100.1	102.9	103.4	103.9	104.1
103.2	103.5	103.8	104.0	104.5
104.4	104.8	104.9	105.2	106.0
104.8	104.9	105.1	106.4	107.5
105.0	105.4	105.7	106.7	108.9

TEXTURE.STA (STA stands for Statistics)

MIN	MAX	MEAN	STDV
100.1	108.9	104.764	1.652

Grids that are common in GIS include digital elevation and land cover grids. You will learn about tools for grid analysis in Chapter 8.

IMAGE THEMES

Images are special grids typically derived from some remote sensing device like a satellite sensor, a digital camera, or a desktop scanner. Examples of images commonly used in remote sensing include digital orthos, satellite imagery, and scanned maps.

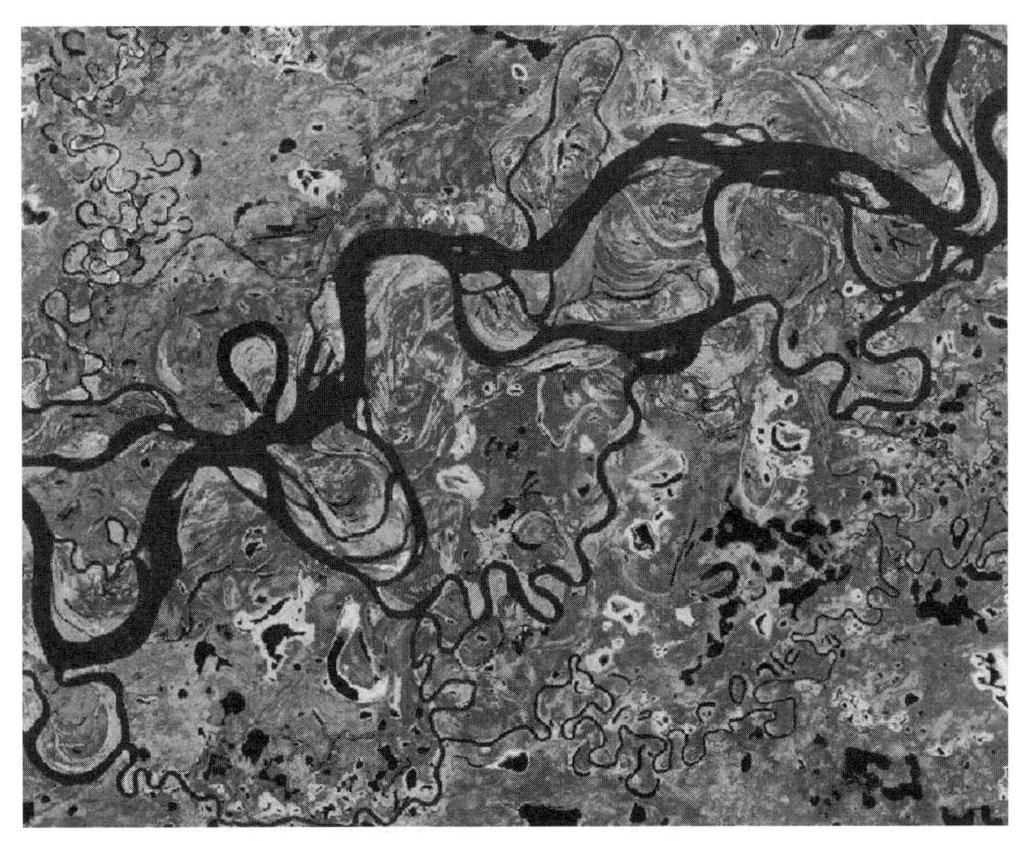

Example satellite image of Yukon River floodplain, Alaska.

You will learn more about image analysis tools in Chapter 9.

TOOLS FOR MANAGING GIS FEATURES

There are several generic tools that we will start with in this chapter that are applicable to managing points, lines, polygons, grids, and images. They are as follows:

- LIST—List the contents of any GIS table.
- COPY—Makes a new copy theme from any point, line, polygon, or grid theme.
- APPEND—Appends 2 or more point, line, or polygon themes.
- KILL—Deletes a user-specified point, line, polygon, or grid theme.
- RENAME—Renames a user-specified point, line, polygon, or grid theme.
- DESCRIBE—Tells the user information about a point, line, polygon, grid, or image theme.

Information produced by DESCRIBE operation:

Point Theme	Line Theme	Polygon Theme	Grid/Image Theme
Number of Points	Number of Arcs and Nodes	Number of Polygons, Arcs, and Nodes	Number of Rows Number of Columns
Size of Attribute Storage	Size of Attribute Storage	Size of Attribute Storage	Data Type (Integer, Floating Point)
Theme Extent: xmin,xmax,ymin,ymax	Theme Extent: xmin,xmax,ymin,ymax	Theme Extent: xmin,xmax,ymin,ymax	Theme Extent: xmin,xmax,ymin,ymax
Spatial Index (Yes or No)	Spatial Index (Yes or No)	Spatial Index Polygon Topology (Yes or No)	Value min,max,mean, standard deviation
GIS Projection Information	GIS Projection Information	GIS Projection Information	GIS Projection Information

TOOLS FOR BUILDING ATTRIBUTE TABLES

Sometimes you will obtain a GIS theme that does not have an attribute table associated with it. The following generic tools can be used for creating tables associated with your themes.

BUILD

- **Builds an attribute table for a point, line, or polygon theme.**

As an example, imagine that you have the following theme.

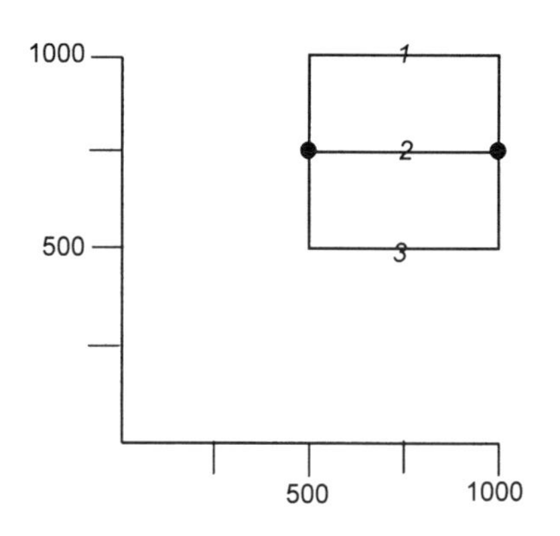

At this point, the GIS knows that you have arcs#1, 2, and 3. You need to use **BUILD** to tell the GIS to determine the polygons and their spatial relationships (called *topology*). After building, you would have the following:

DESCRIBE Theme

Feature Class	Number of Features	Topology?	
Points	3	1	
Lines	3	2	
Polygons	2	Yes	
Xmin	**Ymin**	**Xmax**	**Ymax**
500.000	500.000	1000.000	1000.000

LIST Theme.PAT

Area	Perimeter	Theme#	Theme-ID
-250000.000	2000.000	1	0
125000.000	1500.000	2	0
125000.000	1500.000	3	0

BUILDVAT

- **Builds a value attribute table for an integer grid theme.**

As an example, imagine that you have the following grid theme.

Grid

16	16	27	27	27
16	16	27	27	27
12	12	18	27	27
12	12	12	18	27
12	12	12	12	18

LIST Grid.VAT
file not found

BUILDVAT Grid
LIST Grid.VAT

Value	Count
12	9
18	3
16	4
27	9

BUILDSTA

- **Builds a statistics table for a grid theme.**

As an example, imagine that you have the following grid theme.

Grid

100	101	102	103	104
100	101	103	103	104
101	102	103	104	106
102	103	104	105	106
104	104	105	106	107

LIST Grid.STA
file not found

BUIDSTA Grid
LIST Grid.VAT

MIN	MAX	MEAN	STDV
100.000	107.000	103.320	1.870

EXERCISES

1) You have the following line theme:

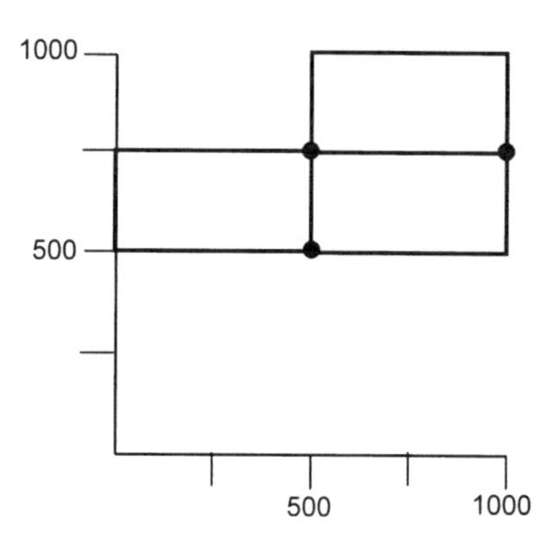

Fill in the following information from using the **DESCRIBE** operation.

Number of Arcs	Number of Nodes	XMIN	YMIN	XMAX	YMAX

2) You have the following points. Point 100 is a well that is 200 feet deep, on parcel 356, while point 101 is a well that is 300 feet deep on parcel 387. Fill in the correct values for the point attribute table associated with this theme.

+100

+101

Well.PAT

Area	Perimeter	Well#	Well-ID	Depth	Parcel

3) You have the following lines. Arc#1 has a length of 100 meters.
Fill in the correct values for the arc attribute table associated with this theme.

1 8 2 7 3 6 4
MAIN STREET
FIRST AVENUE EAST 4
SECOND AVENUE EAST 5
1 2 3
MAPLE AVENUE
5 6 7 8

Fnode#	Tnode#	Length	Street#	Street-ID	Name
			1	1	
			2	2	
			3	3	
			4	4	
			5	5	
			6	6	
			7	7	
			8	8	

4) You have the following polygons. These parcels have the following tax information:

	Parcel 203	Parcel 208	Parcel 209
Property Tax	$1,500	$3,000	$2,000
Owner	Mr. John Smith PO Box 100 Fairbanks, AK	Ms. Jane Doe PO Box 75 Deerville, CA	Mr. Roger Rabbit 235 Rabbit Hole Dr Carrotville, AK

Fill in the correct values for the polygon attribute table associated with this theme.

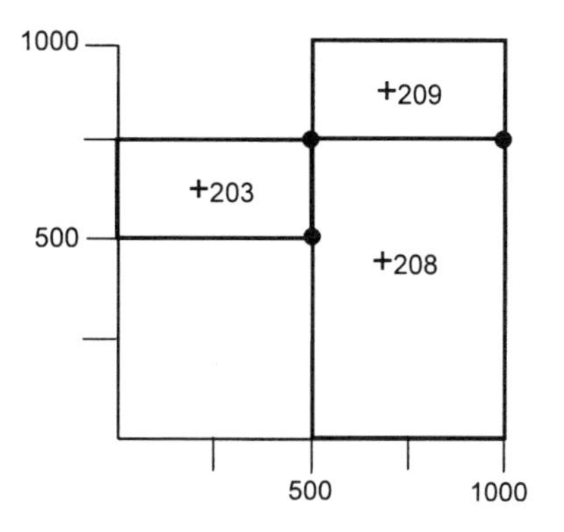

Area	Perimeter	Parcel#	Parcel-ID	Tax	Name	Street	City	State

5) You have the following grid of slope classes. Build a Value Attribute Table for your grid.

Slope_Class Grid:

0	0	1	1	2	2	2	2	2	1
0	0	1	1	2	2	2	2	2	1
0	0	1	1	2	2	2	2	2	1
1	1	1	1	2	2	2	2	2	1
1	1	1	1	2	2	2	2	2	1
1	1	1	1	2	2	2	2	2	1
1	1	1	1	2	2	2	2	2	1
1	1	1	1	2	2	2	2	2	0
1	1	1	1	2	2	2	2	0	0
1	1	1	1	2	2	2	2	0	0

Slope_Class Value Attribute Table:

6) You have the following grid of soil pH. Build a Statistics Table for your grid.

Soil_PH

6.6	6.3	6.2	5.9
6.5	6.7	6.2	6.1
6.5	6.6	6.4	6.3
6.6	6.5	6.5	6.4

Soil_PH Statistics Table

Chapter 2

GIS Tabular Analysis

INTRODUCTION

GIS Information about spatial features is typically stored in tables using a database management system. Typically the databases are stored as spreadsheets with each *row* or *record* corresponding to one feature such as a point, line, or polygon. Each column in the table corresponds to a feature attribute. The table columns are typically called *fields* or *items*.

Each column in a table typically has the following characteristics:

1) *Item Name*. The item name is simply the name of the table column.
2) *Item Type*. The item types most commonly used are binary integer (B), floating point (F), character (C), and date (D). Examples of binary integer items include categorical attributes such as soil texture class, vegetation class, or road surface type. Examples of floating point items include quantitative values such soil pH, tree diameter, or road length. Examples of character items include names such as soil order, plant genus/species, or street name.
3) *Item Width*. This refers to the number of bytes required to store each item. The most basic storage unit for computers is a *Bit* (or Binary Digit). A bit has two possible states, either a 0 or 1. Eight bits together make up a *Byte*.

With one byte, you could represent any integer ranging from 0 to 255:

2^7	2^6	2^5	2^4	2^3	2^2	2^1	2^0
128	64	32	16	8	4	2	1

For example:

1	1	1	1	1	1	1	1	=255

0	0	0	0	0	0	0	0	=0

What integer values would the following bytes represent?

1	0	0	0	0	0	1	0	= ?

0	1	0	1	0	0	0	1	= ?

The first byte would represent $2^7 + 2^1 = 128 + 2 = 130$. The second byte would represent $2^6 + 2^4 + 2^0 = 64 + 4 + 1 = 69$.

The Character type uses a scheme of coding called the American Standard Code for Information Interchange or ASCII to code each character using one byte per character. The ASCII character set is as follows (0 through 31, 128 through 255 are special characters such as bell, carriage return, line feed, escape, control, and so on).

Integer	ASCII Char	Integer	ASCII Char	Integer	ASCII Char	Integer	ASCII Char	Integer	ASCII Char
32		51	3	70	F	89	Y	108	l
33	!	52	4	71	G	90	Z	109	m
34	"	53	5	72	H	91	[	110	n
35	#	54	6	73	I	92	\	111	o
36	$	55	7	74	J	93	]	112	p
37	%	56	8	75	K	94	^	113	q
38	&	57	9	76	L	95	_	114	r
39	'	58	:	77	M	96	.	115	s
40	(	59	;	78	N	97	a	116	t
41	)	60	<	79	O	98	b	117	u
42	*	61	=	80	P	99	c	118	v
43	+	62	>	81	Q	100	d	119	w
44	,	63	?	82	R	101	e	120	x
45	-	64	@	83	S	102	f	121	y
46	.	65	A	84	T	103	g	122	z
47	/	66	B	85	U	104	h	123	{
48	0	67	C	86	V	105	i	124	|
49	1	68	D	87	W	106	j	125	}
50	2	69	E	88	X	107	k	126	~

What do the following three bytes represent as ASCII characters?

0	1	0	0	0	1	1	1	= ?
0	1	0	0	1	0	0	1	= ?
0	1	0	1	0	0	1	1	= ?

The first byte has a decimal equivalent of 64+4+2+1 = 71, which has an ASCII code of G. The second byte has a decimal equivalent of 64+8+1 = 73, which has an ASCII code of I. And the last byte has a decimal equivalent of 64+16+2+1 = 83, which has an ASCII code of S.

Character items require one byte per character since the ASCII coding scheme is typically used to store characters. For example, if you want to store up to 10 characters in an item called NAME, it would require an item width of 10 bytes.

Since date items are stored as YYYYMMDD, the item width for a date type is 8 bytes (one byte for each value in the date).

You can store binary integers as either 2-byte or 4-byte integers.

What would be the range of possible values for a 2-byte (16-bit) integer?

+/-	2^0	2^1	2^2	2^3	2^4	2^5	2^6	2^7	2^8	2^9	2^{10}	2^{11}	2^{12}	2^{13}	2^{14}	= 32,767

The range of values for a 4-byte (32-bit) integer attribute would be +/- 2,147,483,647 Floating point items are stored using either 4- or 8-bytes. A 4-byte width is single-precision real (7 digits of precision), and 8 bytes is double precision (14 digits of precision).

4. *Output Width.* This refers to the number of number of characters used to display the item value. For example, you would need an output width of 13 to list the attribute values of Quaking Aspen (a character attribute value), or 3.14159265359 (a floating point attribute value).
5. *No. of Decimals.* The number of digits to the right of the decimal place for a floating point item.

As an example, imagine that we have a floating point item called PI (3.1415926536) which is stored as follows. The value displayed will depend on how you originally defined the item width, type, and output parameters:

Item Name	Item Width	Output Width	Item Type	No. of Decimals	Value Displayed
PI	4	12	F	5	3.14159
PI	4	3	F	5	***
PI	4	12	F	10	3.1415927410
PI	8	12	F	10	3.1415926536
PI	4	10	B	--	3

Notice that in the second example, *** tells the user that the output format is not appropriate for the attribute value. In this example, 3.145926536 cannot be displayed with 5 digits to the right of the decimal and an output width of only 3 characters. Notice also that the fourth example stores PI with correct precision since the attribute has an item width of 8 bytes. The third example has an item width of 4 bytes which allows for 7 significant digits . . . thus the value displayed beyond 3.141592 . . . (beyond the 7th significant digit) is incorrect. In the last example, the value of 3.14 . . . is stored as a binary integer and therefore all numeric information beyond the decimal point is lost.

SELECTING TABLE RECORDS

To select records, you must first select the table that contains the information you are interested in. Once you have selected your table, you can specify records to select using the following query tools and a logical expression. Logical expressions are questions composed of components called operands, operators, and connectors.

Operands are items or values such as an item name (for examle NEST-ID), a numeric value (for example, 3.14), or a character value (for example, 'Bald Eagle')

Operators allow you to ask questions regarding your operands.

For example, NEST-ID > 100

The following are common GIS expression operators:

Operator	Meaning
EQ or =	equal to
NE or <>	not equal to
GE or >=	greater than or equal to
LE or <=	less than or equal to
GT or >	greater than
LT or <	less than
CN	contains the characters (example: Name CN 'Joe')

NC	Does not contain the characters
IN	In a set of values (example: Soil_Type IN {1,5,7})

Connectors are used to connect simple logical expressions to compound logical expressions. For example: soil_type in {1,5,7} and texture cn 'Silt Loam'.

The following are common GIS expression connectors:

Connector	Meaning
AND	For the condition to be evaluated as true, the logical expressions on both sides of the AND must be true
OR	For the condition to be evaluated as true, the logical expression on one or the other side of the OR must be true
XOR	For the condition to be evaluated as true, the logical condition on one and only one side of the XOR must be true. If both logical expressions are true or both are false, the condition will be evaluated as false.

The following query commands are typically available to build your record selection with.

RESELECT—allows you to ***reduce*** your selected set of records by issuing selection criteria using a logical expression.
ASELECT—allows you to ***add*** records to your selected set of records.
NSELECT—replaces the currently selected records with those ***not*** selected.

As an example, imagine you have a table of soil attributes as follows:

TEXTURE	DRAINAGE	DEPTH
'Silt'	2	50
'Silty Loam'	2	15
'Silt'	2	25
'Silt'	3	50
'Silt'	3	50
'Silty Loam'	2	25
'Silty Loam'	3	50
'Sandy Loam'	1	99
'Silty Loam'	3	5.9
'Silt'	2	50
'Silty Loam'	3	25
'Silty Loam'	2	25

RESELECT DRAINAGE GT 1 AND DEPTH IN {15,25,99} would select the following records:

TEXTURE	DRAINAGE	DEPTH
'Silt'	2	50
'Silty Loam'	2	15
'Silt'	2	25
'Silt'	3	50
'Silt'	3	50
'Silty Loam'	2	25
'Silty Loam'	3	50
'Sandy Loam'	1	99
'Silty Loam'	3	5.9
'Silt'	2	50
'Silty Loam'	3	25
'Silty Loam'	2	25

NSELECT
ASELECT TEXTURE CN 'Silt' would first select the records not currently selected, and then add to the selection set all the records that contain 'Silt' as a texture attribute . . . thus the following records would be selected. Notice that 'Silty Loam' is included since the characters 'Silt' are in that attribute.

TEXTURE	DRAINAGE	DEPTH
'Silt'	2	50
'Silty Loam'	2	15
'Silt'	2	25
'Silt'	3	50
'Silt'	3	50
'Silty Loam'	2	25
'Silty Loam'	3	50
'Sandy Loam'	1	99
'Silty Loam'	3	5.9
'Silt'	2	50
'Silty Loam'	3	25
'Silty Loam'	2	25

DESCRIPTIVE STATISTICS

STATISTICS—Computes descriptive statistics of user-specified items from selected records. The descriptive statistics include ***frequency, mean, sum, maximum, minimum, standard deviation.***

Imagine that you have the following arc attribute table of hiking trails:

Trail-ID	Length	Wild_class	Difficulty
1	2.5	1	1
2	1.0	1	1
3	5.0	2	3
4	15.0	2	3
5	12.0	2	3
6	0.5	1	1
7	5.0	2	2
8	7.5	3	2
9	27.0	3	3
10	13.0	3	2
11	2.0	2	2
12	2.5	1	1

What values would you get for the following STATISTICS queries?

ASELECT
STATISTICS WILD_CLASS
SUM LENGTH

First all records are selected, then statistics are to be summarized by the **Wild_class** attribute. Finally the sum of the attribute **Length** is requested. The following statistics are returned:

Wild_class	Frequency	Sum_Length
1	4	6.5
2	5	39.0
3	3	47.5

ASELECT
RESELECT LENGTH LT 5 AND DIFFICULTY LE 2
STATISTICS WILD_CLASS
MIN LENGTH

First all records are selected, then records that have length less than 5 and difficulty less than or equal to 2 are selected (5 records). Then these 5 records are summarized by Wild class and the minimum length is computed for each Wild class value:

Wild_class	Frequency	Min-Length
1	4	0.5
2	1	2.0

SUMMARIZING TABLES

FREQUENCY—This GIS tool produces a list of the unique attribute values and their frequency for your selected records.

The FREQUENCY program asks you for two parameters:

1) Which item(s) do you want to be analyzed in terms of unique attribute values? (the *FREQUENCY ITEM*).
2) Which item(s) do you want totals of in your output frequency table? (the *SUMMARY ITEM*).

As a simple example, the following table is generated after requesting **Wild_class** as the frequency item and **Length** as the summary item:

Trail-ID	Length	Wild_class	Difficulty
1	2.5	1	1
2	1.0	1	1
3	5.0	2	3
4	15.0	2	3
5	12.0	2	3
6	0.5	1	1
7	5.0	2	2
8	7.5	3	2
9	27.0	3	3
10	13.0	3	2
11	2.0	2	2
12	2.5	1	1

FREQUENCY ITEMS: WILD_CLASS
SUMMARY ITEMS: LENGTH

Case#	Frequency	Wild_Class	Length
1	4	1	6.5
2	5	2	39.0
3	3	3	47.5

Notice that the table contains the same information we got by using the STATISTICS command.

FREQUENCY is different than STATISTICS, in that FREQUENCY can summarize by many combinations of attributes while STATISTICS can summarize only by a single attribute.

For example, in the next example we summarize by both **Difficulty** and **Wild_Class:**

FREQUENCY ITEMS: DIFFICULTY,WILD_CLASS
SUMMARY ITEMS: LENGTH

Case#	Frequency	Difficulty	Wild_Class	Length
1	4	1	1	6.5
2	2	2	2	7.0
3	2	2	3	20.5
4	3	3	2	32.0
5	1	3	3	27.0

OTHER COMMONLY USED TABULAR TOOLS

VIEWING TABLES

DIR—Lists the tables available in your current workspace.
ITEMS—Lists the item definitions (name, type, input/output width, etc.) for your table.
LIST—Lists the information contained in your table.

MANAGING TABLES

Modifying tables:

The following are commonly used to modify table columns and rows:

ALTER—Used to alter the item characteristics such as item name or output width.
CALCULATE—Assigns new values to an item in all selected records, using an arithmetic expression or string. For example, CALCULATE HECTARES = ACRES / 2.471
REDEFINE—Used to create new items that share column space with existing items. One example would be to redefine a new item called AREA_CODE, from an existing item called PHONE_NUMBER.
SORT—Allows you to sort selected records by specified table item(s).
UPDATE—allows you to interactively type in new values for selected record items.

Adding items, records, and tables:

ADD—Allows you to add records or rows to your table.
ADDITEM—is used to add new items or columns to your table.

Deleting items, records, and tables:

DROPITEM—Deletes any specified items from your table.
PURGE—Deletes the selected records from your table.
KILL—Deletes any user-specified tables.

Exporting tables:

COPY—Copies an existing table to a new table.
UNLOAD—Writes selected table information to ASCII text file.
SAVE—Writes selected table information to binary INFO file.

MERGING TABLES

Tables can be merged if there is a key item or field that is in common with the tables.

JOINITEM—Permanently merges two tables.
RELATE – Temporarily merges two or more tables.

Imagine that you have a huge soils database of 20,000 soil polygons with the following polygon attribute table. You could use the **RELATE** command to temporarily link the soils polygons to the soil texture look-up table:

SOILS Polygon Attribute Table

SOILS-ID	AREA	PERIMETER	TEXTURE
1			2
2			3
3			2
4			2
5			2
6			3
7			3
8			1
9			3
10			2

Texture Look-Up Table

TEXT_CODE	TNAME
1	'Sandy Loam'
2	'Silt'
3	'Silty Loam'
4	'Silty Clay Loam'

The advantage of linking to a look-up table instead of storing TNAME values as polygon attributes is that it is much easier to maintain a small look-up table instead of 20,000 polygon attribute records. For example, if we choose to store the texture name in the polygon attribute table, incorrect attribute values such as 'Silt Loam', 'Silty Clay Loam', 'Sandy Silt' would be more difficult to find and correct compared to the linked look-up table approach.

INDEXING ATTRIBUTES

Imagine that you just picked a gallon of blueberries and wanted to find all the blueberry recipes in a cookbook. You could start on page one and search the entire cookbook, page by page in a sequential manner. However, it would be much more efficient to look for the attribute value 'Blueberry' in your cookbook index. In a similar manner, items can be much more efficiently searched if they have been indexed . . .

INDEXITEM—Creates an attribute index to increase query speed for that item.

TABULAR ANALYSIS EXERCISES

1) You have an attribute table about soil polygons. You run the **statistics** program to create a new table summarizing the area of soil polygons by texture class. The output table is as follows:

Texture	Frequency	Sum-Area
1	21389	3371.357086
2	40987	6671.368010
3	*****	27204.001052
4	81298	20271.315022
5	92381	25244.364040

Why is the **Frequency** for texture class 3 not displayed in the table?

How can you solve the problem so that **Frequency** for texture class 3 is displayed in the table?

2) You have the following arc attribute table:

Stream#	Length	Ownership	Trout_count
1	3371.357086	1	156
2	6671.368010	2	354
3	27204.001052	1	45
4	20271.315022	1	98
5	25244.364040	3	322

Which records would be selected in the following expression:
Tables: **SELECT** stream.aat
Tables: **RESELECT** Ownership in {1,3} AND Trout_count GT 200

Which records would be selected in the following expression:
Tables: **SELECT** stream.aat
Tables: **RESELECT** Ownership in {1,3} OR Trout_count GT 200

3) You have a street arc attribute table containing two attributes: **Speed_Limit** which is the maximum allowable speed in miles per hour and **Length** which is the length of each arc in meters. You add another attribute column called **Time** . How would you calculate the time in minutes it would take to travel across each arc at the maximum speed limit? There are 5280 feet in a mile and 3.281 feet in a meter. You start by adding two new columns: **FT_PER_MIN** , the speed limit expressed in feet per minute and **Length_FT** , the arc length in feet.

CALCULATE FT_PER_MIN = ______________________________
CALCULATE Length_FT = ______________________________
CALCULATE TIME = ______________________________

4) Correct the following logical expression:
RESELECT SPECIES CN 'KING SALMON' OR CN 'SOCKEYE SALMON'

5) Correct the following calculation:
CALCULATE ACRES = HECTARES X 2.471

6) Correct the following calculation:
CALCULATE ACRES = AREA / 43,560

7) Correct the following logical expressions:
RESELECT VEGCODE = 1
CALCULATE SHADECOLOR = 27
RESELECT VEGCODE = 2
CALCULATE SHADECOLOR = 35
RESELECT VEGCODE = 3
CALCULATE SHADECOLOR = 67

8) You have selected a **FOREST.PAT** polygon attribute table. You create a new attribute called **SITE_CLASS** based on an existing **SITE_INDEX** attribute. **SITE_CLASS** of 1 would be any polygon with **SITE_INDEX** less than 50, **SITE_CLASS** of 2 would be any polygon with **SITE_INDEX** between 50 and 75, and **SITE_CLASS** of 3 would be any polygon with a **SITE_INDEX** greater than 75.

Fill in the appropriate TABLES commands to do the following:

________________ **/***Add a new attribute column called Site_class**

________________ **/****Select the forest polygon attribute table**

________________ **/***Select all records with site_index less than 50**

________________ **/***Fill in the Site_class attribute with a value of 1**

________________ **/***Select all records in the table**

________________ **/***Select all records with site_index between 50 and 75**

________________ **/***Fill in the Site_class attribute with a value of 2**

________________ **/***Select all records in the table**

________________ **/***Select all records with site_index greater than 75**

________________ **/***Fill in the Site_class attribute with a value of 3**

________________ **/***Select all records in the table**

________________ **/***Select the universe polygon record**

________________ **/***Fill in the Site_class attribute with a value of 0**

9) You have a point attribute table of waterfowl nests containing the following items:

UNIT	The management unit the nest is in
X-COORD	The GIS X-coordinate of each nest location
Y-COORD	The GIS Y-coordinate of each nest location
NEST-ID	The Identification Number of each nest
SPECIES	The species code (1=mallard, 2=pintail, 3=widgeon,4=green wing teal)
AGECLASS	The age class of the nesting duck (1=first year, 2=older than first year)
CLUTCH_SIZE	The number of eggs in each nest

You want to produce a table with the following information:

Unit	Species	Age Class	Total Number of Eggs	Total Number of Nests
1	1	1	121	12
1	1	2	345	42
1	2	1	32	7
1	2	2	213	19
1	3	1	267	22
2	1	1	465	54
2	1	2	132	12
2	3	1	197	15

What would you use for frequency and summary items to generate this information?
Frequency item(s) ________________ **Summary item(s)** ________________

10) There is a proposal to purchase some land for an experimental forest research site. Your job is to produce a table listing the hectares and percent of area for each vegetation class in this area. You have a vegetation polygon attribute table and another table of vegetation names as follows:

Vegetation Polygon Attribute Table

Area	Veg#	Veg-ID	Size-class	Type
-929,7919.191	1	0	0	0
3447.094	2	101	P	1
7017.024	3	102	S	7
and so on...	and so on...	and so on...	and so on...	and so on...

type_names.tbl

-1	'Cutover'
0 '	'Universe polygon'
1	'Black Spruce'
2	'White Spruce'
3	'Aspen'
4	'Birch'
5	'Open Water'
6	'Willow'
7	'Alder'
8	'Dwarf Birch'
9	'Sedge Meadow'
10	'Calamagrostis Grass'

Fill in the appropriate tools in the following flowchart that would produce a sorted table of total hectares and percent of area for each vegetation class.

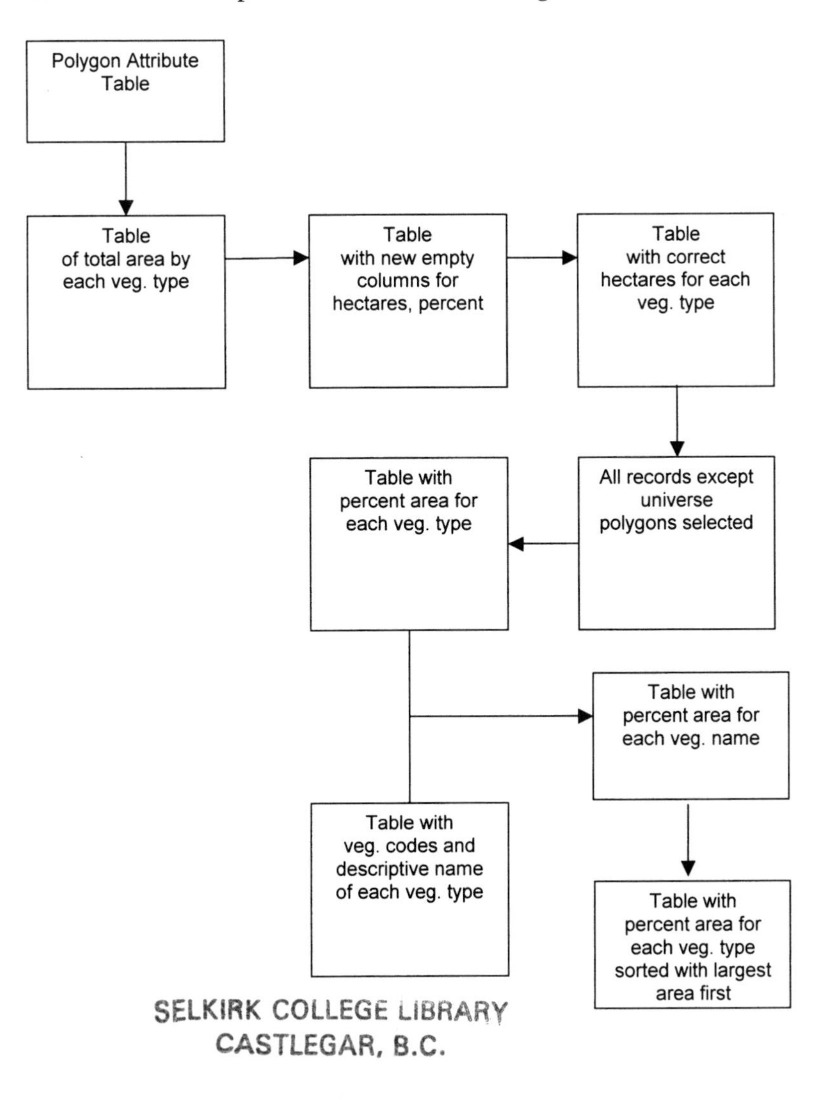

Chapter 3

Point Analysis

INTRODUCTION

A point is a GIS X,Y location that has no area or perimeter. Example point themes include point locations from research plots, radio-collared animals, wells, individual trees, fire hydrants, or utility poles.

The following GIS tools can be used in managing points:

- ADDXY—Add each point X,Y coordinate as new items in the point attribute table.
- NODEPOINT—Create a new point theme from nodes in a line or polygon theme.
- ARCPOINT—Create a new point theme from arc vertices or polygon labels in a line or polygon theme.
- POINTNODE—Transfer attributes from points to the nearest node in a line or polygon theme.

Points can be analyzed using the following GIS tools:

AREA ANALYSIS TOOLS:

- THIESSEN—Create polygons of proximity from points.
- BUFFER—Generate buffer areas of a user-specified distance around points.

DISTANCE ANALYSIS TOOLS:

- NEAR—Compute the nearest distance from points to features in a second point, line, or polygon theme.
- POINTDISTANCE—Compute all point distances between two point themes.

ATTRIBUTE ANALYSIS TOOLS:

- RESELECT—Create a new point theme by selecting points using a logical expression.
- INTERSECT—Transfer polygon attributes to a point theme.

AREA ANALYSIS TOOLS

THIESSEN

- **Converts a point theme into a theme of proximal (or Thiessen) polygons**

Thiessen polygons can be used to apportion a point theme into polygons known as Thiessen or Voronoi polygons. Each polygon will represent only one point's region. Each polygon has the unique property that any location within that polygon is closer to the polygon's source point than to any other points. Let's do the following example.

Step 1) Connect the points 1 to 2, 2 to 3, 3 to 4, and 4 to 1 with dashed lines.
Step 2) Draw the midpoint of each line.
Step 3) Draw a solid line at each midpoint perpendicular to your dashed lines. Each solid line represents the distance halfway between two points. Draw these lines until they merge to create polygons. Erase your dashed lines and points. Build your polygon attribute table.

Input Point Theme

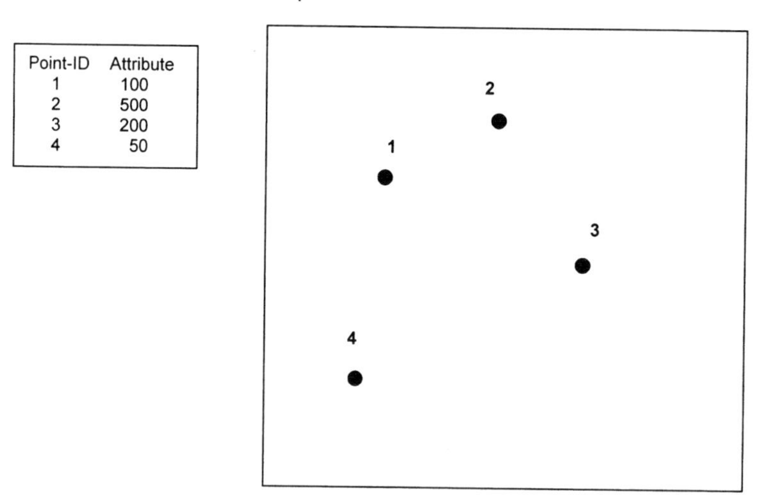

Point-ID	Attribute
1	100
2	500
3	200
4	50

Output Polygon Theme

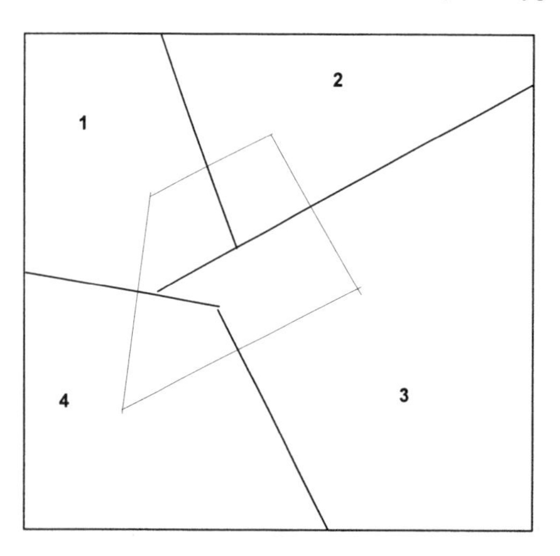

Area	Perimeter	Poly-ID	Attribute
		1	100
		2	500
		3	200
		4	50

BUFFER

- **Creates a buffer polygon of user-specified distance around each point.**

You have two basic options with buffering points.

1) You can buffer all points in your theme by a fixed distance.
2) Or you can use a look-up table to buffer all points in your theme by a variable distance, depending upon each point's attribute value.

Imagine that you have a look-up table called ZONE.LUT and a point attribute table called WELLS.PAT as follows:

ZONE.LUT

Type	Dist
1	10
2	20
3	30

WELLS.PAT

Wells#	Wells-ID	Owner	Type
1	64	1	1
2	45	1	2
3	46	2	1

The following theme results from using TYPE and ZONE.LUT with BUFFER to create variable buffers:

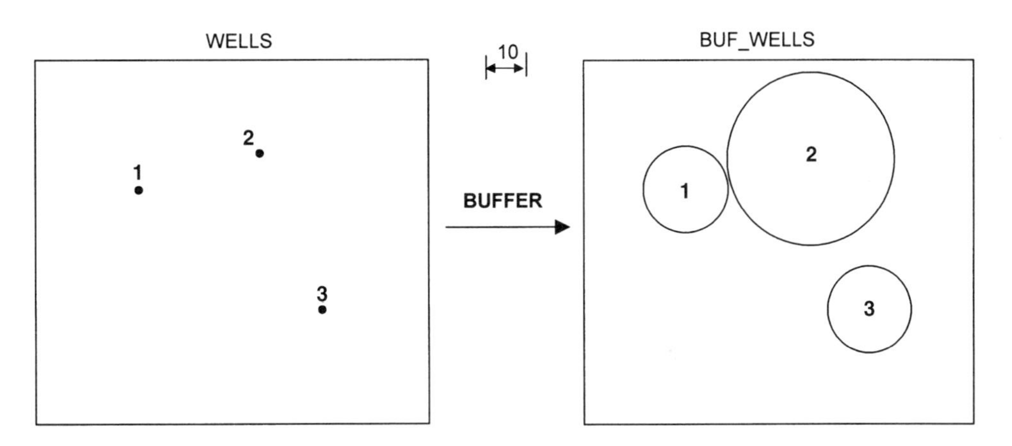

Buf_Well Polygon Attribute Table

Area	Buf_Wells#	Buf_Wells-ID	Inside
-1875.35	1	0	1
312.56	2	1	100
1250.23	3	2	100
312.56	4	3	100

Notice that the output table has a special attribute to indicate whether each polygon is inside a buffer (100) or outside a buffer (1). Also notice that the area of each buffer polygon is not equal to PI * radius^2. For example, for wells 1 and 3, the buffer radius is 10. PI * 10^2 = 314.15926. The GIS has to use many straight arcs to approximate a buffer circle, and thus the area will always be less than a true circular buffer.

Imagine that you change your look-up table as follows

ZONE.LUT

Type	Dist
1	10
2	30
3	30

WELLS.PAT

Wells#	Wells-ID	Owner	Type
1	64	1	1
2	45	1	2
3	46	2	1

The following theme results from using TYPE and ZONE.LUT with BUFFER to create variable buffers:

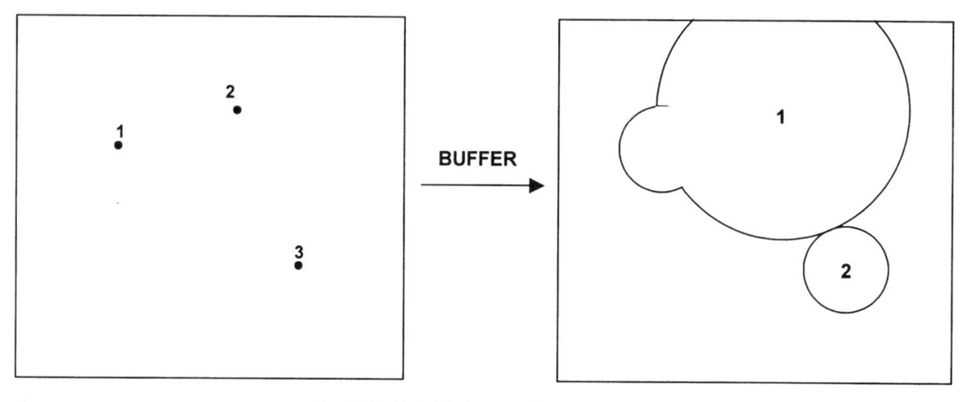

Buf_Wells2 Polygon Attribute Table

Area	Buf_Wells2#	Buf_Wells2-ID	Inside
-1875.35	1	0	1
3058.80	2	1	100
312.56	3	2	100

Notice that if your buffer sizes are relatively large, the number of resulting buffer polygons can be less than the number of original points.

DISTANCE ANALYSIS TOOLS

NEAR

- **Computes the distance from each point in a point theme to the nearest feature in a point, line, or polygon theme.**

For a line or polygon theme, you can specify whether NEAR looks for *only* the nearest arc or polygon ***node*** (NODE option)or whether ***nodes and vertices*** (LINE option) are both searched for. The user also has the option of requesting that the X,Y coordinates of each nearest point be stored in the point attribute table.

Assume the two open dots are points in a point theme. Imagine you ran NEAR with the NODE option using the point theme and the following line theme. The arrows indicate the nearest points selected by the NEAR program:

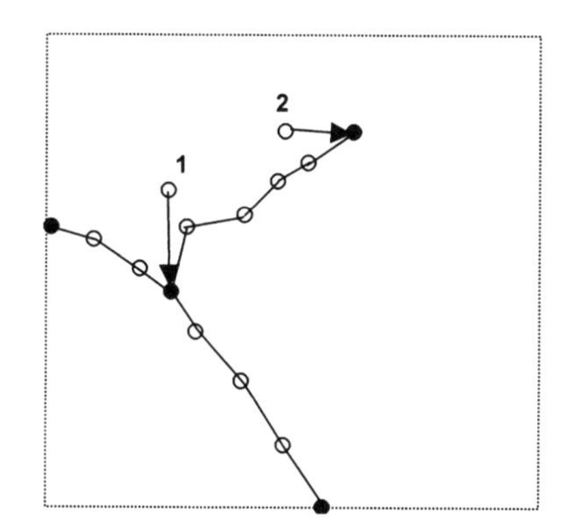

Imagine you ran NEAR with the LINE option using the point theme and the following line theme; the arrows represent the nearest points selected by the NEAR program:

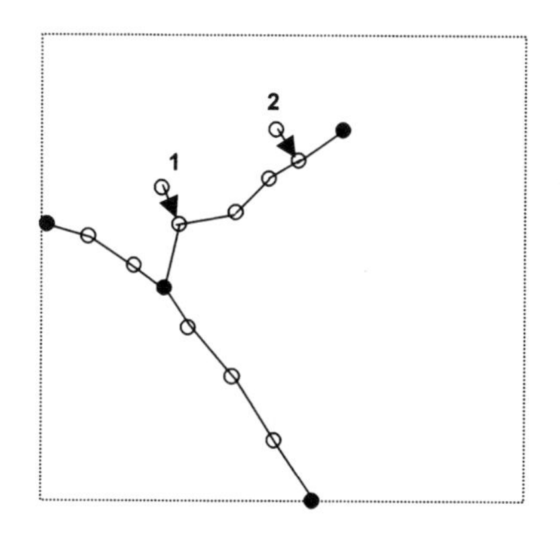

Note: There are two flaws in the NEAR program:

Flaw#1: Imagine you assigned a search radius of 100 meters in the following example. Moose#11 is farther away from any willow polygon than your search radius, and it is therefore assigned a distance of zero.

11
100 m
67

Moose-ID	Attribute	Willow#	**Distance**
1 1	100	0	0.000

Flaw#2: Imagine you assigned a search radius of 1000 meters in the following example. The flaw here is that the GIS computes the distance to the nearest arc, even though the moose is inside a willow polygon.

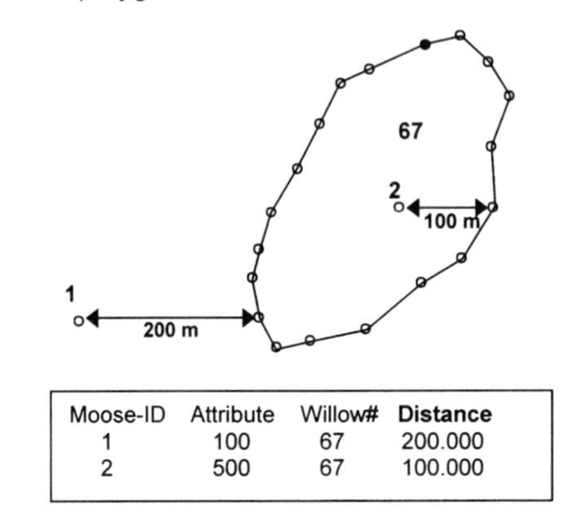

Moose-ID	Attribute	Willow#	Distance
1	100	67	200.000
2	500	67	100.000

POINTDISTANCE

- **Creates a new table that contains the distances between the points in one theme to ALL points in a second theme that are within the specified search radius.**

A distance of zero means that either the two points being compared have exactly the same X,Y coordinates, or the distance is greater than your search radius.

You have two point themes : juvenile mallards and adult mallards:

• = Adults ○ = Juveniles

What would the distance between Adult# 43 and Juvenile#13 be output if you run POINTDISTANCE with a search radius of 200? Assume each square represents a distance of 10 units.

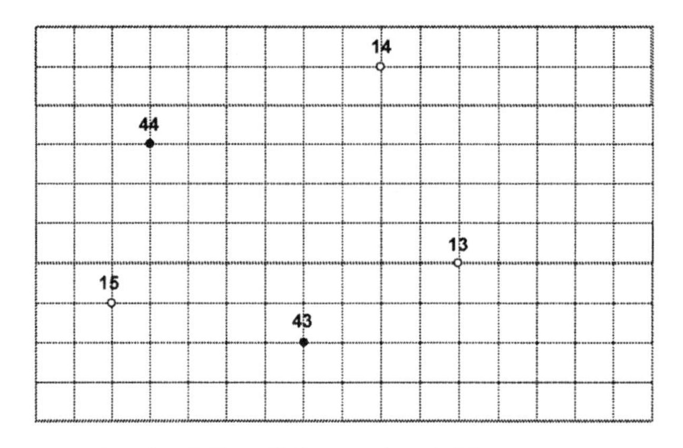

Adult#	Juvenile#	Distance
43	13	?
43	14	
43	15	
44	13	
44	14	
44	15	

In the prior example, Adult# 43 has X,Y coordinates of 70,20 and Juvenile#13 has X,Y coordinates of 110,40. We can use the Pythagorean theorem to compute the straight-line distance between these two locations:

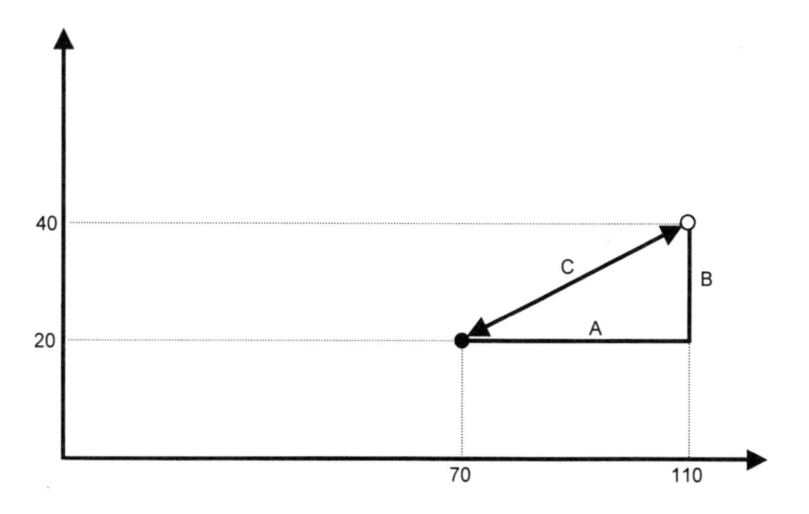

Since the GIS knows the X,Y coordinates of each point, it can easily compute A and B portions of the right triangle: A= 110 – 70 = 40 and B= 40 – 20 = 20

$$C = \text{SQRT } (A^2 + B^2) = \text{SQRT } (40^2 + 20^2) = 44.7$$

Adult#	Juvenile#	Distance
43	13	44.7
43	14	
43	15	
44	13	
44	14	
44	15	

ATTRIBUTE ANALYSIS TOOLS

RESELECT

- **Creates a new point theme by selecting points from existing point theme using user-specified logical expressions.**

Reselect can be used for point, line, and polygon features to select the features you want by asking one or more logical expressions. The following examples use this point theme:

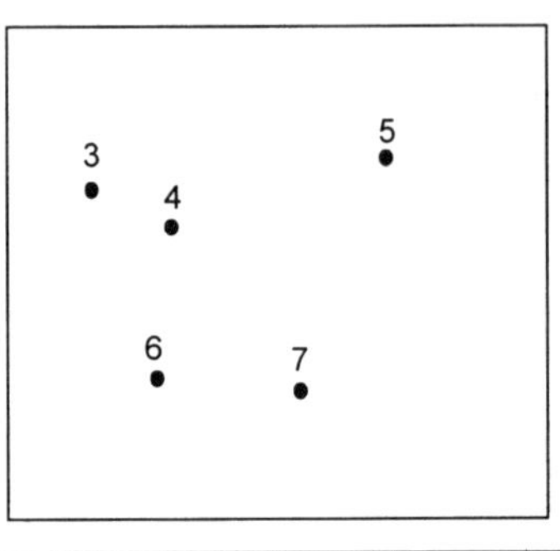

SOILS#	SOILS-ID	PER_SAND	PH	T_NAME	T_CODE
1	3	13	7.45	Silt	3
2	4	17	7.25	Silty Loam	4
3	5	31	6.90	Sandy Loam	2
4	6	29	6.55	Sand	1
5	7	41	6.40	Sand	1

Res PER_SAND gt 15 and PH <= 7.0 /**reselect sandy, acidic soils

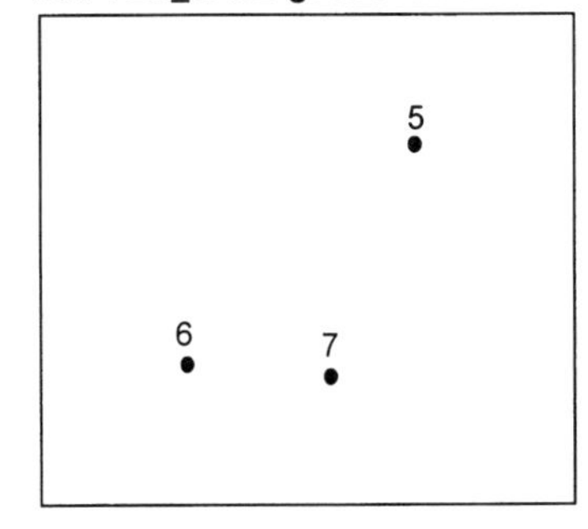

Res T_CODE in { 1,3}
Res T_NAME nc 'Silt'

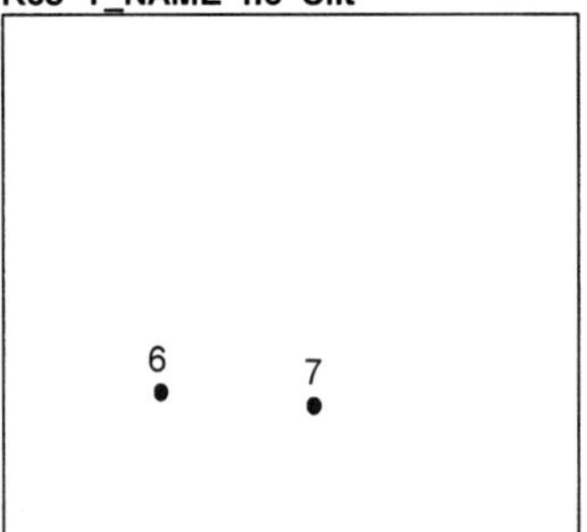

Aselect
Reselect T_CODE in {1,3, 5->8}
Aselect PH > 7

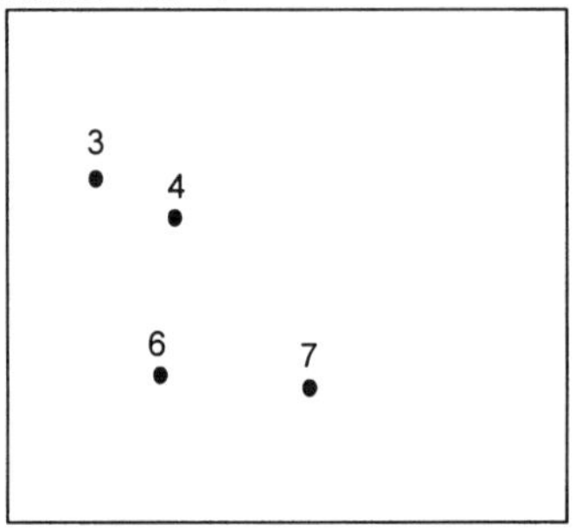

3
4
6
7

INTERSECT

- **For points that intersect with polygons, produce an output point theme containing attributes from both the input point and polygon theme.**

Think of INTERSECT as transferring polygon information to points if the points lie inside the polygons.

SOILS theme

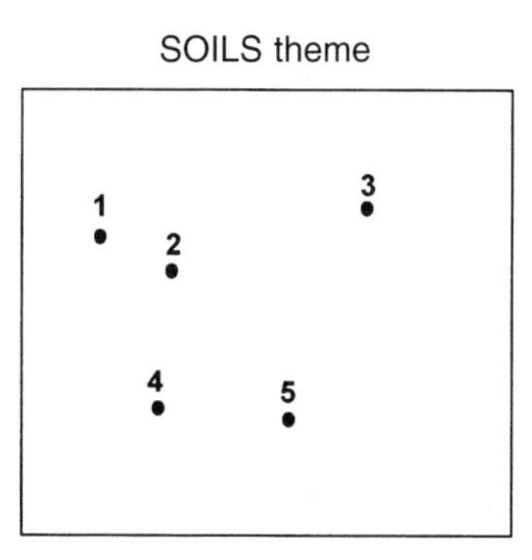

SOILS#	PER_SAND	PH	T_NAME	T_CODE
1	23	7.45	Silt	3
2	14	7.25	Silt	3
3	11	6.90	Sand	1
4	19	6.55	Sand	2
5	21	6.40	Sand	1

SLOPE_CLASS theme

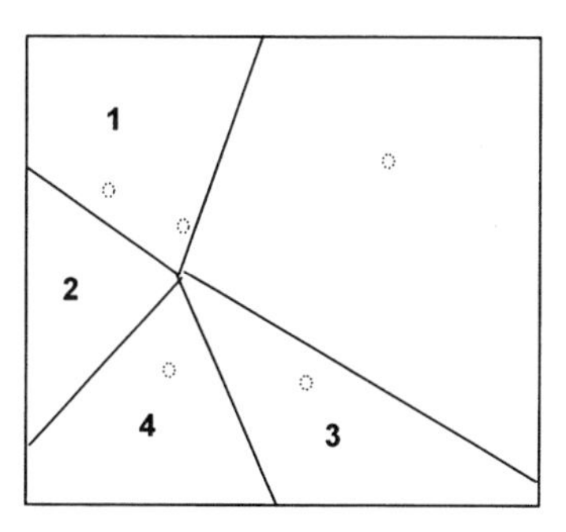

SLOPE_CLASS#	SLOPE_CLASS -ID	GRADIENT	NAME
1	1	23	Very Steep
2	2	14	Moderate
3	3	11	Moderate
4	4	19	Steep

The output theme point attribute table resulting from **INTERSECT** SOILS and SLOPE_CLASS is as follows:

SOIL_SLOPE.PAT

SOIL_ SLOPE#	SOILS#	PER_ SAND	PH	T_NAME	T_CODE	SLOPE_ CLASS#	GRADIENT	NAME
1	1	23	7.45	Silt	3	1	23	Very Steep
2	2	14	7.25	Silt	3	1	23	Very Steep
3	3	11	6.90	Sand	1			
4	4	19	6.55	Sand	2	4	19	Steep
5	5	21	6.40	Sand	1	3	11	Moderate

POINT ANALYSIS EXERCISES

1) You have a point theme of manhole covers and a line theme of streets. You want to determine the street name for each manhole cover.

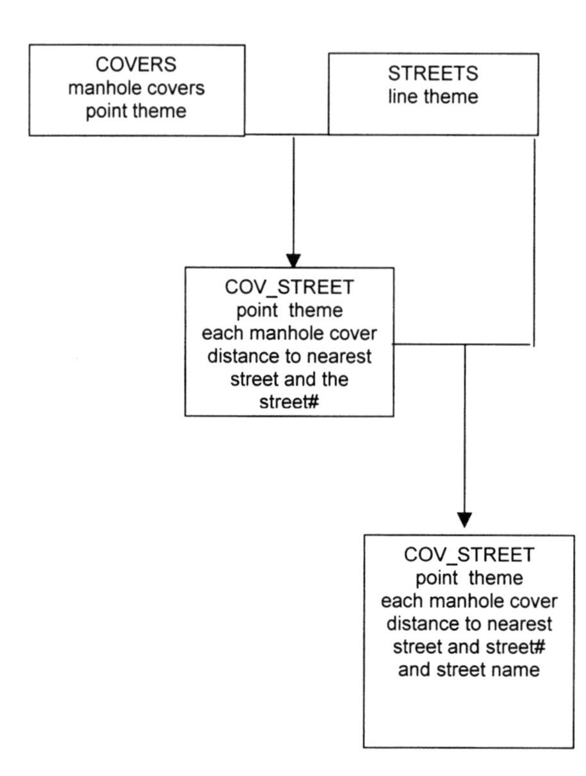

Fill in the following flowchart with the appropriate GIS tools to solve your problem:

2) You have a point theme of wells to sample groundwater. You want to calculate the mean distance for each well to all other wells. Fill in the following flowchart to solve your problem:

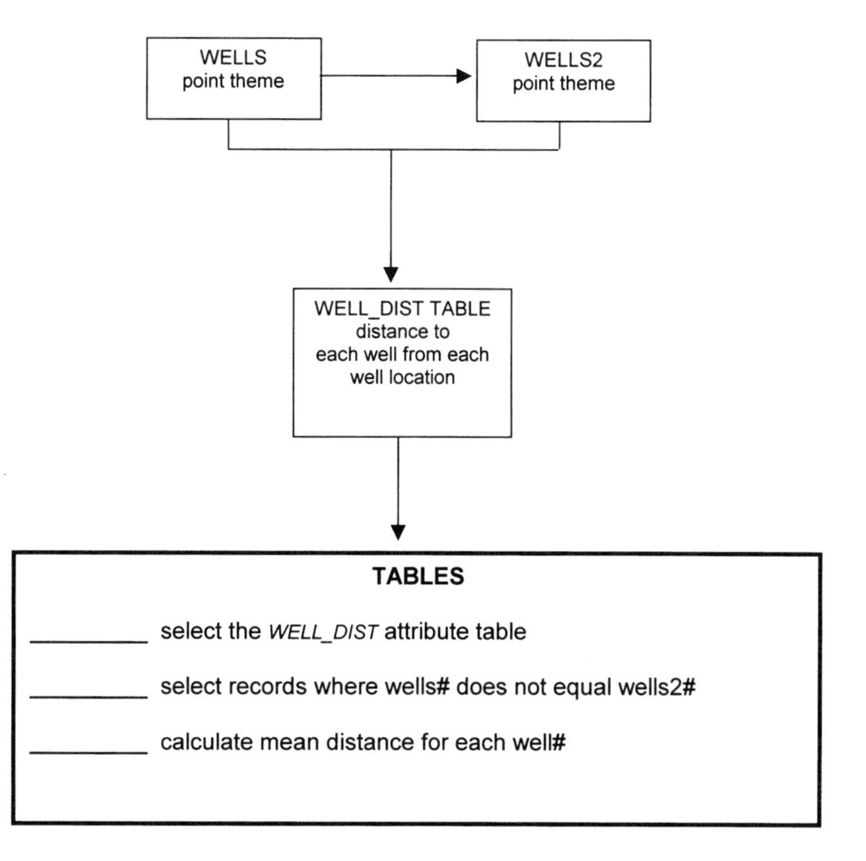

3) You have a point theme of lightning strike locations and a polygon theme of elevation classes. Your goal is to produce a table like the following:

Elevation Zone	Total Area (ha)	Total Lightning Strikes	Lightning Strikes per million ha
1			
2			
3			
4			
5			

Fill in the following flowchart with the appropriate GIS tools to solve your problem:

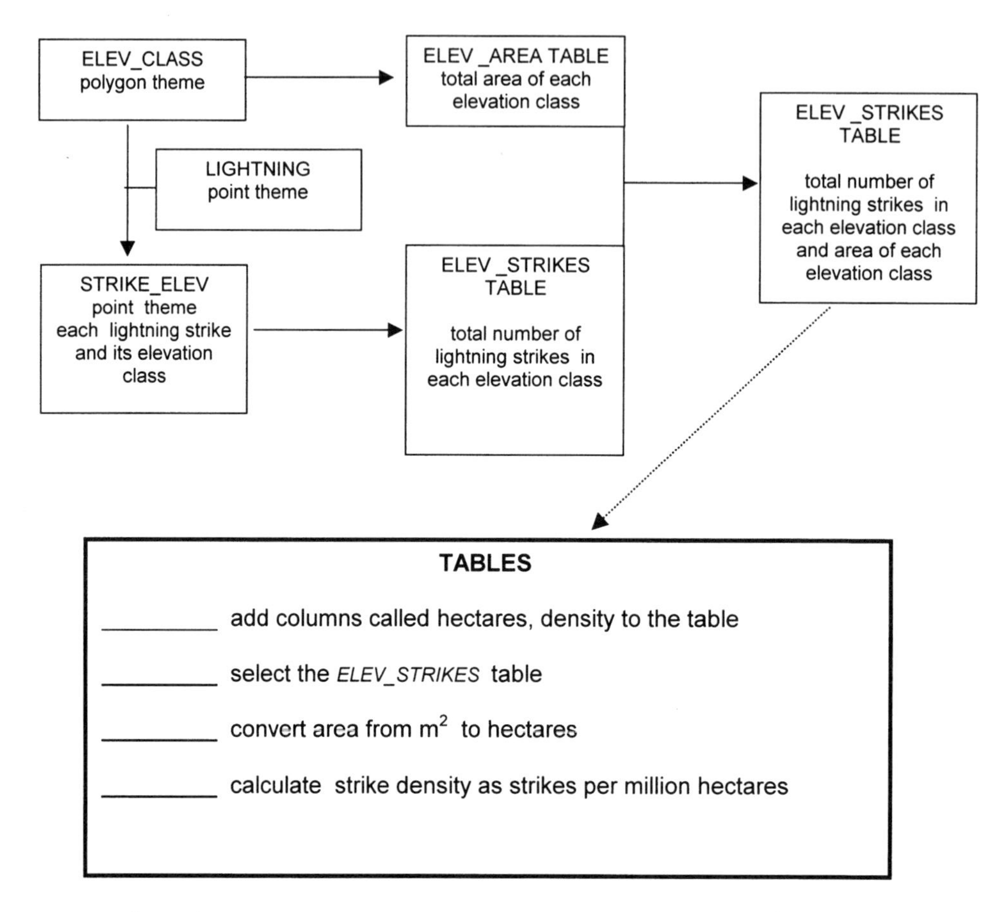

4) You have five GPS survey monuments set up in the Tanana Flats area. Draw five Thiessen polygons associated with these GPS monuments:

• GPS-1
• GPS-2
• GPS-4
• GPS-5
• GPS-3

5) You have a point theme of bufflehead nest box locations along the Chena River. Outline the GIS analysis tools that you would use to select all nest box points that are at least 1 km away from any other nest box.

6) You have a point theme in meters of snowshoe hare locations. You want to know how many of these locations are within 50 meters of a willow polygon . . . .

list hares.pat

HARES#	HARES-ID
1	1
2	2
3	3
4	4
5	5
6	6
7	7
8	8
9	9
10	10

buffer hares buf_50m # # 50 # point
list buf_50m.pat

AREA	PERIMETER	BUF_50M#	BUF_50M-ID	INSIDE
-78123.359	3137.729	1	0	1
7812.336	313.773	2	1	100
7812.336	313.773	3	2	100
7812.336	313.773	4	3	100
7812.336	313.773	5	4	100
7812.336	313.773	6	5	100
7812.336	313.773	7	6	100
7812.336	313.773	8	7	100
7812.336	313.773	9	8	100
7812.336	313.773	10	9	100
7812.336	313.773	11	10	100

list willow.pat

AREA	PERIMETER	WILLOW#	WILLOW-ID
-680190.375	7027.910	1	0
151522.984	1548.660	2	102
392388.719	2466.836	3	103
47269.211	1157.825	4	101
53964.820	990.133	5	104
35044.672	864.457	6	105

intersect buf_50m willow hares_50m
list hares_50m.pat

Area	Perimeter	HARES _50M#	HARES _50M-ID	BUF _50M#	BUF _50M-ID	INSIDE	WILLOW #	WILLOW -ID
-31906.	1735.	1	0	1	0	1	1	0
7812.	313.	2	1	3	2	100	2	102
7812.	313.	3	2	4	3	100	3	103
1293.	175.	4	3	5	4	100	4	101
560.	138.	5	4	5	4	100	4	101
7812.	313.	6	5	6	5	100	3	103
2196.	212.	7	6	7	6	100	5	104
4417.	266.	8	7	11	10	100	6	105

How many of the ten hare locations are within 50 meters of a willow polygon?

7) You have a point coverage of radio-collared moose. The point attributes include Moose-ID, sex, and age. You also have a line coverage of rivers. You want to produce a table as follows:

Moose-ID	**Average Distance to River**

Fill in the following flowchart to produce the above table:

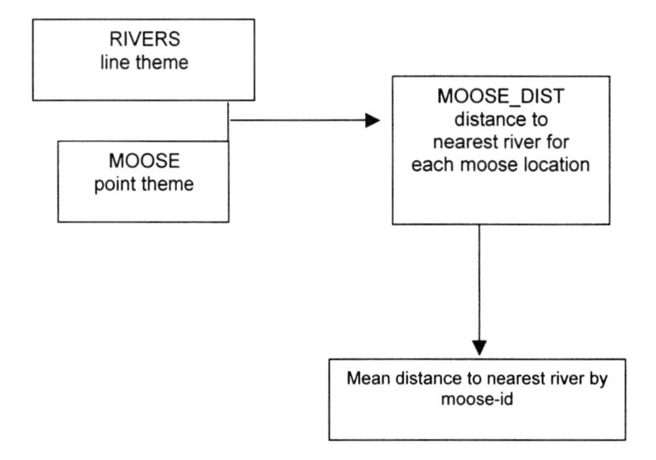

8) You have a point theme of Polar Bear locations stored in longitude/latitude. You want to create a new theme of bear locations north of 70 degrees and between –170 and –100 degrees of latitude. Fill in the following flowchart to produce this new theme.

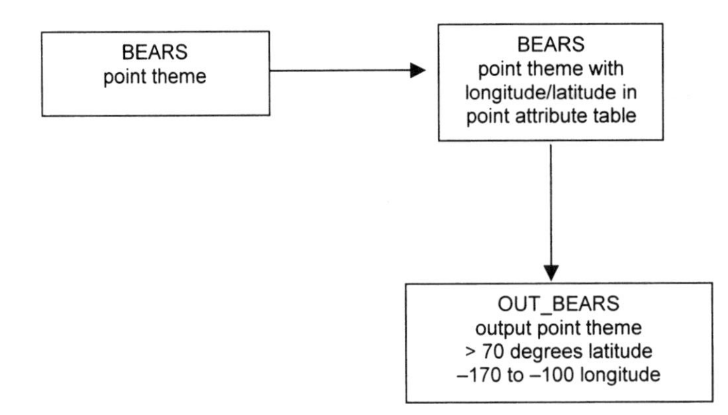

9) You have a theme of endangered plant species locations and a theme of randomly located points. You want to know what percent of plant points are within clay loam soil polygons compared to the percent of random points in clay loam polygons. Fill the following flowchart with the appropriate GIS tools to solve this problem

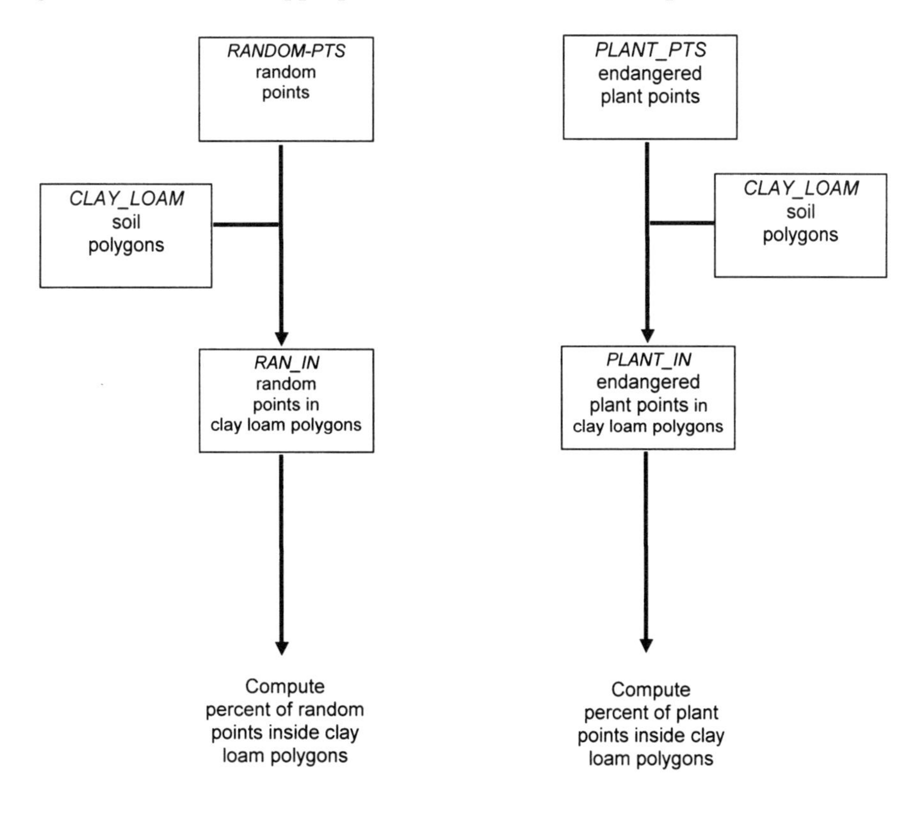

10) You have a theme of endangered plant species locations. You want the mean, minimum, and maximum distance of the plant locations to the nearest clay_loam polygon. Fill the following flowchart with the appropriate GIS tools to solve this problem.

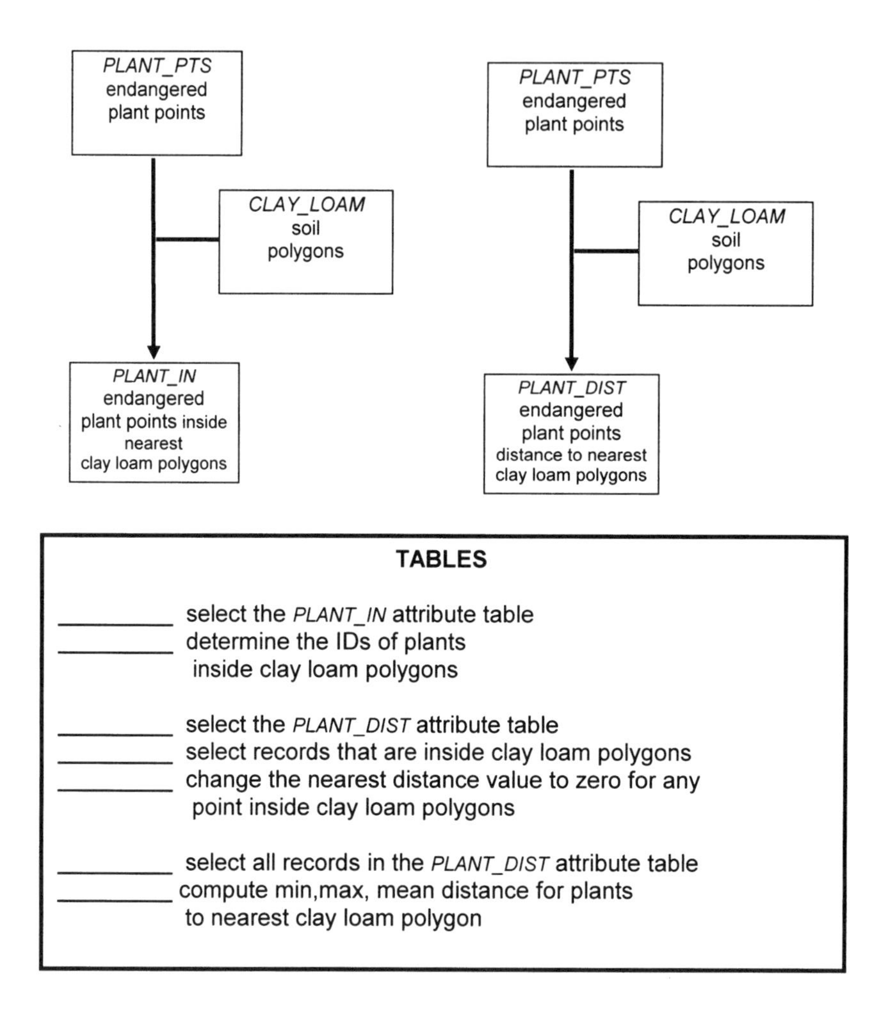

Chapter 4

Line Analysis

INTRODUCTION

A line is a GIS feature that has length but no width. Typical line themes include arcs from roads, streams, hiking trails, pipelines, or contours.

The following GIS tools are commonly used in analyzing lines:

- DISSOLVE—Merges adjacent arcs (by dissolving the node), if they have the same value for a user-specified arc attribute.
- COUNTVERTICES—Add an item in the Arc Attribute Table called VERTICES and fill the item with the total number of vertices (including beginning and ending node) for each arc.
- BUFFER—Generate buffer polygons of a user-specified distance around arcs.
- RESELECT—Create a new line theme by selecting arcs using a logical expression.
- INTERSECT—Transfer the polygon attributes to a line theme.

DISSOLVE

- **Merges adjacent arcs if they have the same attribute value.**

The new Cover-ID of the merged arc will be the lowest Cover-ID of the combined arcs. An #ALL option tells the GIS to merge the adjacent arcs, only if all attributes beyond the Arc-ID are exactly the same.

The following line theme is dissolved based on the item TYPE:

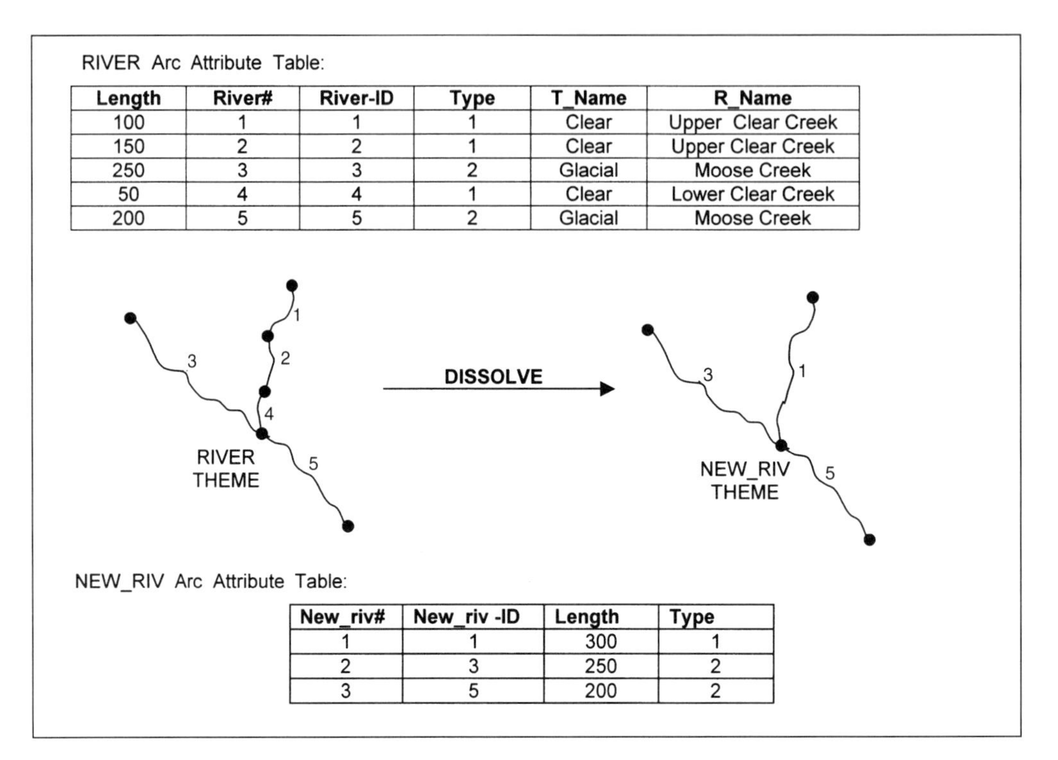

RIVER Arc Attribute Table:

Length	River#	River-ID	Type	T_Name	R_Name
100	1	1	1	Clear	Upper Clear Creek
150	2	2	1	Clear	Upper Clear Creek
250	3	3	2	Glacial	Moose Creek
50	4	4	1	Clear	Lower Clear Creek
200	5	5	2	Glacial	Moose Creek

NEW_RIV Arc Attribute Table:

New_riv#	New_riv -ID	Length	Type
1	1	300	1
2	3	250	2
3	5	200	2

The #ALL option tells **DISSOLVE** to merge the adjacent arcs only if all attributes beyond the Arc-ID are exactly the same.

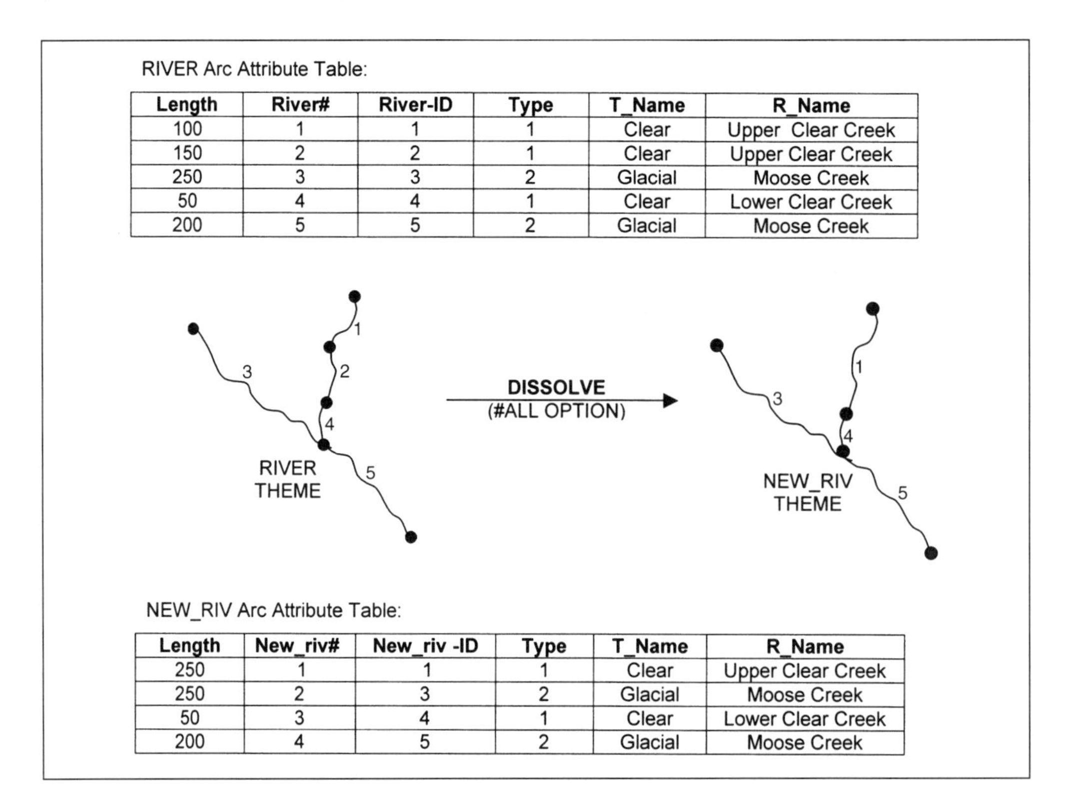

RIVER Arc Attribute Table:

Length	River#	River-ID	Type	T_Name	R_Name
100	1	1	1	Clear	Upper Clear Creek
150	2	2	1	Clear	Upper Clear Creek
250	3	3	2	Glacial	Moose Creek
50	4	4	1	Clear	Lower Clear Creek
200	5	5	2	Glacial	Moose Creek

NEW_RIV Arc Attribute Table:

Length	New_riv#	New_riv -ID	Type	T_Name	R_Name
250	1	1	1	Clear	Upper Clear Creek
250	2	3	2	Glacial	Moose Creek
50	3	4	1	Clear	Lower Clear Creek
200	4	5	2	Glacial	Moose Creek

COUNTVERTICES

- **Counts the number of points in each arc.**

The item called VERTICES is added to the arc attribute table as the total number of points (nodes and vertices) for each arc.

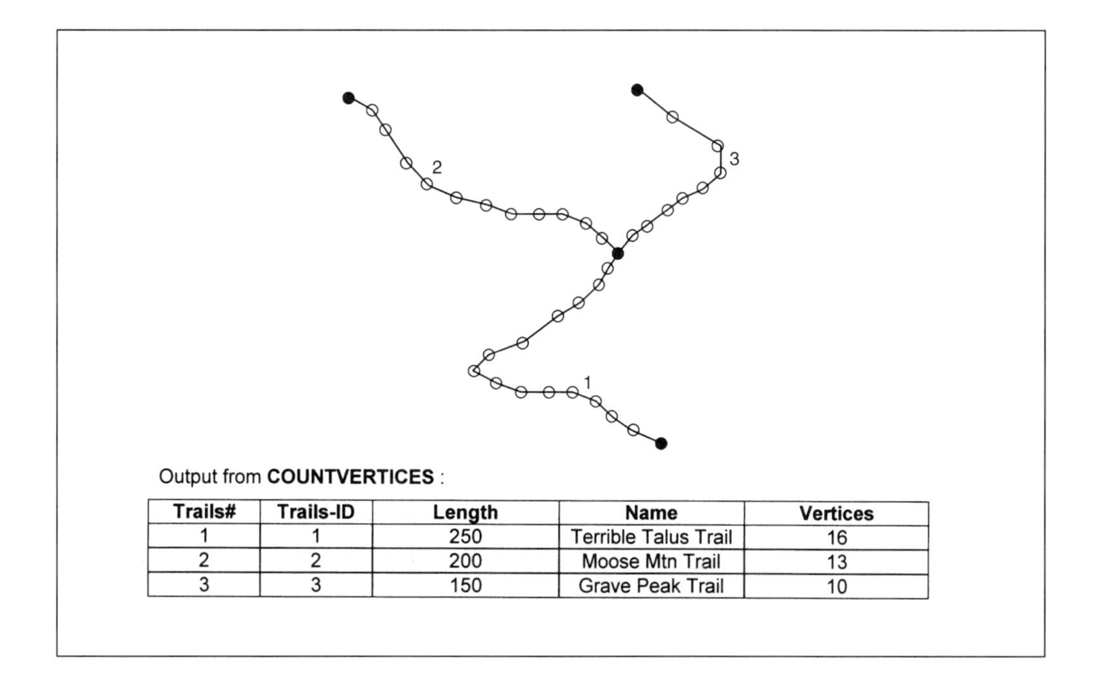

Output from **COUNTVERTICES** :

Trails#	Trails-ID	Length	Name	Vertices
1	1	250	Terrible Talus Trail	16
2	2	200	Moose Mtn Trail	13
3	3	150	Grave Peak Trail	10

BUFFER

- **Create a buffer polygon around each arc.**

When buffering a line theme, you can specify the arc ends to be buffered using a square (FLAT) or semicircle (ROUND) shape. You can also specify whether the buffer should be on both sides of each arc (FULL) , on the topological left side (LEFT) , or on the topological right side (RIGHT) . Like buffering points, you can specify a fixed buffer distance, or by using a look-up table, variable buffers.

Here is an example theme called ROADS and an associated look-up table:

ROADS Polygon Attribute Table

Length	Roads#	Roads -ID	Fnode	Tnode	Surface	Name
	1	1	1	2	1	Winter Road
	2	2	2	3	1	Winter Road
	3	3	4	3	2	Gravel Road
	4	4	3	5	3	Haul Road

ROADS Look-up Table

Surface	Dist
1	50
2	100
3	200

Imagine that you request the following buffers:

Buffer Distance	Buffer Item	End Shape	Buffer Direction	Output Theme
Fixed 100 m	-	Flat	Left	BUF_ROADS1
Fixed 100 m	-	Flat	Right	BUF_ROADS2
Fixed 100 m	-	Round	Full	BUF_ROADS3
Variable (roads.lut)	Surface	Round	Full	BUF_ROADS4

The first step is to figure out the left and right side of each arc based on the location of the from and to nodes:

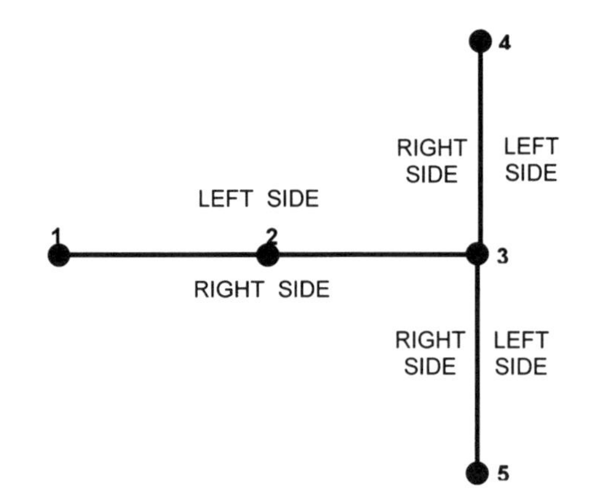

The following polygon themes would be output after buffering:

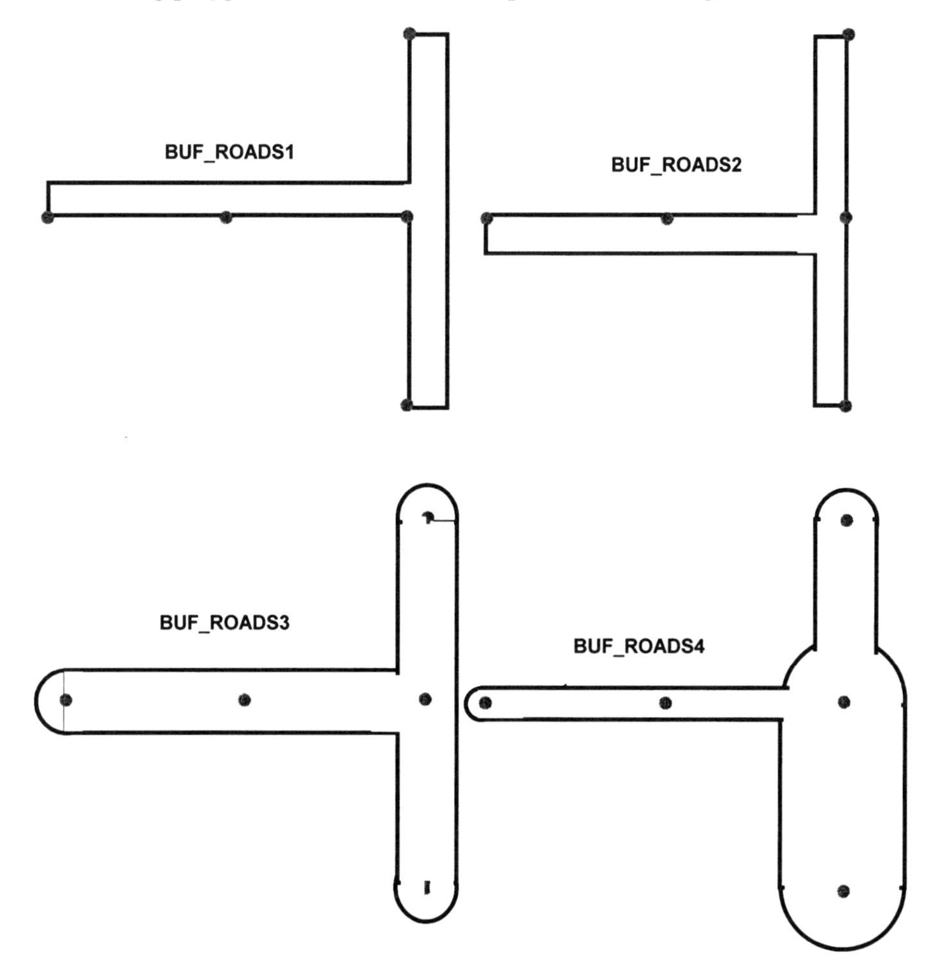

RESELECT

- **Creates a new line theme by selecting arcs based on user-defined logical expression(s).**

In the following example, we use **RESELECT** to create an output theme of short, clearwater streams.

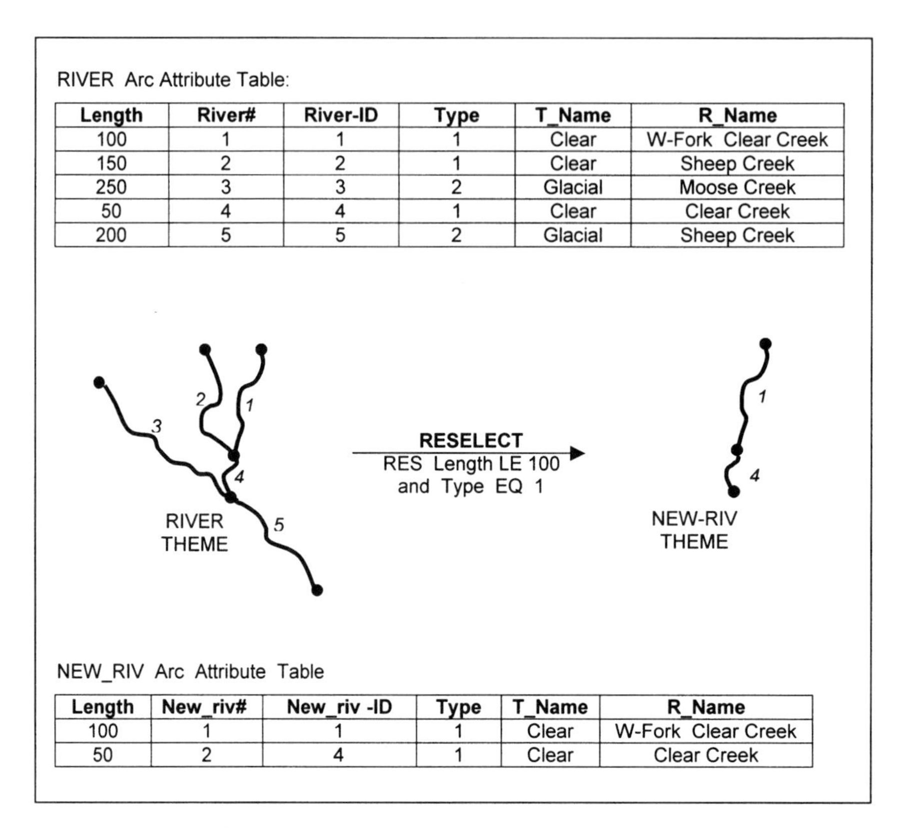

RIVER Arc Attribute Table:

Length	River#	River-ID	Type	T_Name	R_Name
100	1	1	1	Clear	W-Fork Clear Creek
150	2	2	1	Clear	Sheep Creek
250	3	3	2	Glacial	Moose Creek
50	4	4	1	Clear	Clear Creek
200	5	5	2	Glacial	Sheep Creek

NEW_RIV Arc Attribute Table

Length	New_riv#	New_riv -ID	Type	T_Name	R_Name
100	1	1	1	Clear	W-Fork Clear Creek
50	2	4	1	Clear	Clear Creek

INTERSECT

- **Creates a new line theme by intersecting arcs with a polygon theme and then adding the polygon attributes to each arc.**

Imagine the following example:

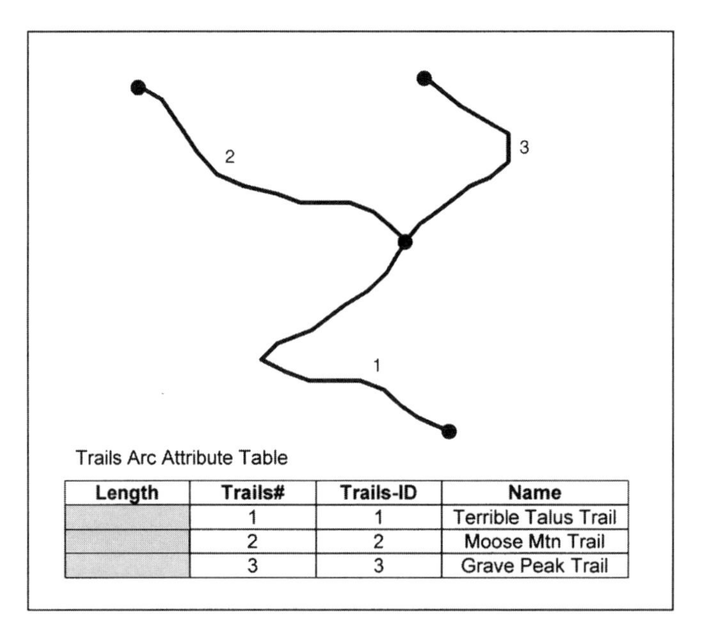

Trails Arc Attribute Table

Length	Trails#	Trails-ID	Name
	1	1	Terrible Talus Trail
	2	2	Moose Mtn Trail
	3	3	Grave Peak Trail

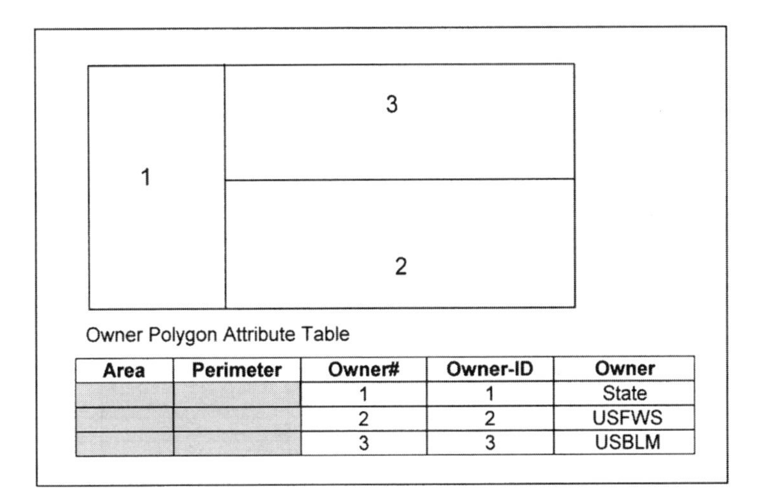

Owner Polygon Attribute Table

Area	Perimeter	Owner#	Owner-ID	Owner
		1	1	State
		2	2	USFWS
		3	3	USBLM

The following theme results from intersecting these two themes:

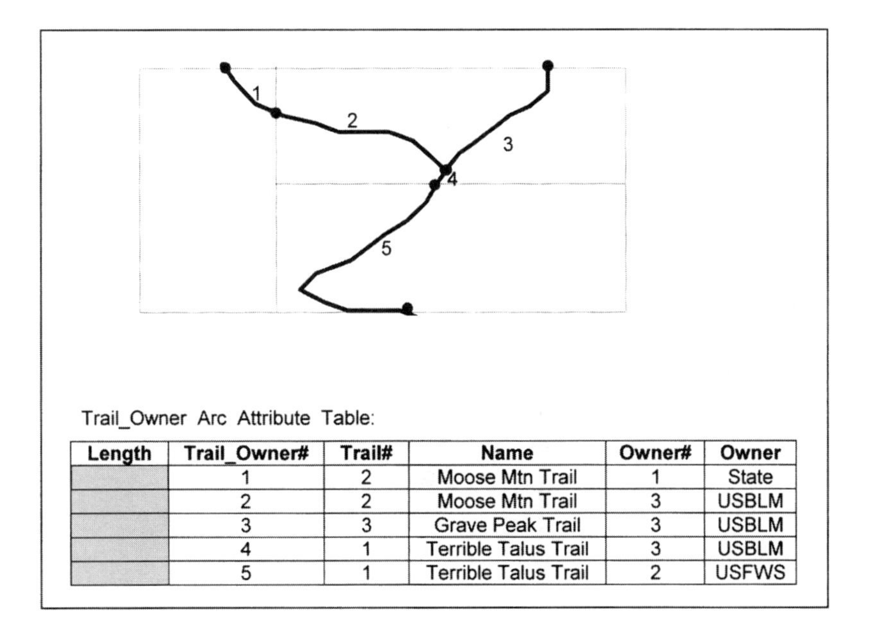

Trail_Owner Arc Attribute Table:

Length	Trail_Owner#	Trail#	Name	Owner#	Owner
	1	2	Moose Mtn Trail	1	State
	2	2	Moose Mtn Trail	3	USBLM
	3	3	Grave Peak Trail	3	USBLM
	4	1	Terrible Talus Trail	3	USBLM
	5	1	Terrible Talus Trail	2	USFWS

Note that you can ***not*** intersect a point and line theme . . . you must intersect either points or lines with a polygon theme. That is because by definition, both points and lines have no area and therefore point in line or line in point intersection is impossible while point in polygon or line in polygon intersection is possible.

LINE ANALYSIS EXERCISES

1) You have the following line theme of pipes. You run **DISSOLVE** using *Pipe_Class* as the dissolve item. What will your output theme look like?

Pipe Arc Attribute Table

Pipe#	Length	Pipe_Class	Diameter	Flow
1	1000	1	36	1500
2	300	1	36	1400
3	700	1	36	1300
4	1000	1	36	1200
5	300	1	36	1100
6	600	1	36	0
7	800	3	12	800
8	1000	2	12	700
9	800	3	10	600
10	800	3	10	600
11	800	3	12	800
12	800	3	12	800
13	1000	2	18	700
14	800	4	4	600

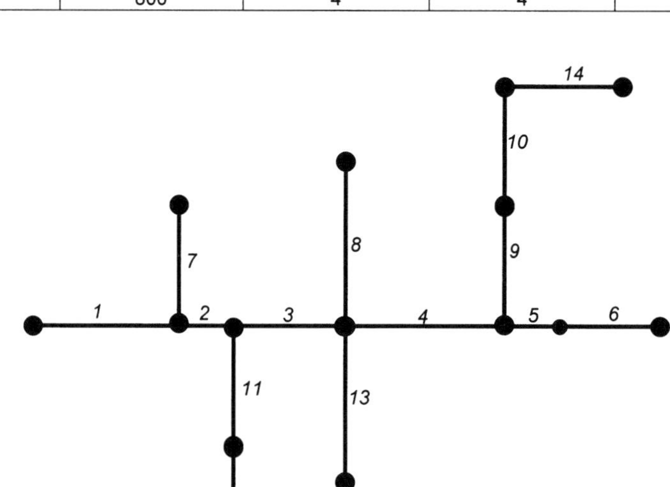

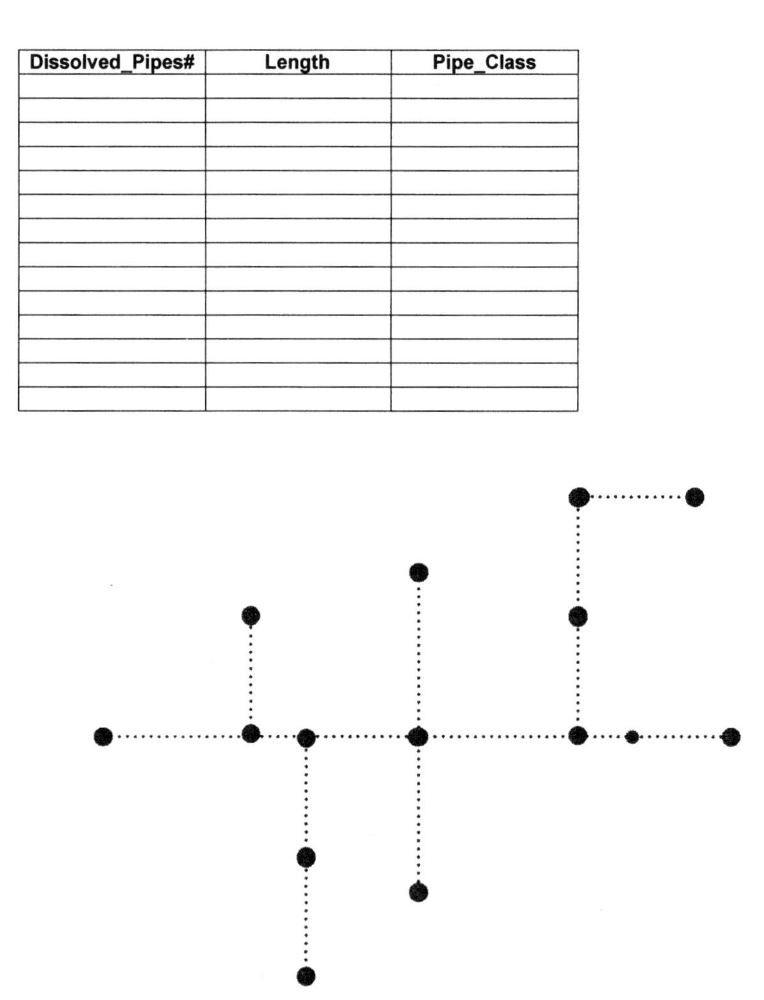

Dissolved_Pipes#	Length	Pipe_Class

2) You have a line theme of roads and a point theme of cabin locations. You want to find all cabins that are within 1 mile of a road. Your GIS coordinate system is in meters.

Fill in the following flowchart to solve the problem:

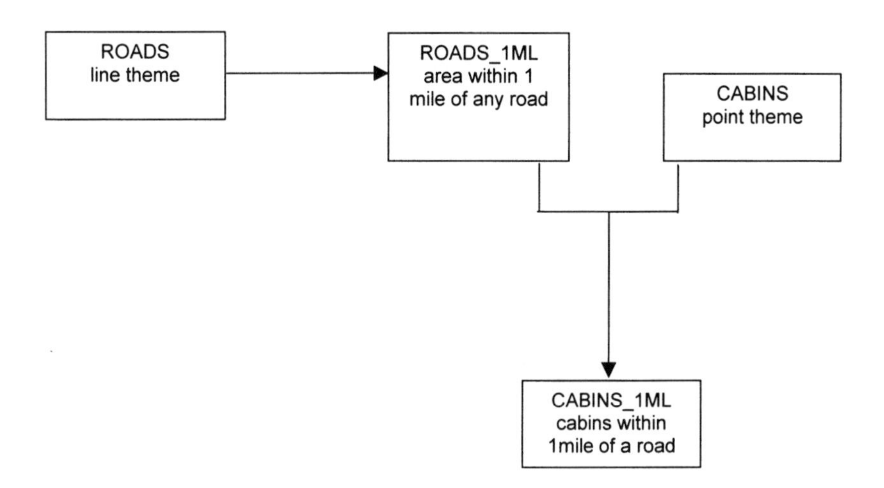

3) You have a line theme of streams with an arc attribute called *King_count* representing the count of king salmon observed along each arc. For each stream, determine the total density of king salmon per mile. Your GIS coordinate system is in meters.

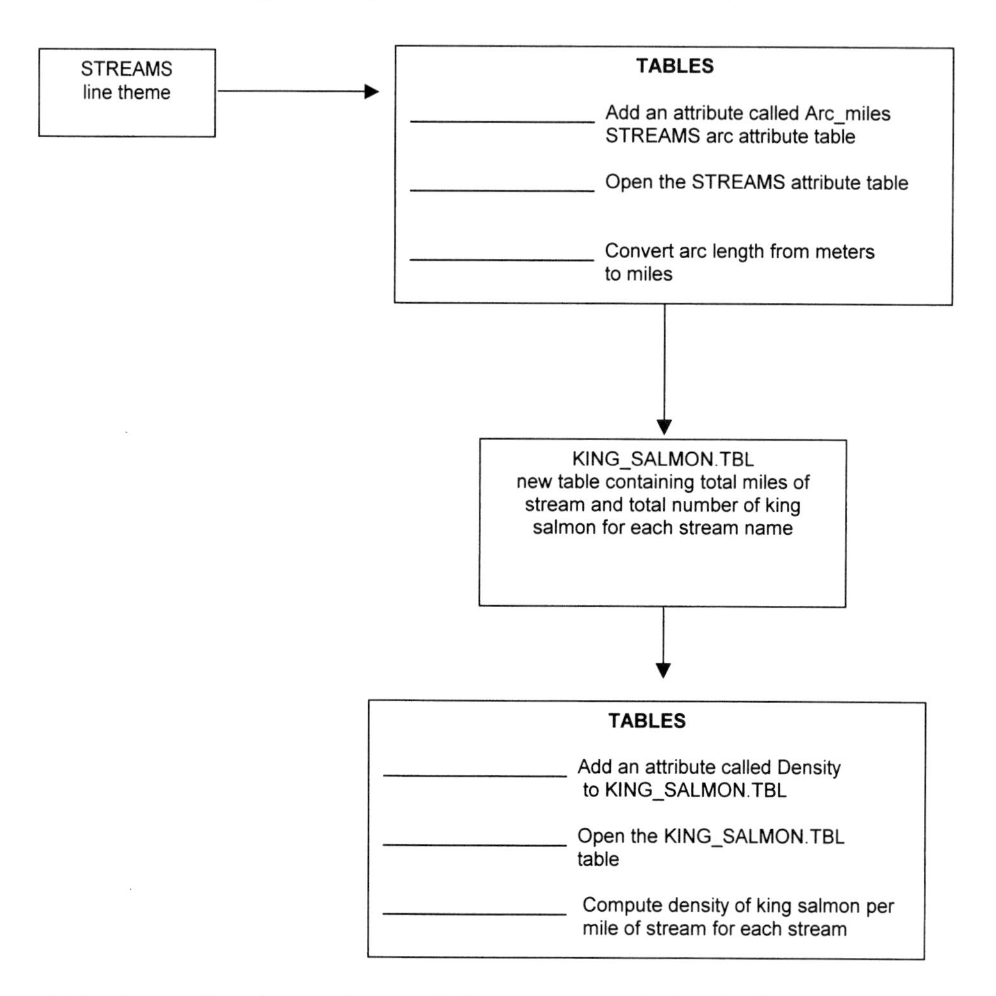

4) You have a line theme where each line represents the border between two vegetation types as follows:

Vegetation Arc Attribute Table

Fnode#	Tnode#	Lpoly#	Rpoly#	Length	Arc#	Arc-ID
		1	2	7409.314	1	2
		3	2	4061.212	2	2
		1	3	231.273	3	3
		1	4	8.255	4	1
		3	1	236.307	5	4
		3	4	4057.586	6	1
		1	4	7073.343	7	1

Vegetation Polygon Attribute Table

Area	Perim.	Vegetation#	Vegetation-ID	Veg_Class	Veg_Name
		1	0	0	
		2	1	3	Pin Oak
		3	2	2	Sweet Gum
		4	3	1	Red Maple

From searching the polygon attribute table, you know there is one stand of Sweet Gum and one stand of Pin Oak in the theme. What is the length of the border between the Sweet Gum and Pin Oak stand? Fill in the appropriate **TABLES** tools to solve the problem:

TABLES	
____________________	Get the arc attribute table
____________________	Find the arc(s) that make up the border between Sweet Gum and Pin Oak stands
____________________	Determine the total length of the border

5) You have a line theme of roads and a point theme of auto accidents. Each arc has a highway code attribute. You want to determine all the accidents that occurred on the Parks Highway (highway code= 2) as a new point theme.

Fill in the following flowchart to solve your GIS problem:

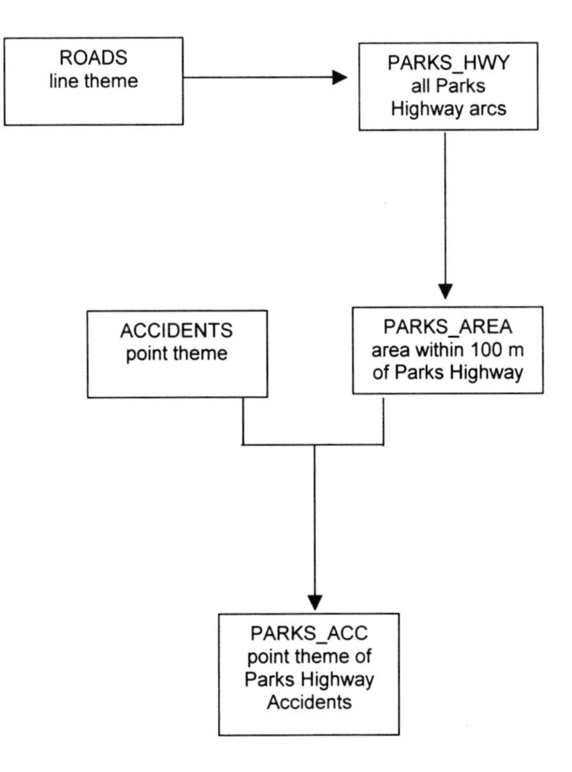

6) You have a polygon theme of parcels and a line theme representing a utility right of way. Each parcel polygon has the owner's phone number and address stored in the polygon attribute table. You want to generate a text file of owner phone numbers and addresses for all parcels within the right of way.

Fill in the following flowchart to solve your GIS problem:

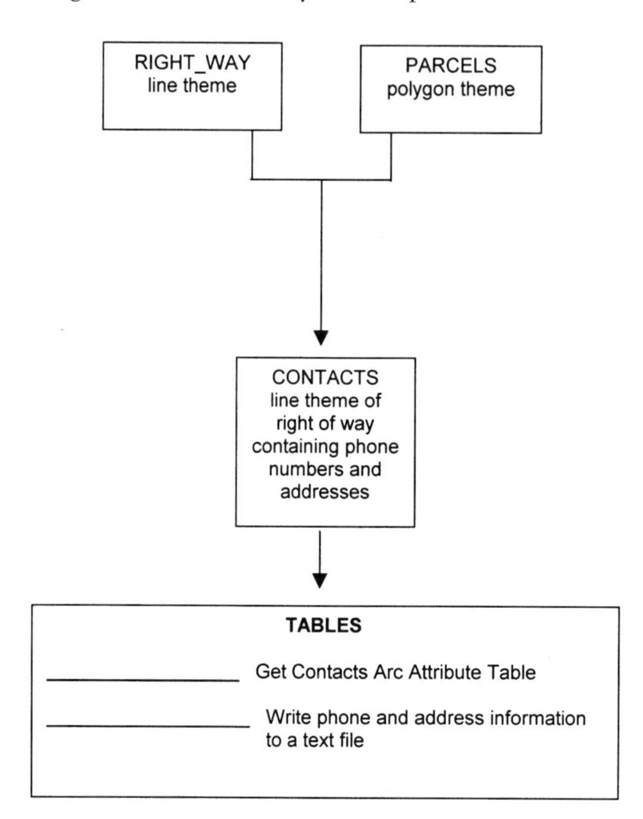

7) You have a line theme of gas pipelines and a polygon theme of ownership. You want to produce a table showing for each ownership, the total length of pipeline by pipe diameter class and by pipe age class.

Fill in the following flowchart to solve your GIS problem:

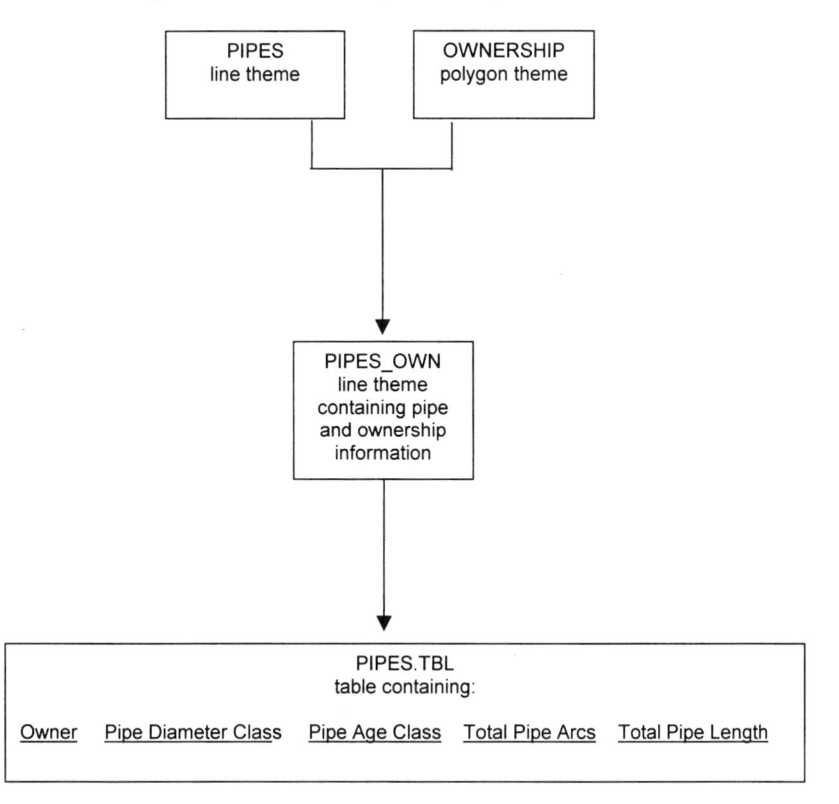

8) You have a line theme of contour lines with an attribute of elevation. Elevations are recorded at contour intervals of 10 meters, ranging from 10 to 980 meters. You want to assign a dashed line to minor contour elevations that are tens of meters (for example: 10,20,30,40, etc.) and a solid line to major contour elevations that are hundreds of meters (for example 100,200,300).

Fill in the following flowchart to solve your GIS problem:

TABLES

____________________ Add a column called *Countour_Class* to the arc attribute table

____________________ Get the arc attribute table

____________________ Get the major elevation contour records

____________________ Assign a *Countour_Class* value of 2

____________________ Get the minor elevation contour records

____________________ Assign a *Countour_Class* value of 1

9) You have a line theme of streets and a point theme of fire hydrant locations. You want to find all areas where the distance between fire hydrants is greater than 1 km and the hydrants are within 10 meters of a street.

Fill in the following flowchart to solve the problem:

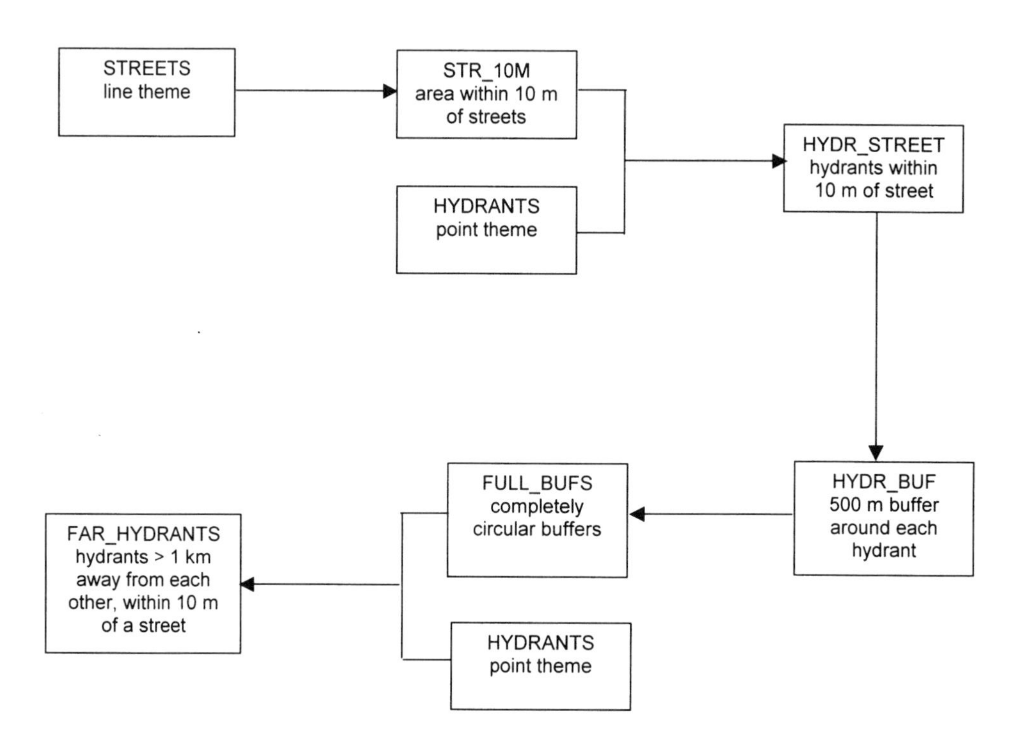

10) A grizzly bear biologist wants you to use a line theme of streams and a point theme of bears to produce the following three tables. The GIS coordinate system is in meters. Your job is to figure out the total area within 1 km of each stream and the total number of bears within these 1 km areas

Stream Name	**Number of Bears** (within 1 km of stream)	**Area (ha)** (within 1 km of stream)
Clear Creek		

Stream Name	**Number of Bears** (within 1 km of stream)	**Area (ha)** (within 1 km of stream)
Moose River		

Stream Name	**Number of Bears** (within 1 km of stream)	**Area (ha)** (within 1 km of stream)
Rapid River		

Fill in the appropriate tools to solve the problem on the next page:

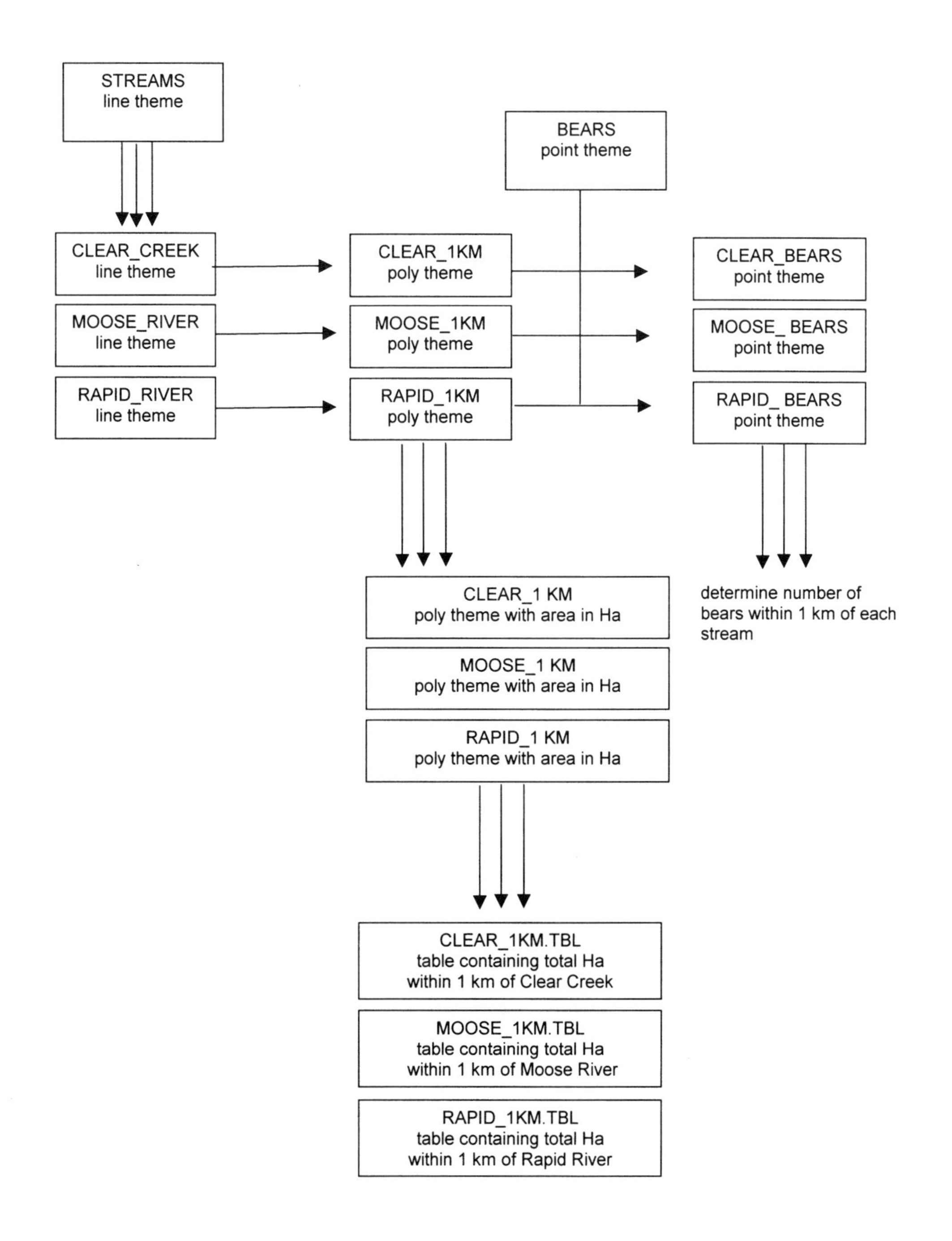
STREAMS
line theme
BEARS
point theme
CLEAR_CREEK
line theme
CLEAR_1KM
poly theme
CLEAR_BEARS
point theme
MOOSE_RIVER
line theme
MOOSE_1KM
poly theme
MOOSE_ BEARS
point theme
RAPID_RIVER
line theme
RAPID_1KM
poly theme
RAPID_ BEARS
point theme
CLEAR_1 KM
poly theme with area in Ha
determine number of
bears within 1 km of each
stream
MOOSE_1 KM
poly theme with area in Ha
RAPID_1 KM
poly theme with area in Ha
CLEAR_1KM.TBL
table containing total Ha
within 1 km of Clear Creek
MOOSE_1KM.TBL
table containing total Ha
within 1 km of Moose River
RAPID_1KM.TBL
table containing total Ha
within 1 km of Rapid River

Chapter 5

Network Analysis

INTRODUCTION

Network analysis is a special type of line analysis involving a set of interconnected lines. Typical networks include themes such as roads, streams, hiking trails, and pipelines. Network Analysis can be used to answer at least four types of questions:

1) **Address Geocoding.** Address geocoding is the process of taking addresses and estimating their locations in your GIS coordinate system. This is done by relating an address database to your GIS theme. Example applications include displaying a house address on a GIS street view for delivery of some product, generating driving directions to a given address, or displaying customer locations in a GIS view from a list of customer addresses.
2) **Optimal Routing.** Optimal routing is the process of delineating the best route to get from one location to one or more locations. The "best route" could be the shortest, the quickest, or the most esthetic, depending on the GIS user's preference for defining "best." Example applications include determining the quickest way to get from a fire station to a fire location, determining the shortest route to divert water lines given a closed junction, or determining the most economic delivery route to several stops.
3) **Finding Closest Facilities.** This is a special type of optimal routing problem where you are trying to find the closest points to a given location. Typically the points are called *facilities* and the given location is called an *event* location. Example applications include determining which two fire stations would have the best response time to a reported fire event, determining the best ambulance station to respond to an accident report event, or finding ten houses for sale that are closest to a day care center.
4) **Resource Allocation.** Resource allocation is the allocation of resources from supply centers to customers on a network. Example applications include a sand and gravel center supplying road sanding after a snowstorm, a school supplying seats to students, a well supplying water to an irrigation network, etc. In resource allocation, typically resources are allocated across the network until either resources are exhausted, or demand across the network is satisfied.

ADDRESS GEOCODING

Address geocoding requires a table of addresses and theme that contains attributes that can be used to match to the table of addresses. Address geocoding most often involves a line theme like streets, but can also be used with point (*e.g.*, light pole IDs), or polygon (*e.g.*, parcels) themes.

Building addresses are most commonly used with street themes in address geocoding. Typical street attributes might include ***prefix*** (e.g., East, North, Old), ***street_name*** (e.g., 5th, Maple, Broadway), ***type*** (e.g., street, avenue, way, boulevard) and ***suffix*** (e.g., NW).

Each street arc typically contains address information such as the beginning and ending addresses on the left and right side of the street. How does the GIS know which side of the street is right and which side is left? Each arc has a beginning node and an ending node. Topological left and right sides are inferred from these nodes. For example:

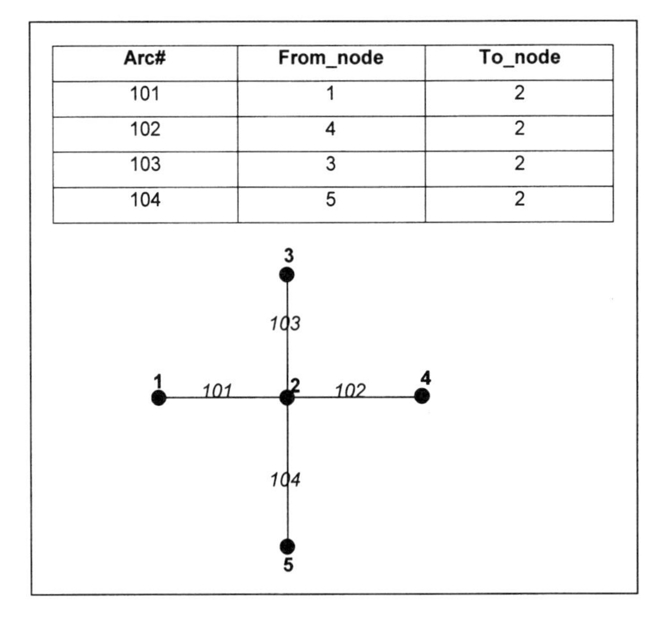

Arc#	From_node	To_node
101	1	2
102	4	2
103	3	2
104	5	2

The to- and from- nodes allow the GIS to delineate the following right and left sides of the streets:

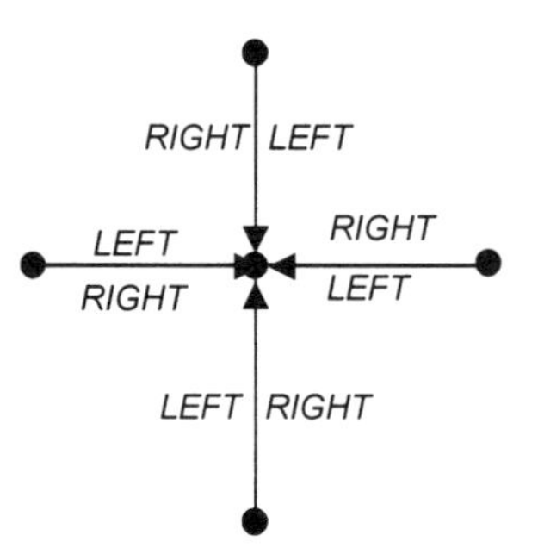

The GIS typically would have address information for the beginning and ending of each arc. For example,

Arc#	Length	Left_From	Left_To	Right_From	Right_To	Prefix	Name	Type	Suffix
101	2700	100	1800	101	1799	West	Main	Street	
102	2200	3201	1801	3200	1800	East	Main	Street	
103	440	300	348	301	349		Third	Ave.	North
104	520	399	351	400	350		Third	Ave.	South

Imagine that you got a report of a broken gas line in a building at 2906 East Main Street. The GIS can estimate the location, first by determining which arc contains the Prefix/Name/Type/Suffix. Once the GIS determines that this is Arc 102, the 2906 address can be estimated using linear interpolation. Since it is an even-number address, it is on the right side of the arc. The address ranges from 3200 to 1800, (a range of 1400) along the right side of the arc. The length of the arc is 2200 meters. The address of 2906 can be geocoded as

[(Max_address - address.)/ range] * arc length

= [(3200 – 2906) /1400] * 2200 = 462 meters along the arc

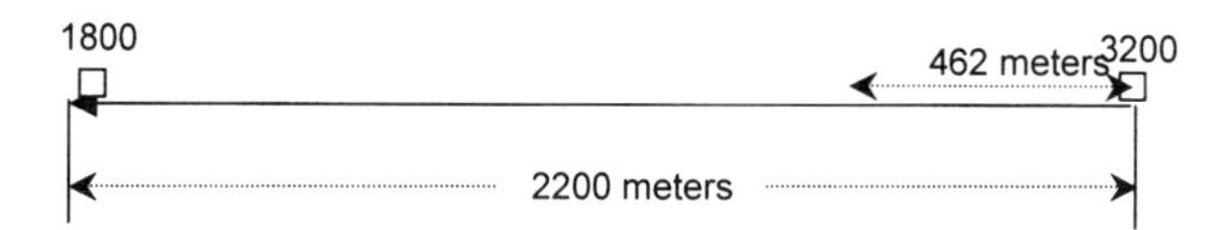

In the previous example, the address perfectly matched the address attributes of an arc. What would happen if the address did not perfectly match? For example, it might be reported as 2906 E. Main Street, or 2906 East Main, or 2906 E. Main St. Most GIS allow the user to specify parameters such as spelling sensitivitiy and minimum match score. A simple example might be, start with a perfect score of 100 and then deduct –25 points for address Prefix/Name/Type/Suffix that does not match any street address attributes. In the above example, 2906 E. Main Street would score 75, 2906 East Main would score 75, and 2906 E. Main St. would score 50. In real GIS applications, the matching score rules are much more sophisticated. The GIS user can specify the minimum acceptable matching score. The higher the minimum matching score, the higher the risk of failure to match all addresses entered for geocoding. The lower the minimum matching score, the higher the risk for some incorrect address geocoding.

OPTIMAL ROUTING

Optimal routing is the process of delineating the best route (or path) to get from one location to one or more locations. It is not usually feasible to test all possible paths that exist in a network. Instead, a pathfinding algorithm is used. Perhaps the most commonly used pathfinding algorithm is the Dijkstra algorithm, first published by E. W. Dijkstra. This algorithm is explained below with a simple example.

Imagine that we have the following network of hiking trails. We want to find the quickest path to get from node 1 (trailhead) to node 7 (tent site). The time to hike each trail varies depending upon trail conditions. Trails 3,5,and 6 are relatively quick hikes.

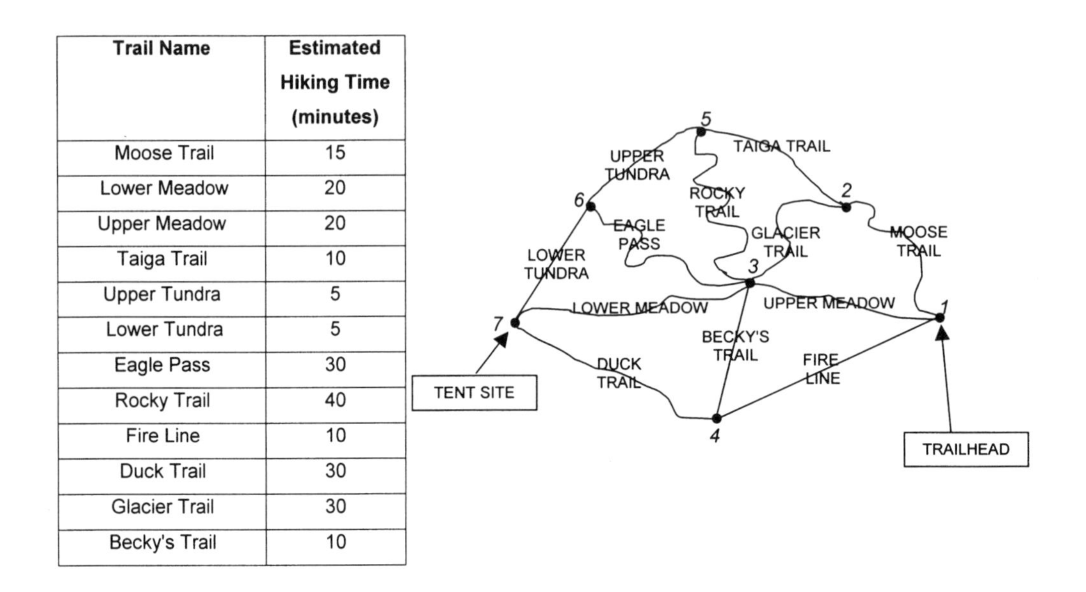

Trail Name	Estimated Hiking Time (minutes)
Moose Trail	15
Lower Meadow	20
Upper Meadow	20
Taiga Trail	10
Upper Tundra	5
Lower Tundra	5
Eagle Pass	30
Rocky Trail	40
Fire Line	10
Duck Trail	30
Glacier Trail	30
Becky's Trail	10

We build two tables, one of nodes that have already been processed, and one of adjacent nodes to process. We start at the trailhead, node#1:

Processed Nodes

Node	Cumulative Cost	Previous Node
1	0	none

Adjacent Nodes

Node	Cumulative Cost	Previous Node
2	15	1
4	10*	1

We then pick the adjacent node with the least cumulative cost (hiking to node#4 will take 10 minutes), and add that node to the processed nodes list.

Processed Nodes

Node	Cumulative Cost	Previous Node
1	0	none
4	10	1

Adjacent Nodes

Node	Cumulative Cost	Previous Node
2	15	1

. . . and scan the nodes adjacent to your latest processed node (4) . . . and add them to the Adjacent Nodes list. There are 3 nodes adjacent to node#4: 1, 3, and 7. Node#1 is already in the processed nodes list, so it is excluded from the adjacent nodes list. The cumulative cost for node#3 is the cost of hiking to node#4 (10 minutes), plus the cost of hiking from node#4 to node#3 (10 minutes). The cumulative cost to get to node#7 is 10 minutes + 30 minutes.

Adjacent Nodes

Node	Cumulative Cost	Previous Node
2	15*	1
3	20	4
7	40	4

We then pick the adjacent node with the least cumulative cost (node#2 with a cost of 15), and add it to the processed nodes list.

Processed Nodes

Node	Cumulative Cost	Previous Node
1	0	none
4	10	1
2	15	1

We then scan the nodes adjacent to your latest processed node (2) . . . and add them to the Adjacent Nodes list. The nodes adjacent to node#2 are nodes 1,3, and 5. Node#1 is already in the processed nodes list, so it is excluded from the adjacent nodes list. The cumulative cost for node 3 is the cost to hike to node#2 (15 minutes) plus the cost to hike from node#2 to node#3 (30 minutes). The cumulative cost to get to node#5 is the time to hike to node#2 (15 minutes) plus the time to hike from node#2 to node#5 (10 minutes).

Adjacent Nodes

Node	Cumulative Cost	Previous Node
3	20*	4
7	40	4
3	45	2
5	25	2

We then pick the adjacent node with the least cumulative cost, and add it to the processed nodes list.

Processed Nodes

Node	Cumulative Cost	Previous Node
1	0	none
4	10	1
2	15	1
3	20	4

Next, scan the nodes adjacent to your latest processed node (3) . . . and add them to the Adjacent Nodes list. The nodes adjacent to node#3 include 1,2,4,5,6,and 7. Nodes

1,2,3, and 4 are already in the processed nodes list and are therefore excluded from the adjacent nodes list. The cumulative cost to hike to node#3 is 20 minutes. To go all the way to node#5 will take 20 + 40 = 60 minutes. To go to node# 3 and then node#6 will take 20 + 30 = 50 minutes. And to hike to node#3 and then node#7 will take 20 + 20 = 40 minutes.

Adjacent Nodes

Node	Cumulative Cost	Previous Node
5	25*	2
7	40	4
5	60	3
6	50	3
7	40	3

We then pick the adjacent node with the least cumulative cost, and add it to the processed nodes list.

Processed Nodes

Node	Cumulative Cost	Previous Node
1	0	none
4	10	1
2	15	1
3	20	4
5	25	2

Scan the nodes adjacent to your latest processed node (5) . . . and add them to the Adjacent Nodes list. The nodes adjacent to node#5 include 2,3, and 6, Nodes 2 and 3 are already in the processed nodes list and therefore are excluded from the adjacent nodes list. The cumulative cost to hike to node 6 is 25 minutes to hike to node#5 and then 5 minutes to hike to node#6, totaling 30 minutes.

Adjacent Nodes

Node	Cumulative Cost	Previous Node
7	40	4
6	50	3
7	40	3
6	30*	5

We then pick the adjacent node with the least cumulative cost, and add it to the processed nodes list.

Processed Nodes

Node	Cumulative Cost	Previous Node
1	0	none
4	10	1
2	15	1
3	20	4
5	25	2
6	30	5

Scan the nodes adjacent to your latest processed node (6) . . . and add them to the Adjacent Nodes list. The nodes adjacent to node 6 include nodes 3,5, and 7. Nodes 3 and 5 are already in the processed nodes list and are therefore excluded from the adjacent nodes list. The cumulative cost to hike to node#7 through node#6 is 5 + 30 = 35 minutes.

Adjacent Nodes

Node	Cumulative Cost	Previous Node
7	40	4
7	40	3
7	35*	6

We then pick the adjacent node with the least cumulative cost and add that to the processed nodes list.

Processed Nodes

Node	Cumulative Cost	Previous Node
1	0	none
4	10	1
2	15	1
3	20	4
5	25	2
6	30	5
7	35	6

All seven nodes are now in the processed nodes list, therefore the processing stops. The quickest route to get to node#7 will take 35 minutes and from the processed nodes list, it is

Node#7 ← Node#6 ← Node#5 ← Node#2 ← Node#1

We can ask the GIS to display this optimal route in our hiking trail network view.

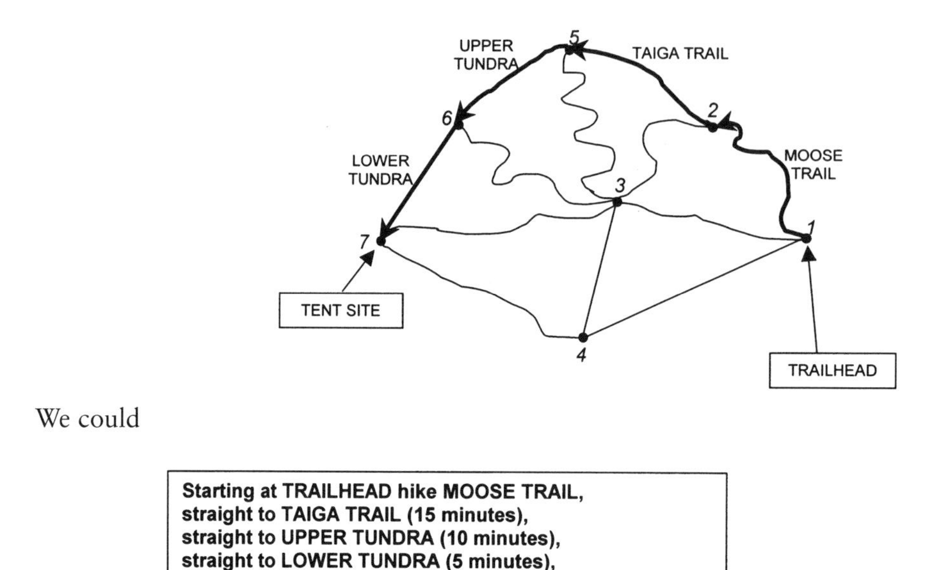

We could

Starting at TRAILHEAD hike MOOSE TRAIL, straight to TAIGA TRAIL (15 minutes), straight to UPPER TUNDRA (10 minutes), straight to LOWER TUNDRA (5 minutes), ending at TENT SITE (5 minutes).

Once the nodes have been processed, all the optimal routes to node#1 from any node are solved. However, there may also be more than one optimal route that is not reflected in the analysis. For example, from the processed nodes list, the optimal route to hike from node#3 is **Node#3** ← **Node#4** ← **Node#1** which would take 20 minutes. The direct **Node#1** → **Node#3** route would also take 20 minutes.

In this simple example, there are relatively few possible routes, so as a check you could try all possible routes. This is generally not possible with complex networks typical of real-life applications.

Possible Routes from Trailhead to Tent Site

Route	Estimated Hiking Time (minutes)
1→2→5→6→7 =	35
1→3→7	40
1→4→3→7	40
1→4→7	40
1→4→3→6→7	55
1→3→6→7	55
1→3→4→7	60
1→2→3→7	65
1→3→5→6→7	70
1→3→2→5→6→7	70
1→4→3→5→6→7	70
1→2→3→6→7	80
1→2→5→6→3→7	80
1→2→5→3→7	85
1→2→3→5→6→7	95

FINDING CLOSEST FACILITIES

This is a special type of optimal routing problem where you are trying to find the closest points to a given location. Typically the points are called *facilities* and the given location is called an *event* location. Example applications include determining which two fire stations would have the best response time to a reported fire event, determining the best ambulance station to respond to an accident report event, or finding ten houses for sale that are closest to a day care center.

As example, imagine that a fire is reported 1610 East Willow Street and we want to determine the 2 fire stations that would have the best response time to that event location. In this simple example, we have a streets theme with "slow" streets that take 3 minutes to travel, and "faster" streets that take 1 minute to travel.

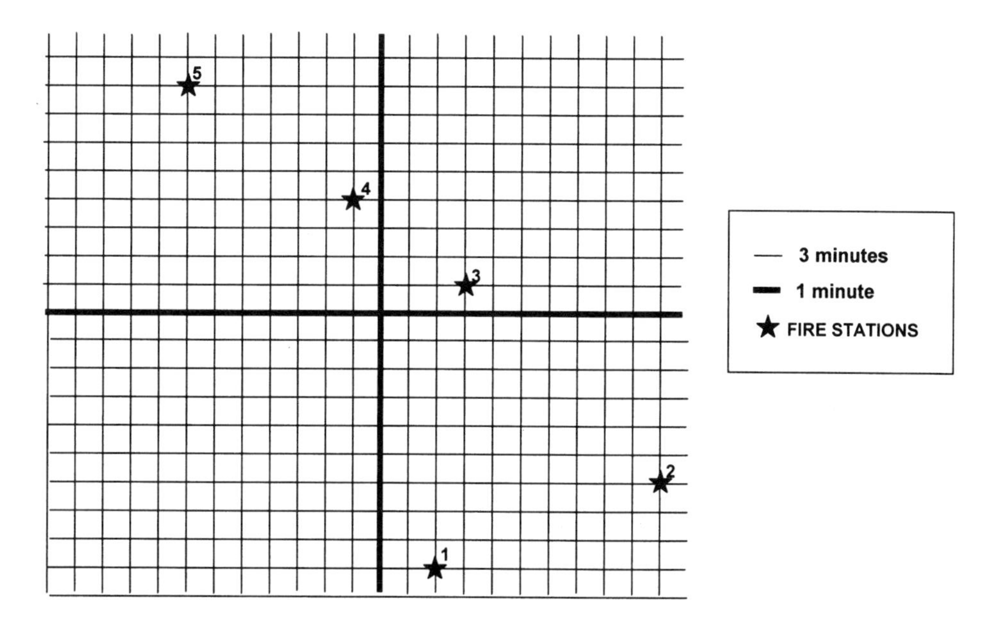

When a fire address is reported, it is geocoded to a street location and then the GIS can compute the optimal path from each fire station to the fire location.

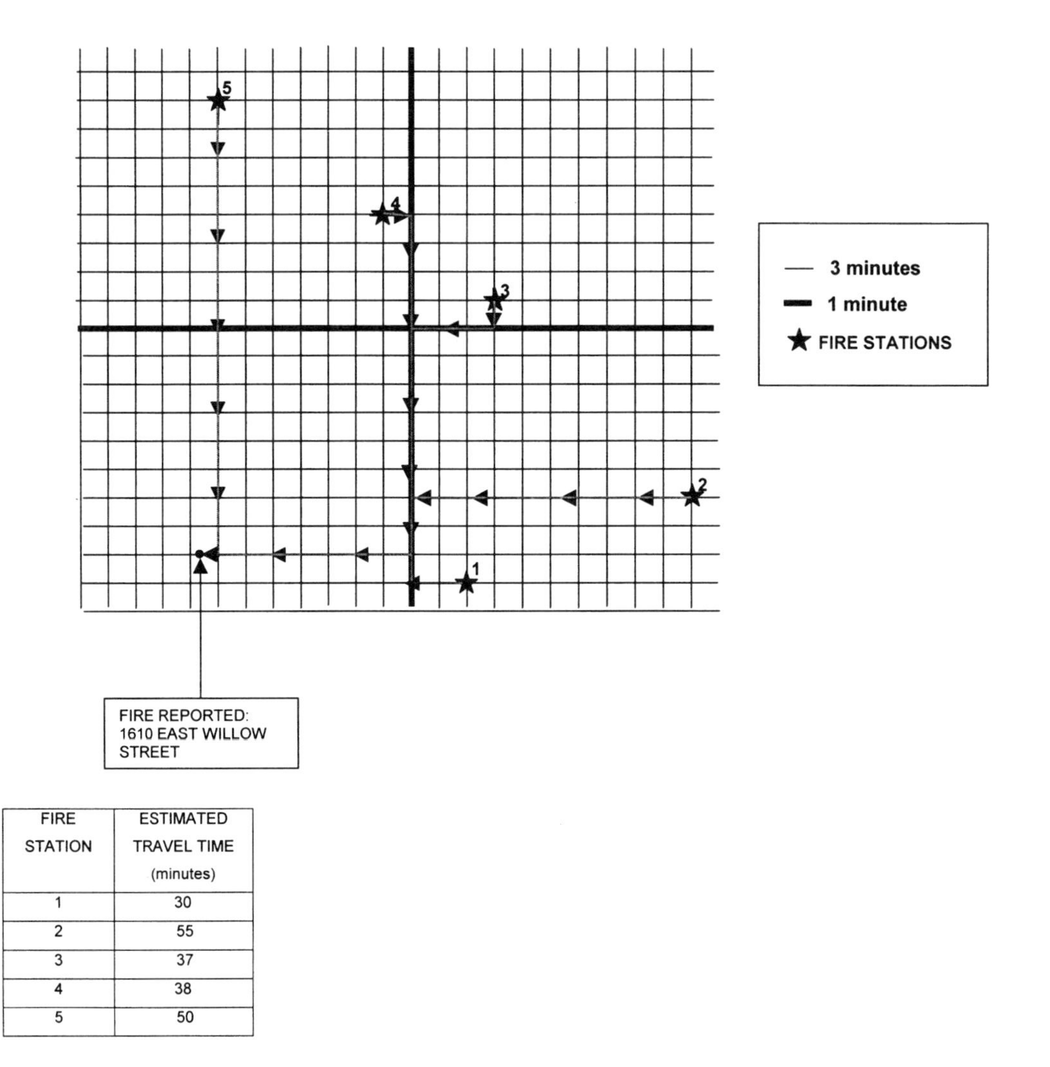

FIRE STATION	ESTIMATED TRAVEL TIME (minutes)
1	30
2	55
3	37
4	38
5	50

Typically the location of facilities are fixed and exist as a GIS point theme. Event locations can be entered into the network in several ways. In dispatching, typically someone calls in an address of the event and it is located by address geocoding. In other applications, the event location may be an existing GIS point theme. For example, a realtor may have a point theme of day care centers, and a point theme of houses currently for sale. She then selects one day care center from a neighborhood a client is interested in, and then runs network analysis to determine the ten closest houses that are for sale. Event locations can also be entered by graphically creating an event point theme in the GIS view.

RESOURCE ALLOCATION

Resource allocation or service area delineation is the allocation of some resource from supply centers to users in the network. Examples include a sand and gravel center supplying plowing and sanding of roads after a snowstorm, a school supplying seats to students, a well supplying water to an irrigation network, etc. In resource allocation, typically resources are allocated across the network until either resources are exhausted, or demand across the network is satisfied.

Imagine that a new recycling center is being proposed. Your job is to analyze two alternative locations. What is the area within a 1 km drive of each proposed location?

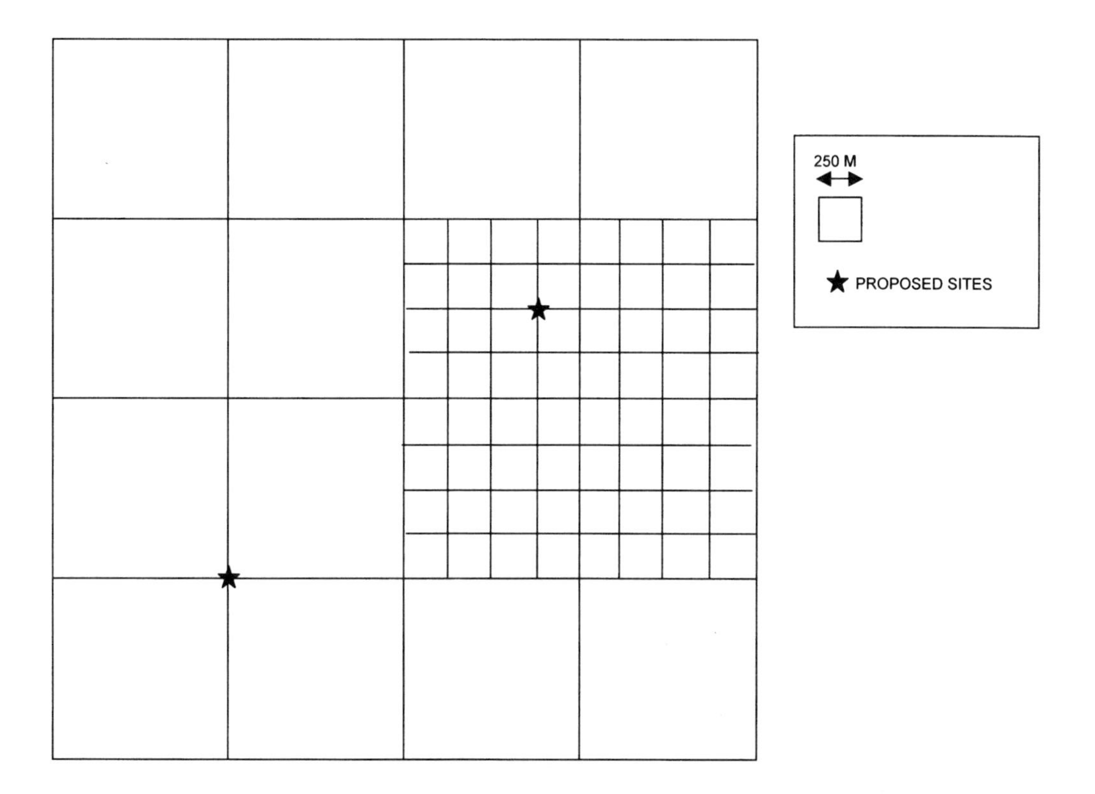

The first step is to start at each service center and go out along the network until the limit of 1km is met. This is the *service network* of each proposed recycle center.

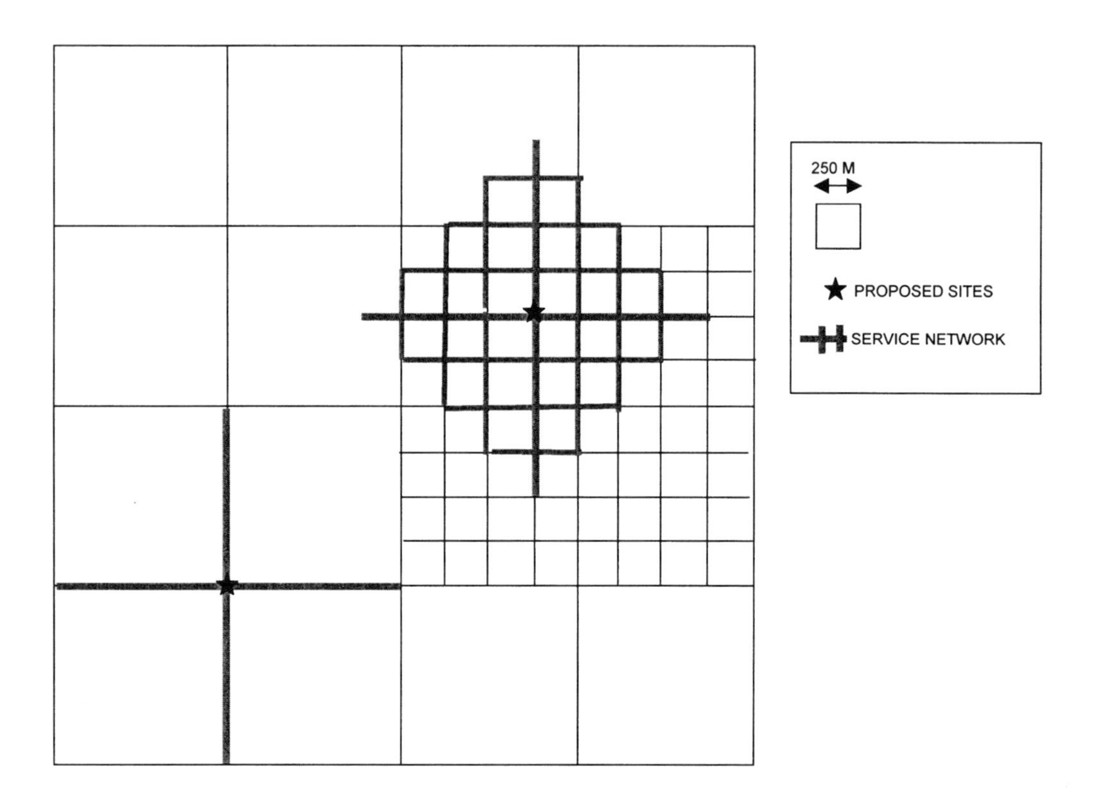

In some applications, you might want a polygon theme of *service areas* corresponding to your service networks. An example might be where you want to select all houses from a point theme that are within each service area, so that you can generate mailing labels from each house address. Service areas or service polygons are delineated by connecting the outside nodes of each service network as follows:

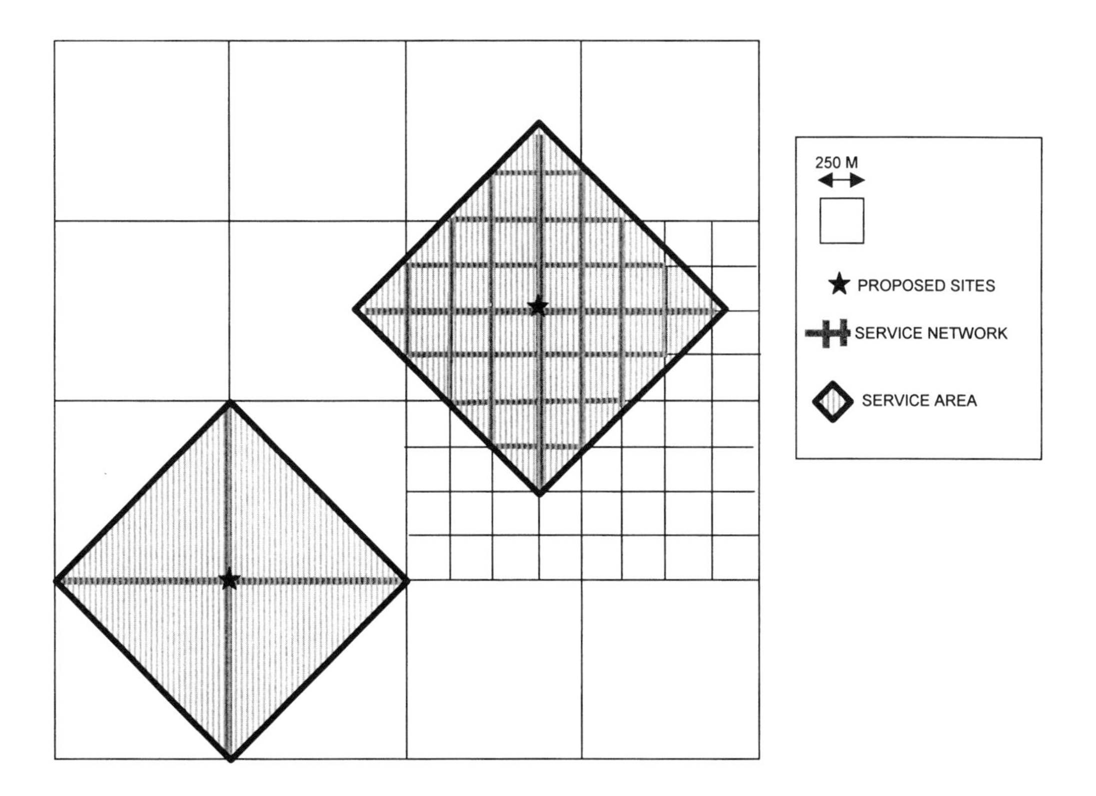

In some applications, the resource centers may have a supply constraint. For example, schools can only hold a maximum number of students, a reservoir can only hold some maximum quantity of water, and so on. In our simple recycling center example, imagine that each center has the maximum capacity of processing 6,000 units per week. And assume that each household generates 10 units of recyclable waste per week. If our streets theme has an attribute containing the number of households within each arc, we could model this capacity constraint. In this simple example, we will assume 50 households per 250 m of arc in neighborhood A, and 10 households per 250 m of arc in neighborhood B. We could allocate along the network up to 1 km from each center, until the center capacity has been reached.

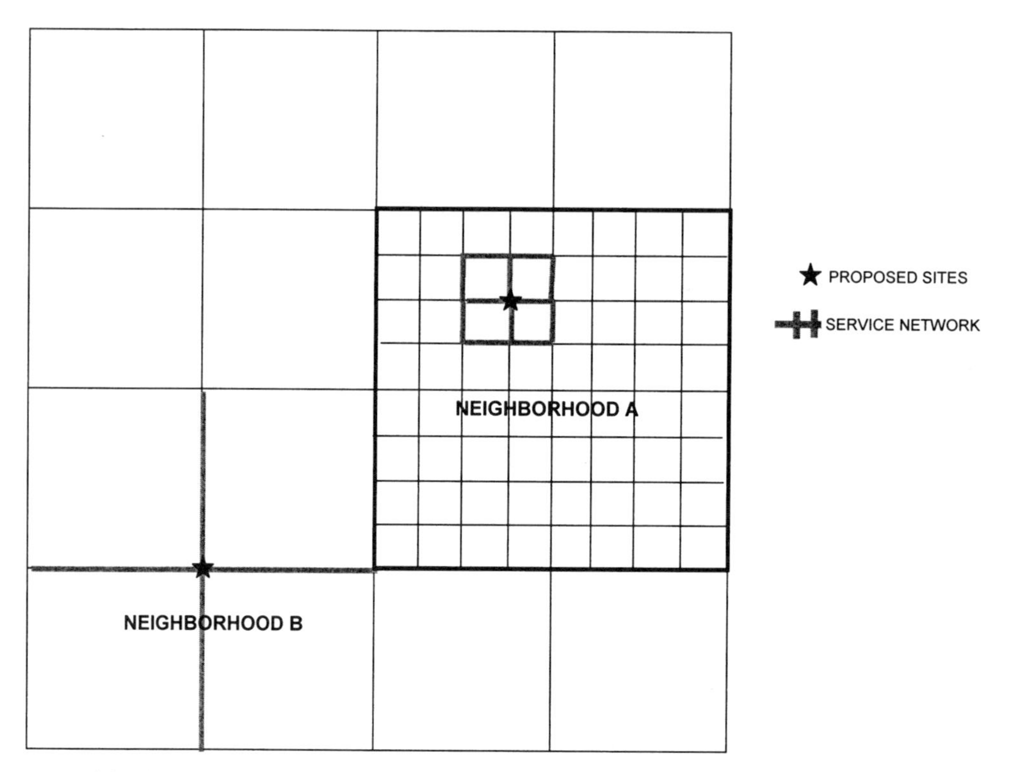

Neighborhood A is more densely populated and the center reaches capacity at 600 households (or 600 X 10 = 6,000 units per week). Neighborhood B is less densely populated and reaches the 1 km distance limit while 160 households supply 1,600 units per week and therefore the capacity is not met for allocation of households to this facility.

TURN TABLES IN NETWORKS

So far we have dealt with relatively simple networks. Complexities such as stop signs, U-turns, one-way streets, red lights, and so on, can easily be incorporated in networks. One way of incorporating such complexities is by using *turn tables*.

A turn table can be used to assign impedance values at nodes. Left turns are assigned positive angles, while right turns are assigned negative angles. For example, a stop sign or red light could be modeled as follows:

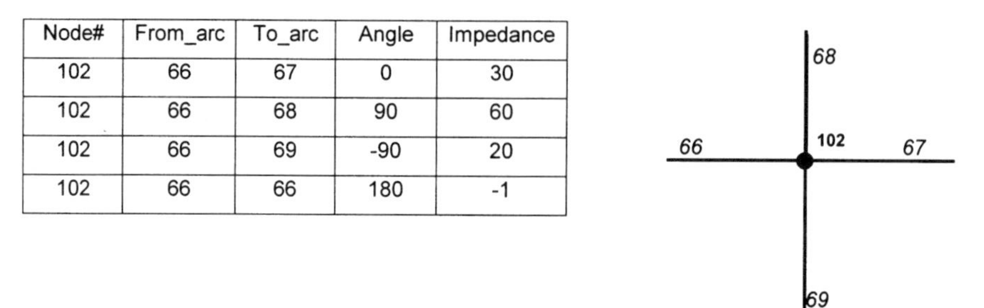

Node#	From_arc	To_arc	Angle	Impedance
102	66	67	0	30
102	66	68	90	60
102	66	69	-90	20
102	66	66	180	-1

In the above example, to go straight from arc 66 to arc 67 would take 30 seconds. A left turn from arc 66 to 68 would take 60 seconds while a right turn would take 20 seconds. A U-turn on arc 66 is not allowed and therefore is assigned a negative impedance value.

If a turn is not represented in a turntable, it will simply have no travel cost associated with it. Turntables can also be used to model impossible turns such as a freeway overpass at nodes 101 and 102.

Node#	From_arc	To_arc	Angle	Impedance	Turn
101	67	66	0	0	STRAIGHT
101	67	68	-90	-1	RIGHT
101	67	69	90	-1	LEFT
101	67	67	180	-1	U-TURN
101	66	67	0	-1	STRAIGHT
101	66	69	-90	-1	RIGHT
101	66	68	90	-1	LEFT
101	66	66	180	-1	U-TURN
102	78	79	0	0	STRAIGHT
102	78	70	-90	-1	RIGHT
102	78	69	90	-1	LEFT
102	78	78	180	-1	U-TURN
102	79	78	0	-1	STRAIGHT
102	79	69	-90	-1	RIGHT
102	79	70	90	-1	LEFT
102	79	79	180	-1	U-TURN

NETWORK ANALYSIS EXERCISES

1) You have a report of a fire at 1721 First Avenue West. Draw the location of that address on the following streets theme.

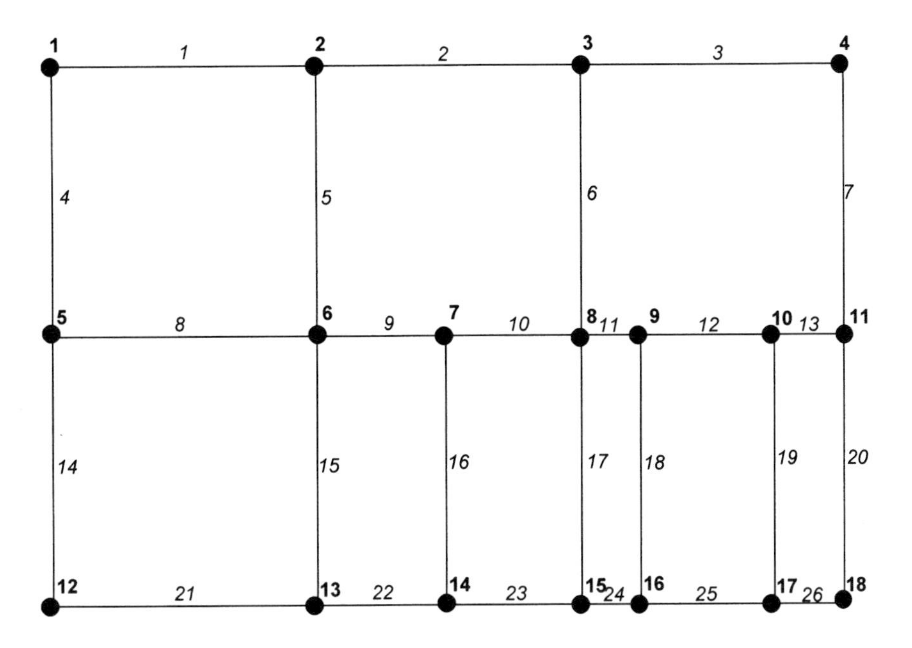

Arc#	Length	F_node	T_node	Left_from	Left_to	Right_from	Right_to	Name	Type	Suffix
1	400	1	2	200	300	201	299	Main	Street	
2	400	2	3	302	400	301	399	Main	Street	
3	400	3	4	402	500	401	499	Main	Street	
4	400	5	1	100	2	99	1	Third	Avenue	
5	400	6	2	100	2	99	1	Second	Avenue	West
6	400	8	3	100	2	99	1	First	Avenue	West
7	400	11	4	100	2	99	1	Center	Street	
8	400	6	5	1814	1998	1813	1997	First	Avenue	West
9	200	7	6	1744	1812	1743	1811	First	Avenue	West
10	200	8	7	1692	1742	1691	1741	First	Avenue	West
11	100	9	8	1612	1690	1611	1689	First	Avenue	West
12	200	10	11	1552	1610	1551	1609	First	Avenue	West
13	100	11	10	1500	1550	1549	1551	First	Avenue	West
14	400	12	5	200	102	199	101	Third	Avenue	West
15	400	13	6	200	102	199	101	Second	Avenue	West
16	400	14	7	692	742	691	741	Spruce	Street	
17	400	15	8	200	102	199	101	First	Avenue	West
18	400	16	9	552	610	551	609	Aspen	Street	
19	400	17	10	744	812	743	811	Birch	Street	
20	400	18	11	200	102	199	101	Center	Street	
21	400	12	13	1900	1950	1949	1951	Second	Avenue	West
22	200	13	14	1952	2010	1951	2009	Second	Avenue	West
23	200	14	15	2012	2090	2011	2089	Second	Avenue	West
24	100	15	16	2092	2142	2091	2141	Second	Avenue	West
25	200	16	17	2144	2212	2143	2111	Second	Avenue	West
26	100	17	18	2114	2298	2113	2297	Second	Avenue	West

2) Given the following network theme, determine the path that is the quickest to get from point A to point B.

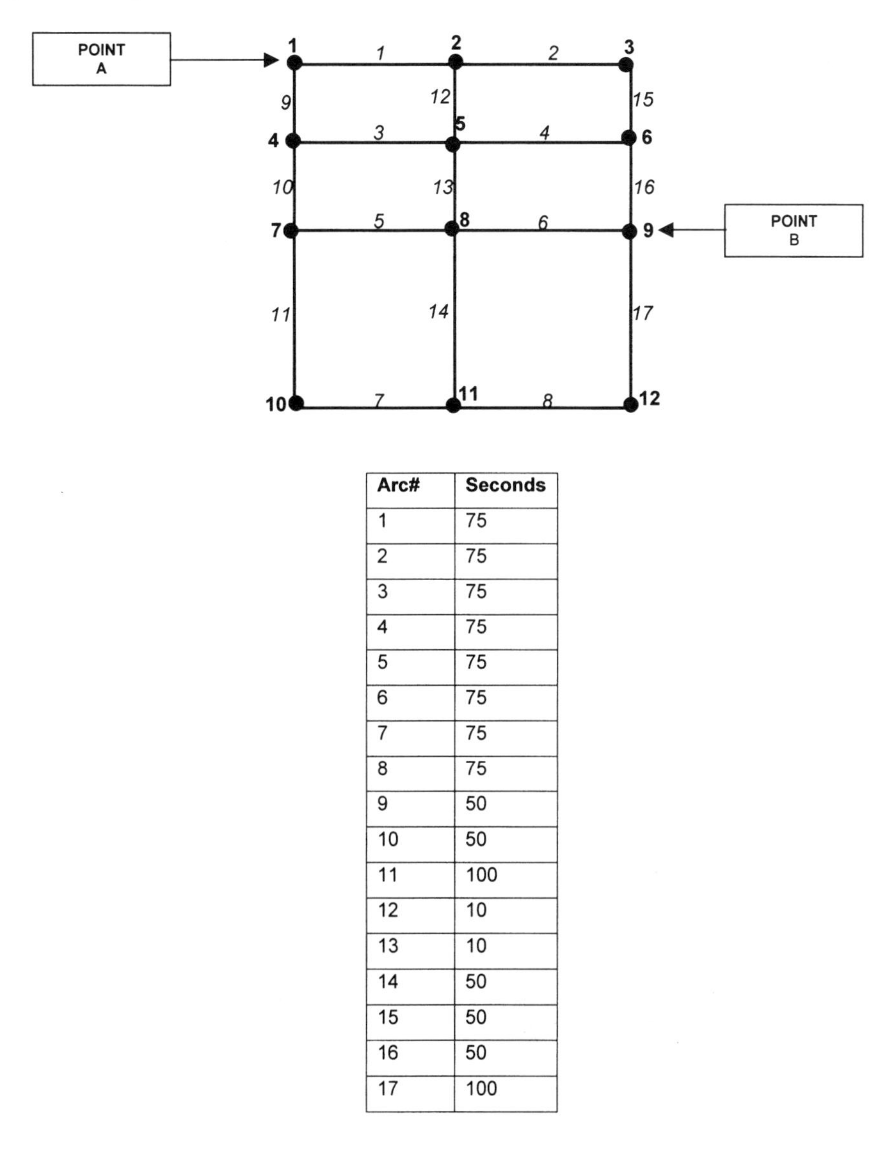

Arc#	Seconds
1	75
2	75
3	75
4	75
5	75
6	75
7	75
8	75
9	50
10	50
11	100
12	10
13	10
14	50
15	50
16	50
17	100

3) You are managing wilderness cabins on a network of backcountry ski trails. The average skiing time per 100 meters of trail for various slopes is as follows:

Slope Class	- Down Slope Time (seconds per 100 m)	+ Up Slope Time (seconds per 100 m)
Level	40	40
0-5 percent	30	60
5-15 percent	20	120
>15 percent	15	200

Arc#	Length_meters	From_node	To_node	Slope	Time_secs	Back_secs
1	1100	1	2	0		
2	1000	2	5	0		
3	1200	6	5	-15		
4	1800	7	6	+15		
5	1100	3	7	0		
6	1000	1	3	0		
7	1400	4	3	-18		
8	2200	4	1	-20		
9	1300	4	5	-10		
10	1200	4	6	-25		
11	2000	4	7	-12		

Delineate service network that represents all trails that are within a half hour from the Colorado Creek Cabin.

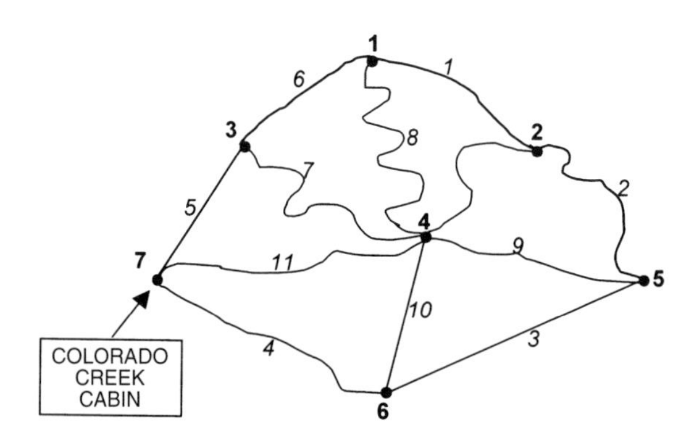

Chapter 6

Dynamic Segmentation

INTRODUCTION

Dynamic Segmentation is a special type of line analysis. Arcs are segmented dynamically and temporarily for the analysis. Dynamic segmentation is especially useful in situations when arcs are used for many different applications. For example, imagine that you have an arc from a streets theme:

The street maintenance department wants to segment the street based on pavement type.

The police department wants to segment the street based on speed limits.

And the fire department wants to segment the street based on hydrant locations.

You can see that the arc becomes quite segmented and complex with these three applications . . .

An alternative is to maintain the original streets theme and use dynamic segmentation for each application. Dynamic segmentation has another advantage: it does ***not*** require X,Y coordinates of events that are used to segment an arc. You can use events such as accidents with distance from an intersection, interpretive stops with highway mile markers, and so on. This type of linear data is called route-measure formatted data.

ROUTES AND SECTIONS

A *route* is a collection of arc segments or *sections* which have attributes. A route does not necessarily have to start or end at a node. The following might be an example of a bus route.

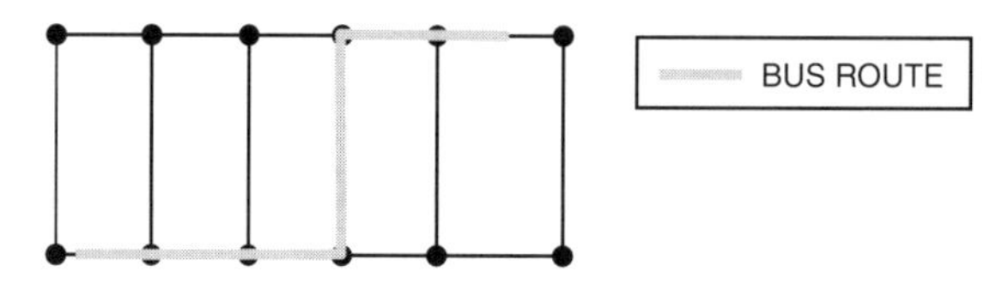

And a collection of routes is called a route system. The following is an example of a route system composed of three bus routes.

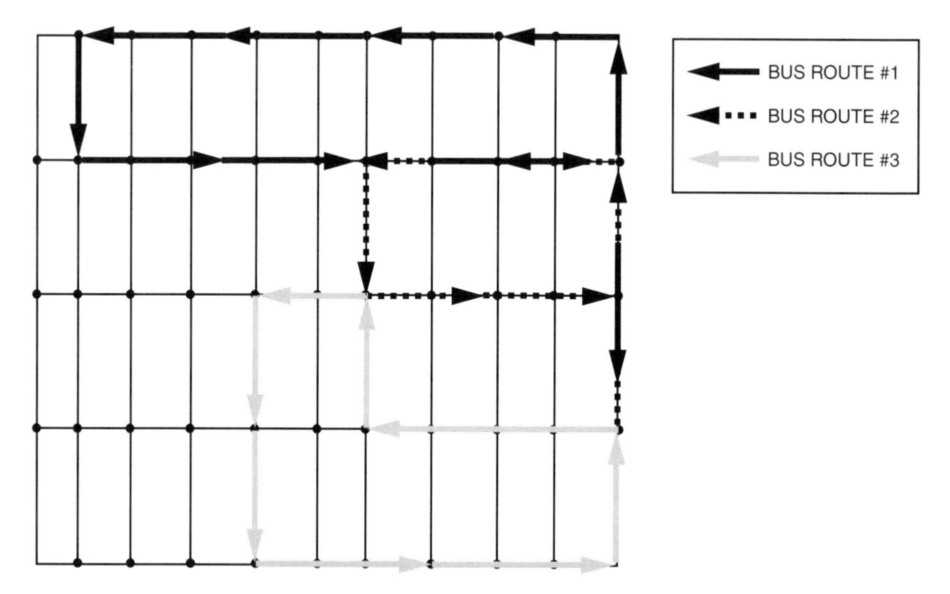

Each route is composed of one or more sections. For example, assume you have a road route 19 that has three sections: dirt from mile 0 to mile 1.5, gravel from mile 1.5 to 3.5, and paved from mile 3.5 to 10. This could be represented as a route with the following section attributes:

ARC #5 ARC #6

Route Attribute Table

Route#	Route-ID
1	19

Section Attribute Table

Arc#	Routelink#	F-Meas	T-Meas	F-Pos	T-Pos	Section#	Surface
5	1	0	1.5	0	33..33	1	Dirt
5	1	1.5	3.5	33.33	77.77	2	Gravel
5	1	3.5	4.5	77.77	100	3	Paved
6	1	4.5	10.0	0	100	4	Paved

where the **From-** and **To-** measure attributes are the miles at the start and end of each section. Corresponding to these attributes are the **From-** and **To-** positions which are the percentage of the arc. For example, Arc#5 is 4.5 miles long. Section#1 is from the start of Arc#5 (0) to 33.33 percent (1.5 / 4.5) of Arc#5 at mile 1.5.

EVENT TYPES

An event describes a location or a section along a route. Examples from transportation include traffic accident event locations, road surface types for each road section, or speed limit along each road section. There are three types of events, depending on how measures are used to locate the events. The three event types are point, continuous, and linear.

Point events are events that occur at a precise point location on a route. Examples include the location of fire hydrants on a street route or the location of stops along a nature trail route. An example event table for nature trail stops might be as follows:

Example Point Event Table

Route-ID	Dist_from_trailhead	Description
1	10	Welcome stop
1	245	Snag stop
1	320	Blind at pond
2	15	Welcome stop
2	125	Cave entrance

Continuous events represent the locations where a change occurs along a route. Examples include the location where the speed limit changes on a street route or the location where the class of white water changes along a river route. An example event table for a white-water river might be as follows:

Example Continuous Event Table

Route-ID	To_Distance	Rapids_class
1	10	3
1	16	2
1	22	1
1	35	2
1	50	1

From the start of the route until distance 10, the river is class 3. From distance 10 to distance 16, the river is class 2, and so on. Typically the last record of the continuous event table represents the end of the route. In the above example, the river is class 1 from distance 35 to distance 50.

Linear events represent the from- and to- locations along a route. Therefore they can be discontinuous. Examples include sections of a street route that were paved in 1999, or stretches of a stream route that contain spawning sockeye salmon. An example linear event table for a stream route might be as follows:

Example Linear Event Table

Route-ID	From	To	Salmon_Count	Stream
1	10	15	13	Lower Clear Creek
1	22	23	12	Upper Clear Creek
2	14	40	31	Upper Clear Creek
2	45	57	42	Red Riffle1
2	58	70	21	Red Riffle2

All events must have at least two attribute fields. First, there must be a key field that identifies which route each event belongs to. Second, there must be a measure field that identifies the location of the event along the route.

CREATING EVENTS

There are three basic ways to create events, 1) tabular entry, 2) creating point events from a point theme, and 3) creating linear or continuous events from a polygon theme.

Creating events by tabular entry can be done either by using the GIS tables' utilities or an external database management system. The basic process is to create an event table, define the appropriate columns, and then enter the correct data. There will be one record for each event. Once an event table has been created, the key field and measure field are used to link the event table to a route system. This is typically done using the **EVENTSOURCE** tool. The following example establishes the event source relationship between the *rapids.tbl* event table and the *river* route system.

RAPIDS.TBL

Route-ID	To_Distance	Rapids_class
1	10	3
1	16	2
1	22	1
1	35	2
1	80	1

RIVER SECTION TABLE:

Arc#	Routelink#	F-Meas	T-Meas	F-Pos	T-Pos
1	1	0	20	0	100
2	1	20	40	0	100
3	1	40	60	0	100
4	1	60	80	0	100
5	2	80	100	0	100
6	2	100	200	0	100

RIVER ROUTE TABLE:

Route#	Name
1	Upper Float
2	Lower Float

To establish the event source, you would specify the relationships linking the event table to the route system. For example, you would specify the event source type is *continuous*, the event key field is **Route-ID**, while the route key field is **Routelink#**, and the event to-measure field is **To_Distance**. Once the event source has been established, the GIS can compute the locations of the events. For example,

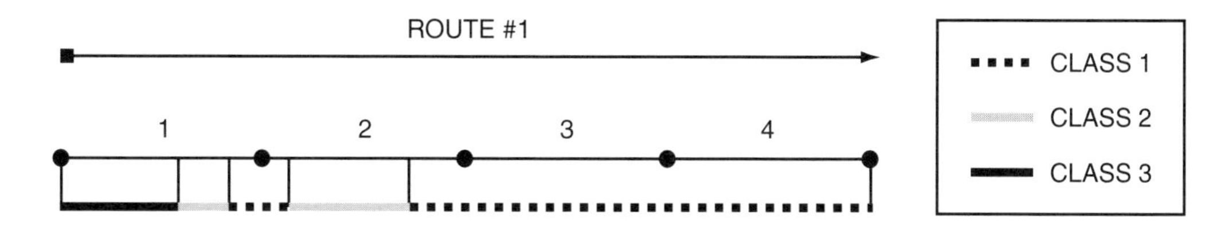

You can create point, linear, or continuous events by creating event tables. You can also create a point event table from a point theme. You could do this using the **ADDROUTEMEASURE** tool. You need to specify the name of your point theme, the name of your route theme and route system, the name of your output event table. You can also specify a search radius if your points are not perfectly aligned with your route system. As an example, imagine that we have a route system for irrigation and a point theme of gate locations and when they were last inspected.

Gate#	Date
1	06/21/2000
2	06/21/2000
3	07/21/1993
4	06/21/2000
5	06/21/2000
6	06/21/2000
7	08/28/1995

IRRIGATION SECTION TABLE:

Arc#	Routelink#	F-Meas	T-Meas	F-Pos	T-Pos
1	1	0	15	0	100
2	1	15	25	0	100
3	1	25	35	0	100
4	1	35	45	0	100
5	1	45	60	0	100
6	1	60	65	0	100
7	2	0	15	0	100
8	3	0	15	0	100
9	3	15	20	0	100
10	4	0	15	0	100
11	5	0	20	0	100
12	6	0	20	0	100

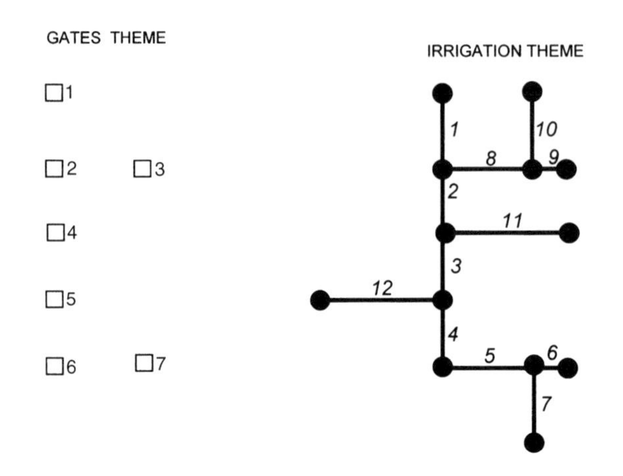

We enter the ADDROUTEMEASURE command and specify *GATES* as the name of our point theme, *IRRIGATION* as the name of our route theme and route system, and *cleaned_gates.tbl* as the name of our output event table. The output event table contains the measures in the route system for each gate. You could also link the date of cleaning from the original gates point attribute table.

cleaned_gates.tbl

Gate#	Route#	Measure	Date
1	1	0	06/21/2000
2	1	15	06/21/2000
4	1	25	06/21/2000
5	1	35	06/21/2000
6	1	45	06/21/2000
7	1	60	08/28/1995
7	2	0	08/28/1995
2	3	0	06/21/2000
3	3	15	07/21/1993
3	4	0	07/21/1993
4	5	0	06/21/2000
5	6	0	07/21/1993

You can also create event tables from a polygon theme. As an example, imagine that you have a line theme of hiking trails and a polygon theme of forest types. You want to create an event table delineating the forest types along the hiking trails. You could do this using the **POLYGONEVENTS** tool. You need to specify the name of your polygon theme, the name of your route theme and route system.

FOREST POLYGON ATTRIBUTE TABLE

Forest#	Type	Name
1	16	Quaking Aspen
2	17	Paper Birch
3	9	White Spruce

TRAILS SECTION TABLE:

Arc#	Routelink#	F-Meas	T-Meas	F-Pos	T-Pos
6	1	0	1000	0	100
7	1	1000	2000	0	100
8	1	2000	3000	0	100

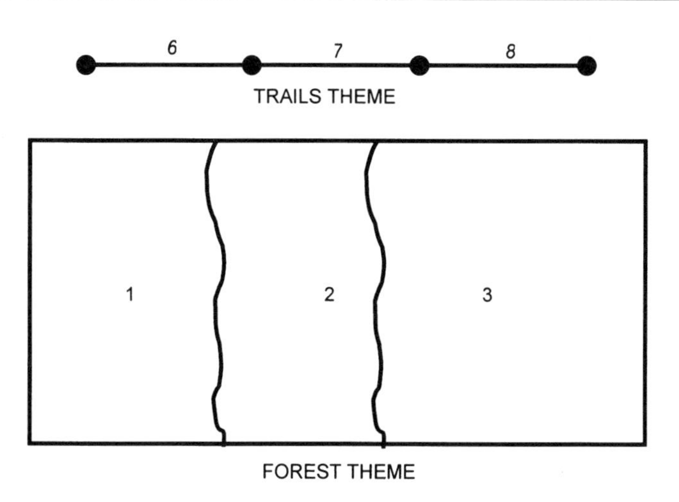

The resulting event table would contain all the information from the polygon theme for each route section:

Route-ID	From	To	Forest#	Type	Name
1	0	800	1	16	Quaking Aspen
1	800	1800	2	17	Paper Birch
1	1800	3000	3	9	White Spruce

ANALYSIS TOOLS

Many of the line analysis tools can be used with routes or sections after you have converted the selected routes or sections to arcs. Important tools that you can use with routes/sections and events are listed in the following table:

Analysis Tools for Working with Routes and Sections

Tool	Function
RESELECT	Select and save user-specified routes or sections to a new theme
ROUTEARC	Converts each route into an arc in the output theme
SECTIONARC	Converts each section into an arc in the output theme
ROUTESTATS	Computes descriptive statistics for routes and sections

Analysis Tools for Working with Events

Tool	Function
EVENTPOINT	Convert point events into a new point theme
EVENTARC	Convert events into a new line theme
OVERLAYEVENTS	Combine two or more event tables
DISSOLVEEVENTS	Combines adjacent records in linear event tables if they have the same value

RESELECT

- **Creates a route system by selecting routes or sections specified by the user**

Imagine that we have a line theme of streams as follows:

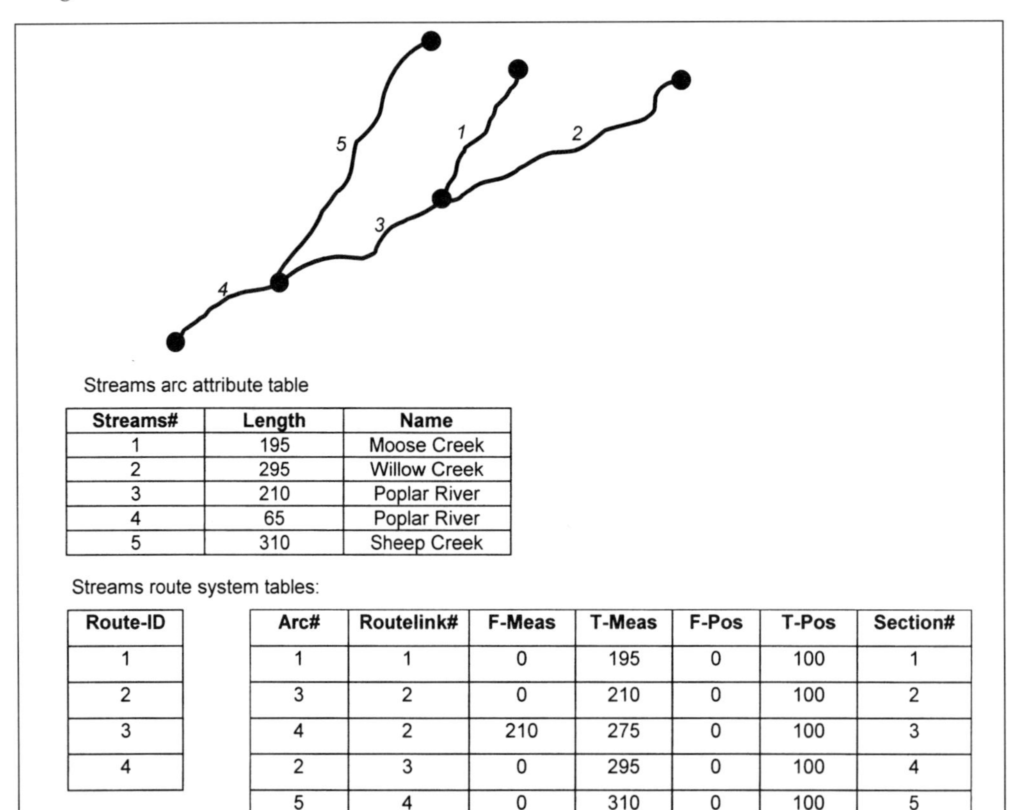

Streams arc attribute table

Streams#	Length	Name
1	195	Moose Creek
2	295	Willow Creek
3	210	Poplar River
4	65	Poplar River
5	310	Sheep Creek

Streams route system tables:

Route-ID
1
2
3
4

Arc#	Routelink#	F-Meas	T-Meas	F-Pos	T-Pos	Section#
1	1	0	195	0	100	1
3	2	0	210	0	100	2
4	2	210	275	0	100	3
2	3	0	295	0	100	4
5	4	0	310	0	100	5

RESELECT ROUTE-ID EQ 2

What would the output theme look like? In this example, route 2 is selected for the output theme . . .

Output arc attribute table

Arc#	Length	Name
3	210	Poplar River
4	65	Poplar River

Output route system tables:

Route-ID
2

Arc#	Routelink#	F-Meas	T-Meas	F-Pos	T-Pos	Section#
3	2	0	210	0	100	1
4	2	210	275	0	100	2

ROUTEARC

- **Convert each route into one arc in the output theme**

Imagine that we have a line theme of streams as follows:

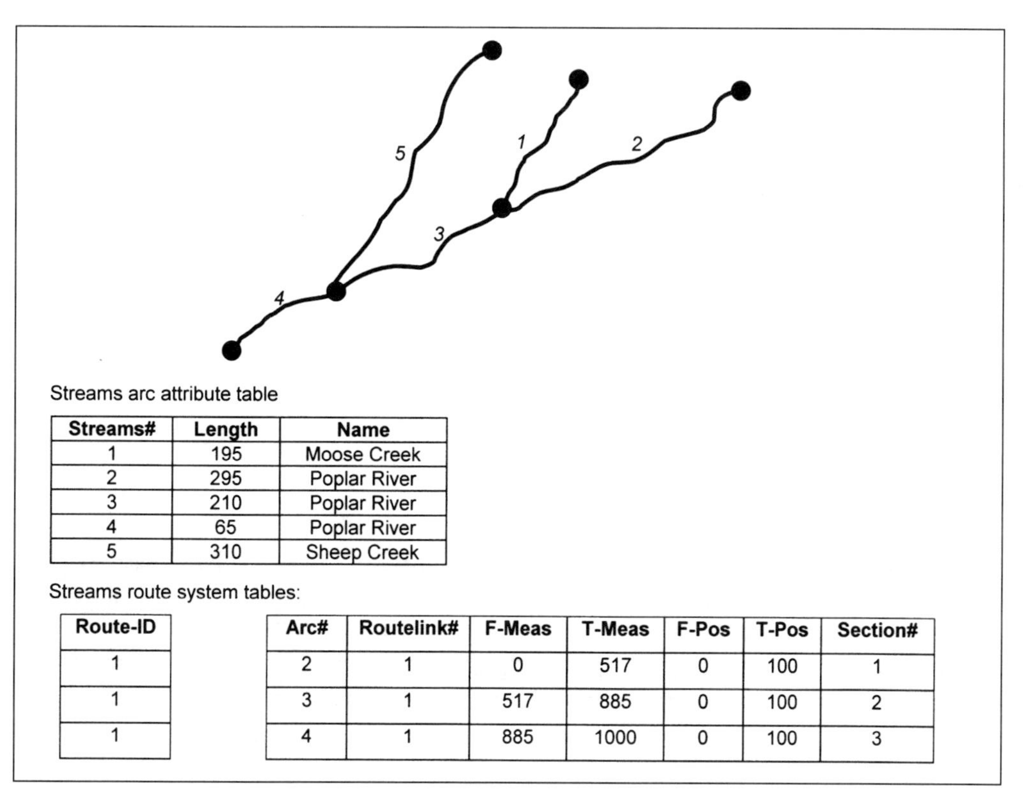

Streams arc attribute table

Streams#	Length	Name
1	195	Moose Creek
2	295	Poplar River
3	210	Poplar River
4	65	Poplar River
5	310	Sheep Creek

Streams route system tables:

Route-ID
1
1
1

Arc#	Routelink#	F-Meas	T-Meas	F-Pos	T-Pos	Section#
2	1	0	517	0	100	1
3	1	517	885	0	100	2
4	1	885	1000	0	100	3

What would the output theme generated by ROUTEARC look like? Route 1 would be output as a single arc . . .

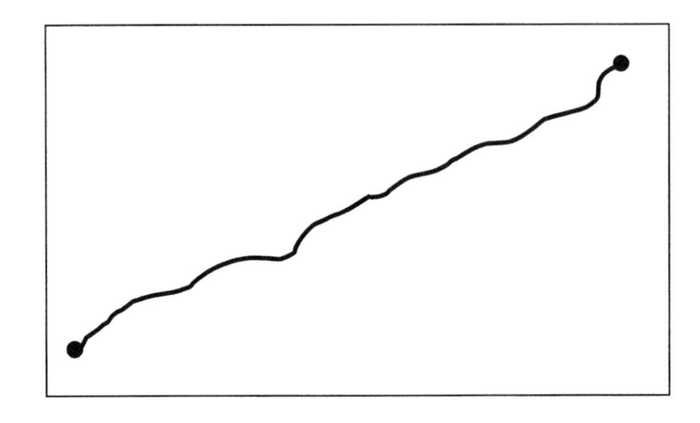

SECTIONARC

- **Convert each section into one arc in the output theme**

Imagine that we have a line theme of streams as follows:

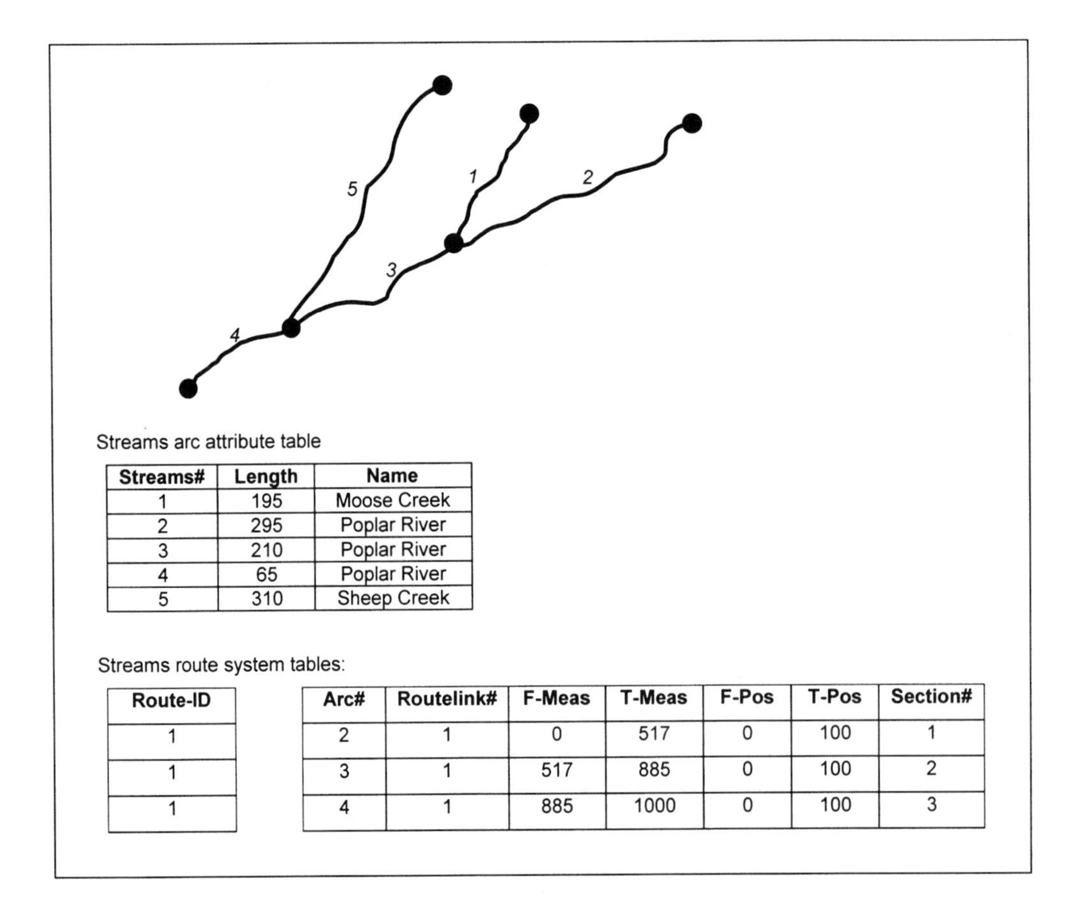

Streams arc attribute table

Streams#	Length	Name
1	195	Moose Creek
2	295	Poplar River
3	210	Poplar River
4	65	Poplar River
5	310	Sheep Creek

Streams route system tables:

Route-ID
1
1
1

Arc#	Routelink#	F-Meas	T-Meas	F-Pos	T-Pos	Section#
2	1	0	517	0	100	1
3	1	517	885	0	100	2
4	1	885	1000	0	100	3

What would the output theme generated by SECTIONARC look like? Each section of route 1 would be output as an arc . . .

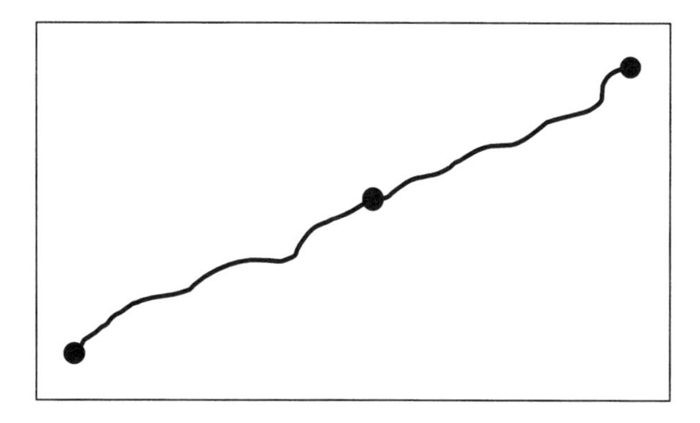

ROUTESTATS

- **Computes descriptive statistics for routes and sections**

The following table describes the types of descriptive statistics that are available.

Route Stats Keyword	**Description**
ARCLENGTH	Total length of each route in GIS coordinate units.
MEASURELENGTH	Total length of each route in route measure units.
LOWMEASURE	The lowest measure value in each route.
HIGHMEASURE	The highest measure value in each route.
FIRSTSECTION	The internal ID number of the first section of each route.
LASTSECTION	The internal ID number of the last section of each route.
NUMSECTIONS	The total number of sections in each route.
Section Stats Keyword	
ARCLENGTH	Total length of each section in GIS coordinate units.
MEASURELENGTH	Total length of each section in route measure units.
DELTAPOSITION	The percent of arc spanned for each section.
RATIO	Measure length / Length in GIS coordinate units for each section.
PREVSECTION	The internal ID number of the section before each section. A negative value indicates the start of an arc.

Imagine that we have a line theme of streams as follows:

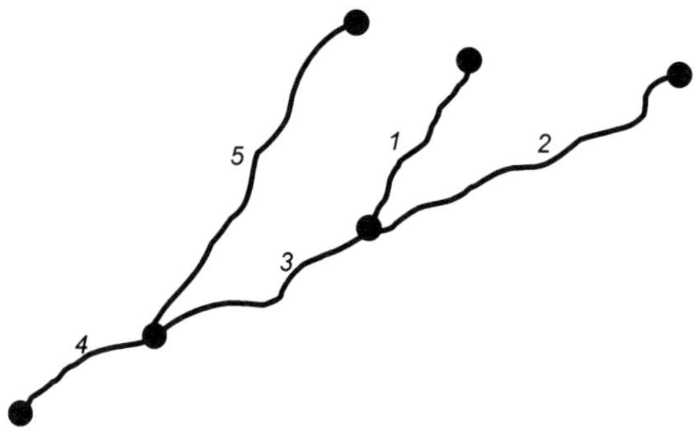

Streams arc attribute table

Streams#	Length	Name
1	195	Moose Creek
2	295	Poplar River
3	210	Poplar River
4	65	Poplar River
5	310	Sheep Creek

Streams route system tables:

Route-ID
1
1
1

Arc#	Routelink#	F-Meas	T-Meas	F-Pos	T-Pos	Section#
2	1	0	517	0	100	1
3	1	517	885	0	100	2
4	1	885	1000	0	100	3

What would the output tables generated by ROUTESTATS contain, if you requested the following descriptive statistics?

RouteStats: ***numsections*** RouteStats: ***arclength*** SectionStats: ***measurelength***

In this example, there is one route. The route statistics requested are the total number of sections in route#1 (3 sections) and the total length of the route in GIS coordinate units (295 + 210 + 65 = 570). The section statistics requested is the length of each section in measure units. Section 1 is 517 units long, section 2 is 885 – 517 = 368, and section 3 is 1000 – 885 = 115 measure units long.

Route#	Numsections	Arclength
1	3	570

Section#	Measurelength
1	517
2	368
3	115

EVENTPOINT

- **Convert point events into a new point theme**

Imagine an example of stops along a nature trail . . .

Point Event Table

Route-ID	Dist_from_trailhead	Description
1	10	Welcome stop
1	245	Snag stop
1	320	Blind at pond
2	15	Welcome stop
2	125	Cave entrance

Trails Section Table

Arc#	Routelink#	F-Meas	T-Meas	F-Pos	T-Pos
1	1	0	195	0	100
2	1	195	310	0	100
3	1	310	575	0	100
4	2	0	295	0	100
4	2	295	510	0	100

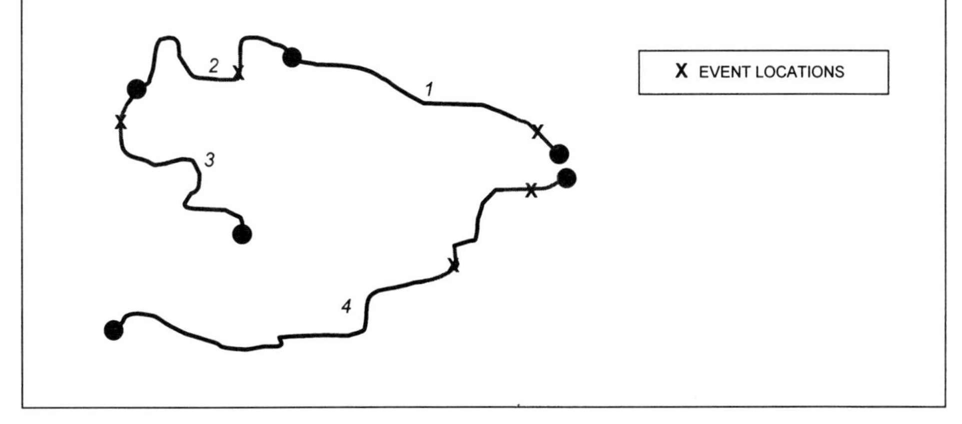

The output point theme would be as follows:

Point Attribute Table

Point#	Route-ID	Dist_from_trailhead	Description
1	1	10	Welcome stop
2	1	245	Snag stop
3	1	320	Blind at pond
4	2	15	Welcome stop
5	2	125	Cave entrance

Output Point Theme:

+2
+3
+1
+4
+5

EVENTARC

- **Convert linear or continuous events into a new line theme**

As an example, we have an event table of rapid class along a route corresponding to a float down a river.

RAPIDS.TBL

Route-ID	To_Distance	Rapids_class
1	10	3
1	16	2
1	22	1
1	35	2
1	80	1

RIVER SECTION TABLE

Arc#	Routelink#	F-Meas	T-Meas	F-Pos	T-Pos	Section#
1	1	0	20	0	100	1
1	1	20	40	0	100	2
1	1	40	60	0	100	3
1	1	60	80	0	100	4

The output theme would contain the following arcs:

1 2 3 4 5

OUTPUT ARC ATTRIBUTE TABLE

Arc#	Route-ID	To_Distance	Rapids_class
1	1	10	3
2	1	16	2
3	1	22	1
4	1	35	2
5	1	80	1

OVERLAYEVENTS

- **Combine two or more event tables**

As an example, imagine that you have two event tables: one of bird observations, and one of vegetation types along a transect.

Vegetation along Transect Event Table

Route#	From	To	Veg_code	Veg_name
1	0	100	2	Oak-Hickory
1	100	300	3	Red Maple
1	300	500	4	Silver Maple
1	500	700	7	Sweet Gum
1	700	1000	8	Birch/Beech/Maple

May 29 Birds Observed at Transect Locations Event Table

Route#	Date	Time	Species	Location
1	05/29/2000	600	Hermit Thrush	250
1	05/29/2000	605	House Wren	300
1	05/29/2000	615	Towhee	400
1	05/29/2000	620	House Wren	550
1	05/29/2000	625	HermitThrush	700
1	05/29/2000	630	Scarlet Tanager	800
1	05/29/2000	635	Northern Oriole	850

There are two possible options to use with OVERLAYEVENTS: *intersect* or *union. Intersect* will return only the events that occur at the same location in all event tables. *Overlay* will return all the events from any location in any event table.

Output Table from OVERLAYEVENTS *(UNION OPTION)*

Route#	From	To	Veg_code	Veg_name	Date	Time	Species
1	0	100	2	Oak-Hickory			
1	100	250	3	Red Maple			
1	250	250	3	Red Maple	05/29/2000	600	Hermit Thrush
1	250	300	3	Red Maple			
1	300	300	3	Red Maple	05/29/2000	605	House Wren
1	300	300	4	Silver Maple	05/29/2000	605	House Wren
1	300	400	4	Silver Maple			
1	400	400	4	Silver Maple	05/29/2000	615	Towhee
1	400	500	4	Silver Maple			
1	500	550	7	Sweet Gum			
1	550	550	7	Sweet Gum	05/29/2000	620	House Wren
1	550	700	7	Sweet Gum			
1	700	700	7	Sweet Gum	05/29/2000	625	Hermit Thrush
1	700	700	8	Birch/Beech/Maple	05/29/2000	625	Hermit Thrush
1	700	800	8	Birch/Beech/Maple			
1	800	800	8	Birch/Beech/Maple	05/29/2000	630	Scarlet Tanager
1	800	850	8	Birch/Beech/Maple			
1	850	850	8	Birch/Beech/Maple	05/29/2000	635	Northern Oriole
1	850	1000	8	Birch/Beech/Maple			

Output Table from OVERLAYEVENTS (*INTERSECT OPTION*)

Route#	Location	Date	Time	Species	Veg_code	Veg_name
1	250	05/29/2000	600	Hermit Thrush	3	Red Maple
1	300	05/29/2000	605	House Wren	3	Red Maple
1	300	05/29/2000	605	House Wren	4	Silver Maple
1	400	05/29/2000	615	Towhee	4	Silver Maple
1	550	05/29/2000	620	House Wren	7	Sweet Gum
1	700	05/29/2000	625	HermitThrush	7	Sweet Gum
1	700	05/29/2000	625	HermitThrush	8	Birch/Beech/Maple
1	800	05/29/2000	630	Scarlet Tanager	8	Birch/Beech/Maple
1	850	05/29/2000	635	Northern Oriole	8	Birch/Beech/Maple

DISSOLVEEVENTS

- **Combines adjacent records in linear event tables if they have the same value**

You already learned about DISSOLVE in the chapter on line analysis. DISSOLVEEVENTS is analogous and can be used to simplify event tables.

As an example, we could use Veg_code as the dissolve item from the following event table, and produce the following output event table by using DISSOLVEEVENTS.

Input Event Table → DISSOLVEEVENTS → Output Event Table

Input Event Table

Route#	From	To	Veg_code
1	0	100	2
1	100	250	3
1	250	250	3
1	250	300	3
1	300	300	3
1	300	300	4
1	300	400	4
1	400	400	4
1	400	500	4
1	500	550	7
1	550	550	7
1	550	700	7
1	700	700	7
1	700	700	8
1	700	800	8
1	800	800	8
1	800	850	8
1	850	850	8
1	850	1000	8

Output Event Table

Route#	From	To	Veg_code
1	0	100	2
1	100	300	3
1	300	500	4
1	500	700	7
1	700	1000	8

DYNAMIC SEGMENTATION EXERCISES

1) You have a route system of irrigation canals as follows. Circle the first section of route#1 and the second section of route#3.

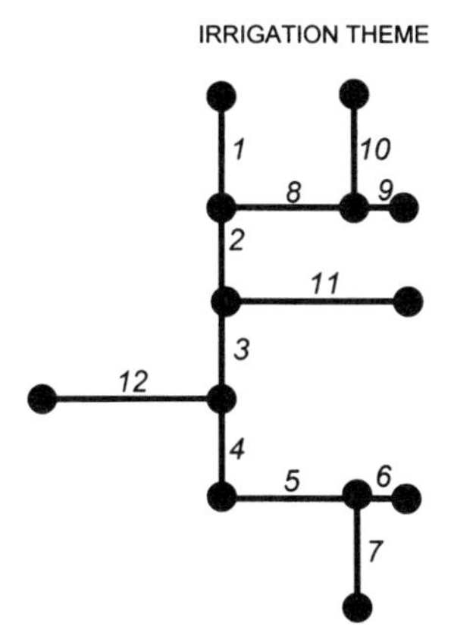

IRRIGATION SECTION TABLE:

Arc#	Routelink#	F-Meas	T-Meas	F-Pos	T-Pos
1	1	10	15	66.66	100
2	1	15	25	0	100
3	1	25	35	0	100
4	1	35	45	0	100
5	1	45	60	0	100
6	1	60	65	0	100
7	2	0	15	0	100
8	3	0	15	0	100
9	3	15	17.5	0	50
10	4	0	15	0	100
11	5	0	20	0	100
12	6	0	20	0	100

2) You have the following point theme of auto accidents and network of streets.

Accidents Point Attribute Table

Accident#	Date	Num_Vehicles
1	01/01/2000	1
2	01/01/2000	3
3	01/02/2000	2
4	01/04/2000	2
5	01/07/2000	1
6	01/07/2000	1
7	01/08/2000	2

Streets Section Table

Arc#	Routelink#	F-Meas	T-Meas	F-Pos	T-Pos
1	1	0	15	0	100
2	1	15	25	0	100
3	1	25	35	0	100
4	1	35	45	0	100
5	1	45	60	0	100
6	1	60	65	0	100
7	2	0	15	0	100
8	3	0	15	0	100
9	3	15	20	0	100
10	4	0	15	0	100
11	5	0	20	0	100
12	6	0	20	0	100

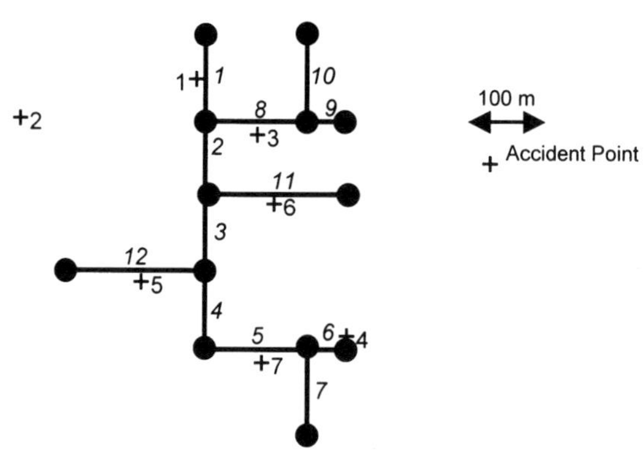

You enter the **ADDROUTEMEASURE** command and specify *Accidents* as the name of your point theme, *Streets* as the name of your route theme and route system, and *acc_events.tbl* as the name of our output event table. You specify a search radius of 100 meters. Fill in your output event table:

acc_events.tbl

Accident#	Route#	Measure	Date	Num_Vehicles

3) You have a route system of streams as follows.

Streams arc attribute table

Streams#	Length	Name
1	195	Moose Creek
2	295	Poplar River
3	210	Poplar River
4	65	Poplar River
5	310	Sheep Creek

Streams route system tables:

Route-ID
1
2
3
4

Arc#	Routelink#	F-Meas	T-Meas	F-Pos	T-Pos	Section#
1	1	0	195	0	100	1
3	2	0	210	0	100	2
4	2	210	275	0	100	3
2	3	0	295	0	100	4
5	4	0	310	0	100	5

Salmon Event Table

Route#	From	To	Sockeye_Count	King_Count
2	100	120	30	0
2	140	160	40	10
2	180	190	20	5
2	190	195	18	6
2	200	210	0	5

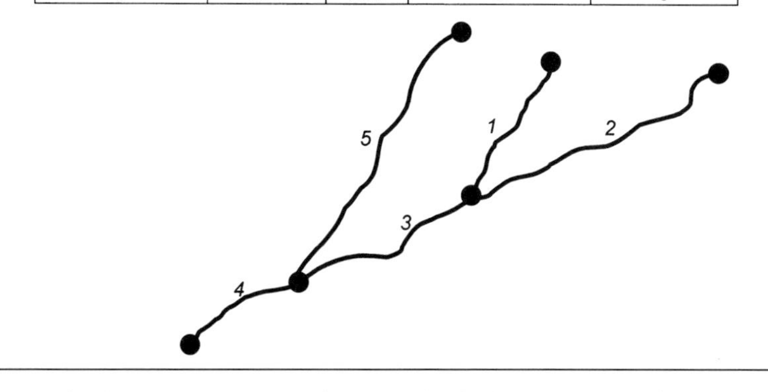

Circle the stream section that has the highest density of Sockeye Salmon.
Circle the stream section that has the highest density of King Salmon.

4) You have a route system of streams as follows. What would be the output if you run **RESELECT** on your streams with the following expression:

RES SECTION# EQ 3

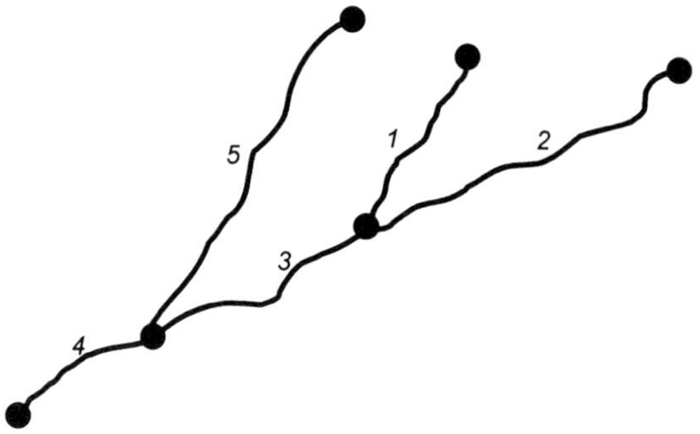

Streams arc attribute table

Streams#	Length	Name
1	195	Moose Creek
2	295	Willow Creek
3	210	Poplar River
4	65	Poplar River
5	310	Sheep Creek

Streams route system tables:

Route-ID
1
2
3
4

Arc#	Routelink#	F-Meas	T-Meas	F-Pos	T-Pos	Section#
1	1	0	195	0	100	1
3	2	0	210	0	100	2
4	2	210	275	0	100	3
2	3	0	295	0	100	4
5	4	0	310	0	100	5

5) You have the following route system. Draw what the output theme would look like if you applied the **ROUTEARC** command to this route system.

Route Attribute Table

Route#	Route-ID
1	19

Section Attribute Table

Arc#	Routelink#	F-Meas	T-Meas	F-Pos	T-Pos	Section#	Surface
5	1	0	1.5	0	33..33	1	Dirt
5	1	1.5	3.5	33.33	77.77	2	Gravel
5	1	3.5	4.5	77.77	100	3	Paved
6	1	4.5	10.0	0	100	4	Paved

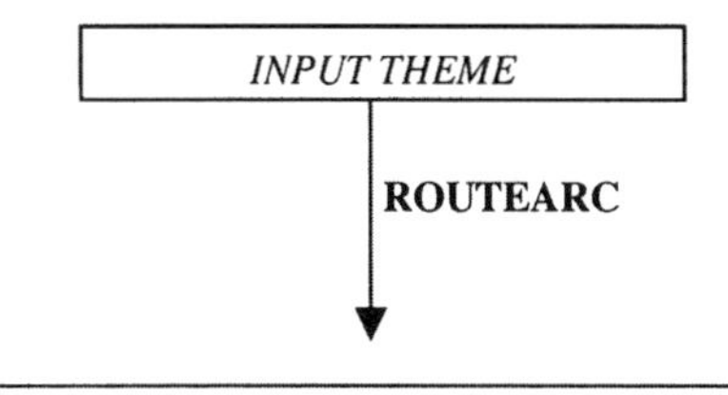

6) You have the following route system. Draw what the output theme would look like if you applied the **SECTIONARC** command to this route system.

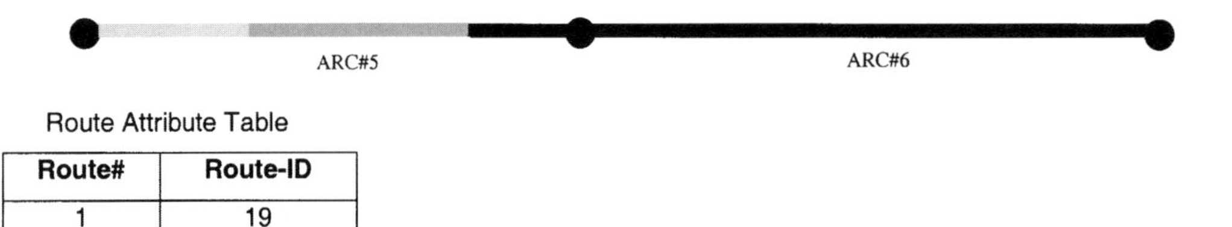

Route Attribute Table

Route#	Route-ID
1	19

Section Attribute Table

Arc#	Routelink#	F-Meas	T-Meas	F-Pos	T-Pos	Section#	Surface
5	1	0	1.5	0	33..33	1	Dirt
5	1	1.5	3.5	33.33	77.77	2	Gravel
5	1	3.5	4.5	77.77	100	3	Paved
6	1	4.5	10.0	0	100	4	Paved

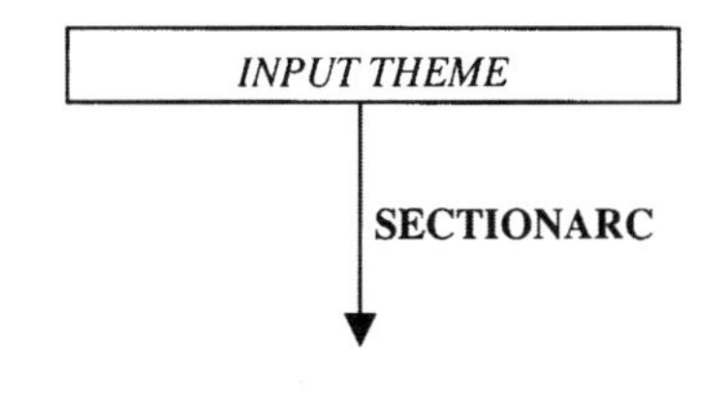

7) You have the following bus route system:

Arc Attribute Table

Arc#	Length
1	500
2	500
3	1000
4	500
5	1000
6	500
7	1000
8	500
9	1000
10	500
11	1000

Bus route system tables:

Route-ID
1
2
3

Arc#	Routelink#	F-Meas	T-Meas	F-Pos	T-Pos	Section#
1	1	0	50	0	100	1
2	1	50	100	0	100	2
3	1	100	200	0	100	3
1	2	0	50	0	100	4
2	2	50	100	0	100	5
4	2	100	150	0	100	6
5	2	150	250	0	100	7
6	2	250	300	0	100	8
7	2	300	350	0	50	9
1	3	0	50	0	100	10
2	3	50	100	0	100	11
8	3	100	150	0	100	12
9	3	150	250	0	100	13
10	3	250	300	0	100	14
11	3	300	325	0	25	15

What would be the results if you run **ROUTESTATS** on your route system with the following options?

RouteStats: ***ARCLENGTH***
RouteStats: ***MEASURELENGTH***

8) You have the following point and linear events along a street route system. Draw the output themes that would result from using **EVENTPOINT** and **EVENTARC**.

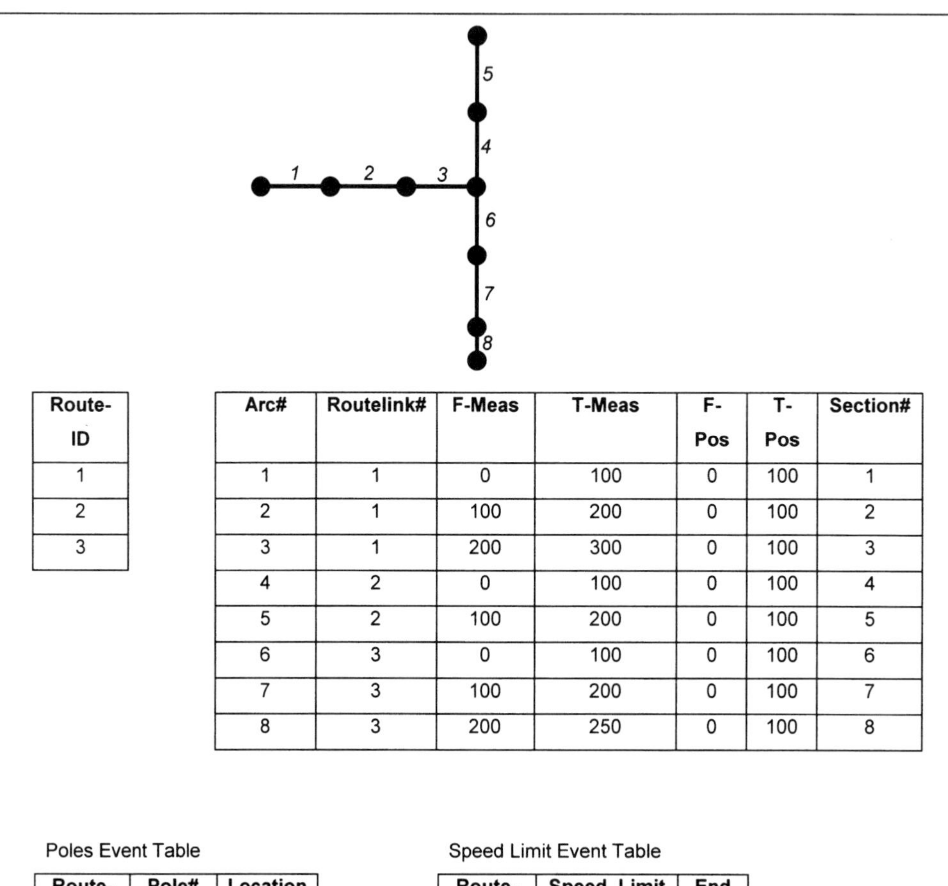

Route-ID
1
2
3

Arc#	Routelink#	F-Meas	T-Meas	F-Pos	T-Pos	Section#
1	1	0	100	0	100	1
2	1	100	200	0	100	2
3	1	200	300	0	100	3
4	2	0	100	0	100	4
5	2	100	200	0	100	5
6	3	0	100	0	100	6
7	3	100	200	0	100	7
8	3	200	250	0	100	8

Poles Event Table

Route-ID	Pole#	Location
1	110	50
1	111	150
2	245	50
2	246	150
3	247	75
3	248	200

Speed Limit Event Table

Route-ID	Speed_Limit	End
1	45	300
2	35	200
3	35	100
3	25	125
3	35	250

Output Point Theme:

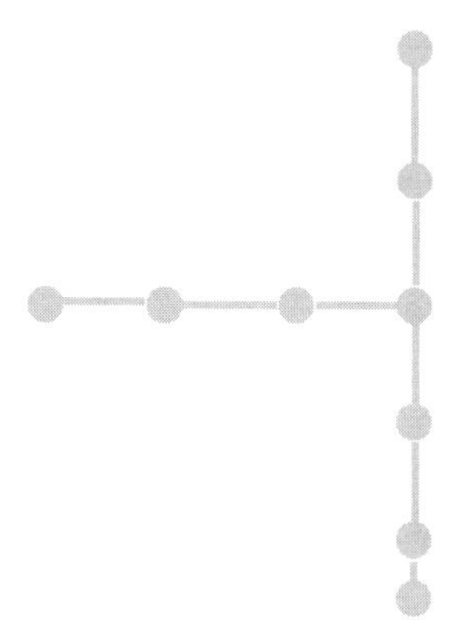

Output Line Theme:

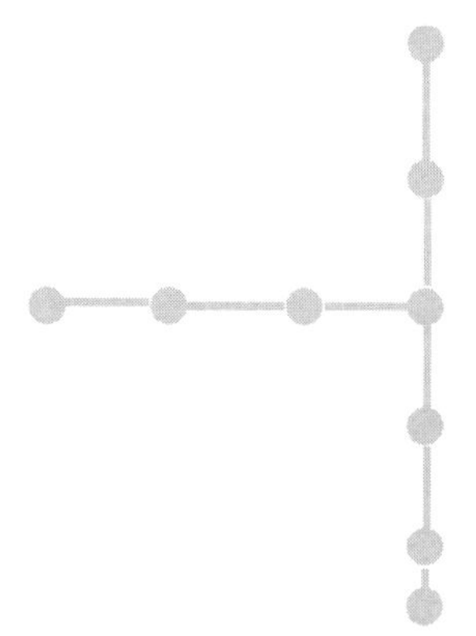

9) You have two event tables of king salmon counts along a stream route. Fill in the output table from executing **OVERLAYEVENTS** with these two event tables.

1999 King Salmon Counts

Route#	From	To	Count_1999
1	60	65	3
1	110	130	18
1	165	175	14

2000 King Salmon Counts

Route#	From	To	2000_Count
1	60	75	1
1	110	130	3
1	165	175	2

Stream Section Table

Arc#	Routelink#	F-Meas	T-Meas	F-Pos	T-Pos
1	1	0	200	0	200

OVERLAYEVENTS OUTPUT TABLE (*UNION OPTION*)

Route#	From	To	Count_1999	Count_2000
1	60	65		
1	65	75		
1	110	130		
1	130	165		
1	165	175		
1	175	175		

OVERLAYEVENTS OUTPUT TABLE (*INTERSECT OPTION*)

Route#	From	To	Count_1999	Count_2000
1	60	65		
1	110	130		
1	165	175		

10) You have the following route system and irrigation event table. Fill in the output event table resulting from using **DISSOLVEEVENTS** with ***Pipe_Type*** as the dissolve item.

Input Route System

Route#	Arclink#	F-Meas	T-Meas	F-Pos	T-Pos
1	1	0	100	0	100
1	2	100	200	0	100
1	3	200	250	0	100
1	4	250	300	0	100
2	1	0	100	0	100
2	2	100	200	0	100
2	5	200	250	0	100
2	6	250	300	0	100
3	1	0	100	0	100
3	2	100	200	0	100
3	7	200	250	0	100
3	8	250	300	0	100

Input Irrigation Pipe Event Table

Route#	From	To	Pipe_Type	Date_Installed
1	0	65	1	07/29/2000
1	65	100	1	07/30/2000
1	100	175	1	08/01/2000
1	175	200	1	08/02/2000
1	200	300	2	08/03/2000
2	0	65	1	07/29/2000
2	65	100	1	07/30/2000
2	100	175	1	08/01/2000
2	175	200	1	08/02/2000
2	200	300	2	08/03/2000
3	0	65	1	07/29/2000
3	65	100	1	07/30/2000
3	100	175	1	08/01/2000
3	175	200	1	08/02/2000
3	200	250	2	08/03/2000
3	250	300	2	08/04/2000

Output Table from DISSOLVEEVENTS

Route#	From	To	Pipe_Type

Chapter 7

Polygon Analysis

INTRODUCTION

A polygon is a GIS feature that has an area and perimeter. Typical polygon themes may include themes such as country or state boundaries, township sections, vegetation polygons, soil mapping units, lakes, parcels, or watersheds.

The following are some GIS tools that you can use in analyzing polygons:

GIS tools already covered in point and line analysis chapters:

- COUNTVERTICES—Count the number of vertices (including nodes) in each polygon and add the count as a polygon attribute called **VERTICES**.
- BUFFER—Generate buffer polygons of a user-specified distance around polygons.
- RESELECT—Create a new polygon theme by selecting polygons using a logical expression.
- DISSOLVE—Merge adjacent polygons if they have the same value for a user-specified polygon attribute.

New tools:

- ELIMINATE— Merges selected polygons with neighboring polygons that have the largest shared border between them. Used to remove "sliver" polygons.
- CLIP—Use a polygon theme as a "cookie-cutter" to cut out features from a point, line, or polygon theme. A ***cut and copy*** operation to produce a single output theme.
- SPLIT—A ***cut and copy*** operation that uses each polygon as a separate "cookie-cutter" to produce *many* output themes.
- ERASE—Eliminate polygons in a theme from the area covered by polygons in a second theme. A ***cut*** operation.
- UPDATE—Replace polygons in a theme with polygons from a second theme. A ***cut and paste*** operation.
- INTERSECT, UNION, IDENTITY—Transfer the polygon attributes to a second polygon theme (spatial join).

DISSOLVE

- **Merges adjacent polygons if they have the same attribute value.**

The output polygons will only contain the item used for the dissolving and no additional attributes. If **#ALL** is used as the dissolve item, then original polygon attributes will be output but **User-IDs** will be altered. The **#ALL** option is typically used following the edge-matching of tiles or map sheets.

In the following example, **MapClass** is used as the dissolve item

Veg_polys Polygon Attribute Table

Area	Perimeter	Veg_polys#	ID	Type	Veg_name	MapClass
-	(universe poly)	1	0			
		2	11	1	Bl. Spruce	1
		3	12	2	W. Spruce	1
		4	13	3	Aspen	2
		5	14	3	Aspen	2
		6	15	1	W. Spruce	1
		7	16	4	P. Birch	2

VEGPOLYS Polygons

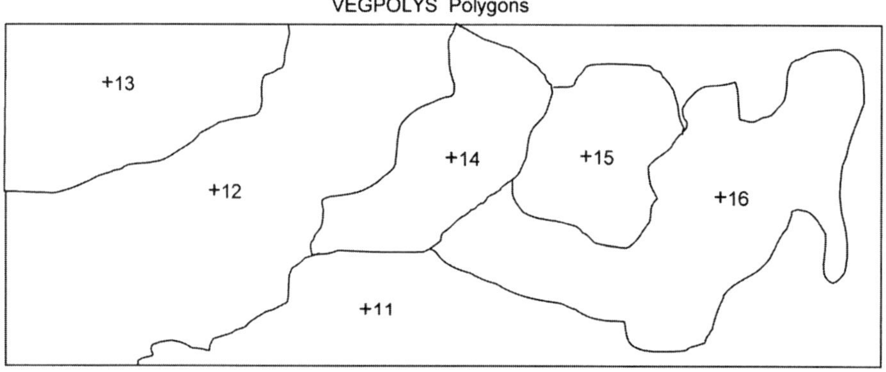

Dissolved Polygons

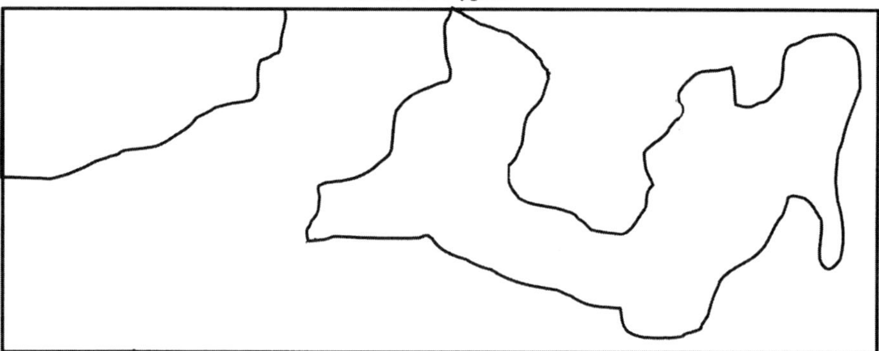

ELIMINATE

- Merges *selected* polygons with neighboring polygons that have a shared border between them.

ELIMINATE is often used to remove "sliver" polygons. You can control merging of selected polygons by using the **BORDER** or **AREA** keywords. The default, **BORDER** merges a selected polygon with a neighboring unselected polygon by dropping an arc and retaining the *longest shared border*. **AREA** drops an arc and retains the polygon with the *largest area*.

In the following example, we use **ELIMINATE** with **RESELECT AREA LT 100 AND AREA GT 0** on the following theme and the **BORDER** option:

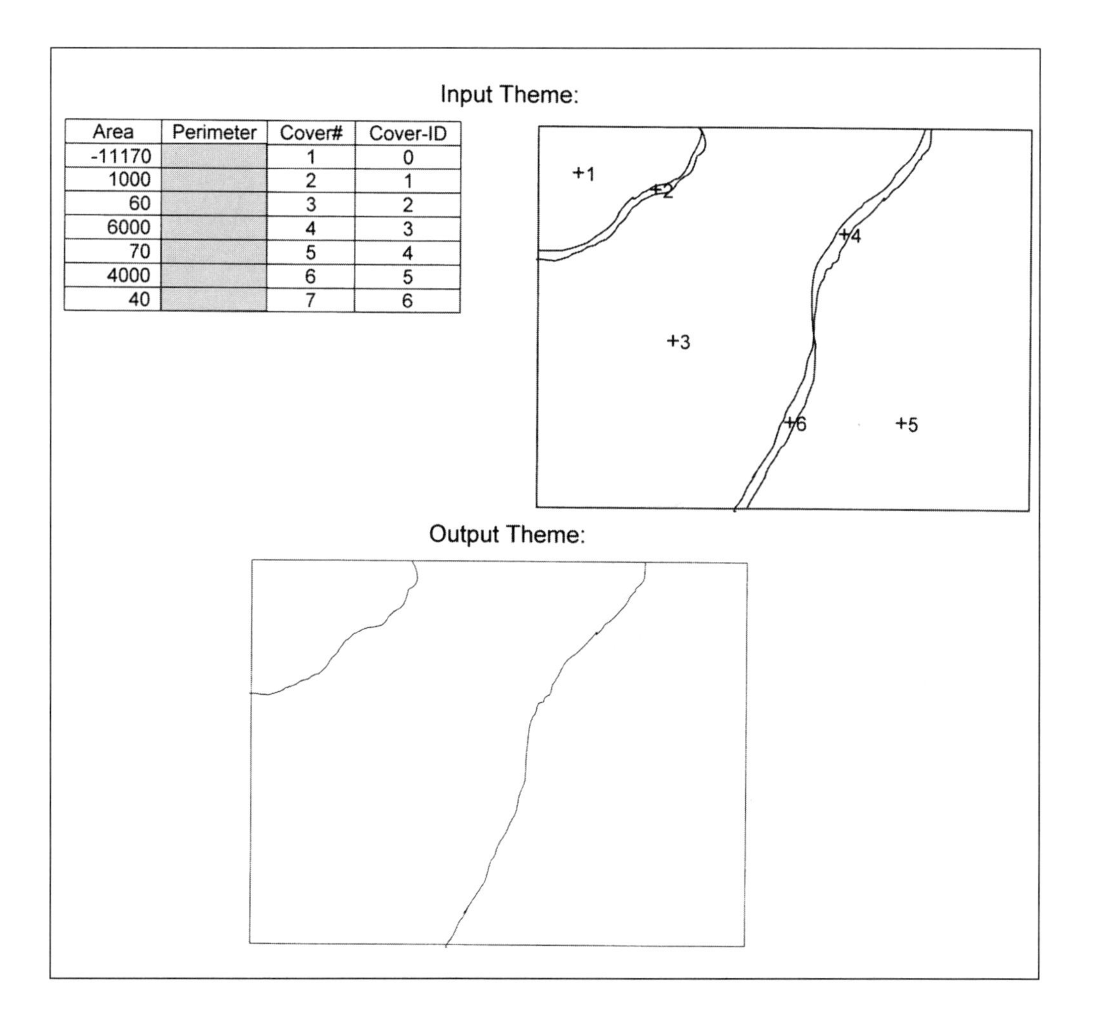

Area	Perimeter	Cover#	Cover-ID
-11170		1	0
1000		2	1
60		3	2
6000		4	3
70		5	4
4000		6	5
40		7	6

CLIP

- Use a polygon theme as a "cookie cutter" to cut out features from a point, line, or polygon theme. A cut and copy operation to produce a *single* output theme. You can clip point, line, or polygon themes.

Imagine that you have a theme containing all lands owned by the state Department of Natural Resources. You want to create a new theme of vegetation polygons within these properties. So you clip **VEGPOLYS** with **DNR_LANDS:**

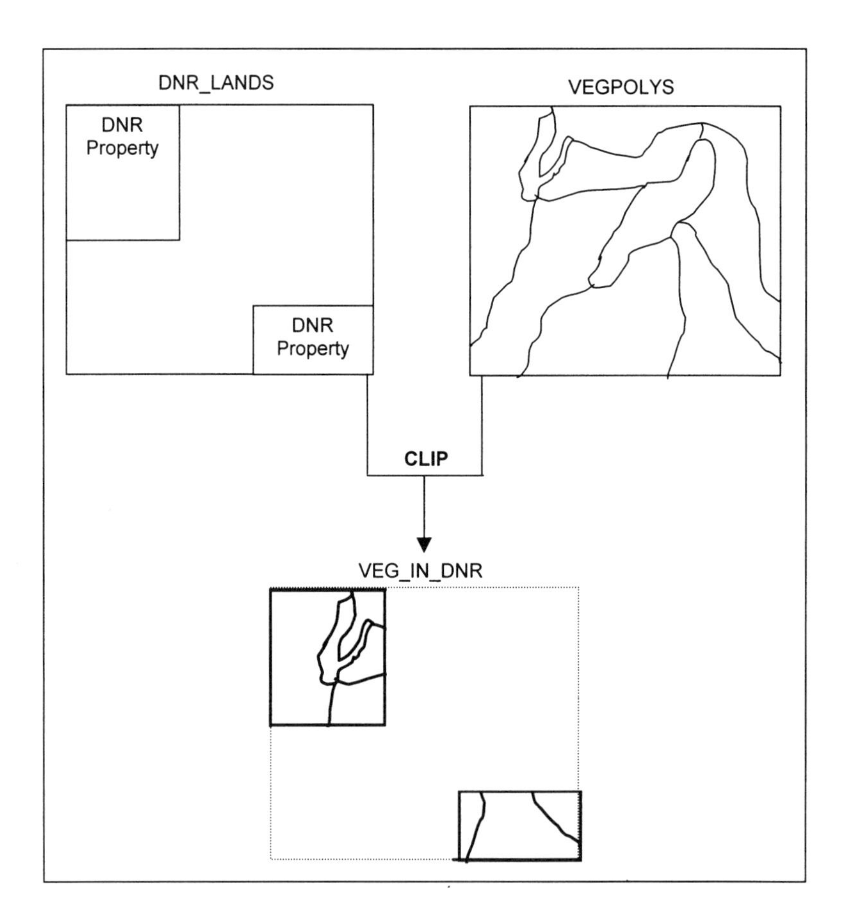

SPLIT

- A cut and copy operation that uses each polygon as a separate "cookie cutter" to produce *many* output themes. You can split point, line, or polygon themes.

Imagine that you have a polygon theme of watershed boundaries and a wells point theme. You want to produce separate point themes of wells inside each watershed 1 and watershed 2.

SPLIT WELLS WATERSHEDS WATERSHED-ID POINT

Enter the 1st theme: **WELLS_SHED1**
Enter item value: **1**

Enter the 2nd theme: **WELLS_SHED2**
Enter item value: **2**

Enter the 3rd theme: **end**

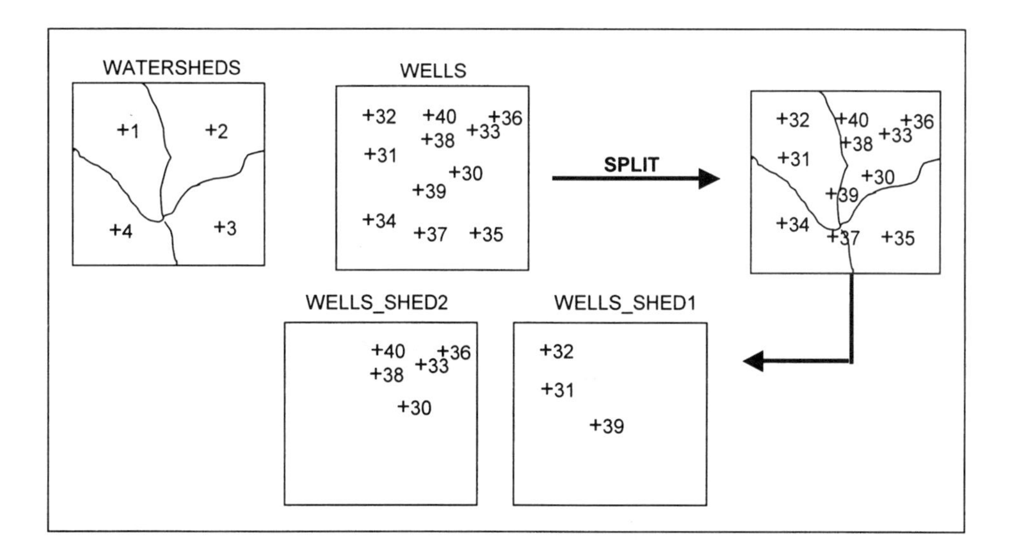

ERASE

- Eliminate polygons in a theme from the area covered by polygons in a second theme. A cut operation that works on point, line, or polygon themes.

Imagine that you have a line theme of winter skiing trails and a polygon theme of wetlands. We create a new theme called **HIKE_TRAILS** representing the trails above water in the summertime.

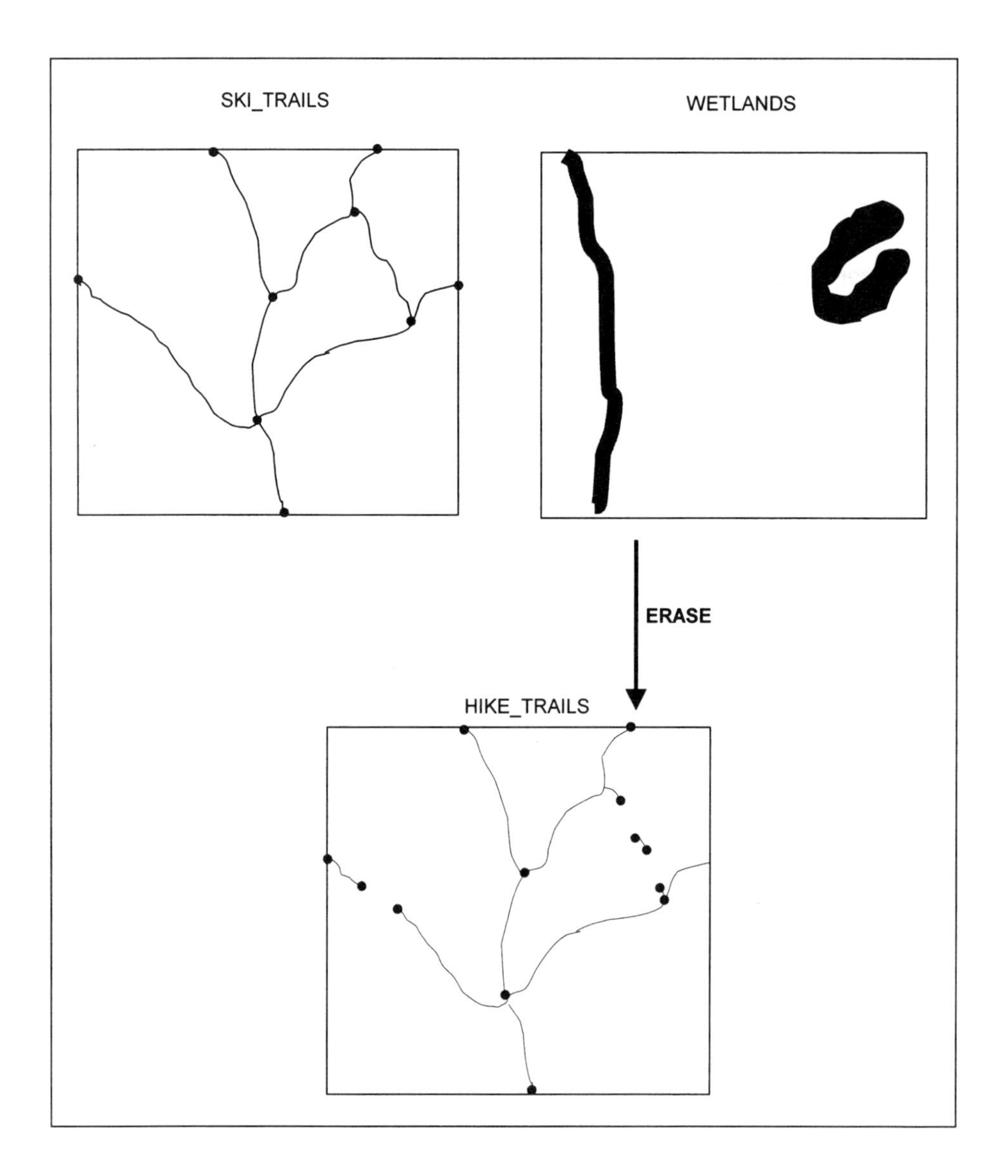

UPDATE

- Replace polygons in a theme with polygons from a second theme. A cut and paste operation. The keyword **KEEPBORDER** or **DROPBORDER** specifies whether or not the outside border of the update theme will be kept after it is inserted into the input theme.

Imagine that you have a polygon theme of vegetation and polygon theme of planted spruce seedlings. You want to update your vegetation theme.

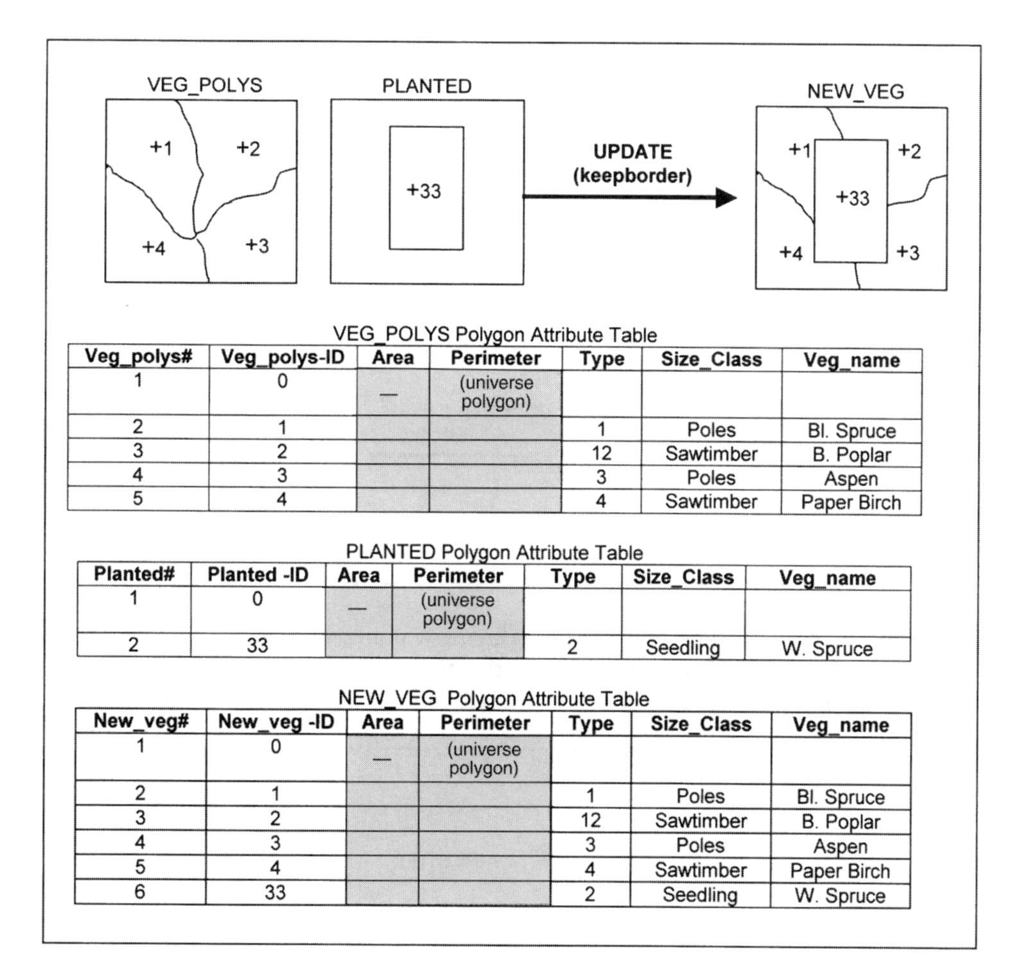

VEG_POLYS Polygon Attribute Table

Veg_polys#	Veg_polys-ID	Area	Perimeter	Type	Size_Class	Veg_name
1	0	—	(universe polygon)			
2	1			1	Poles	Bl. Spruce
3	2			12	Sawtimber	B. Poplar
4	3			3	Poles	Aspen
5	4			4	Sawtimber	Paper Birch

PLANTED Polygon Attribute Table

Planted#	Planted -ID	Area	Perimeter	Type	Size_Class	Veg_name
1	0	—	(universe polygon)			
2	33			2	Seedling	W. Spruce

NEW_VEG Polygon Attribute Table

New_veg#	New_veg -ID	Area	Perimeter	Type	Size_Class	Veg_name
1	0	—	(universe polygon)			
2	1			1	Poles	Bl. Spruce
3	2			12	Sawtimber	B. Poplar
4	3			3	Poles	Aspen
5	4			4	Sawtimber	Paper Birch
6	33			2	Seedling	W. Spruce

SPATIAL JOINS: INTERSECT, UNION, IDENTITY

- Transfer the polygon attributes to a second polygon theme (spatial join).
  You have already learned about the **INTERSECT** tool for intersecting points and lines with polygon themes.

INTERSECT creates a new theme by overlaying a point, line, or polygon theme with an intersecting polygon theme. The output theme contains only the features *inside* the intersecting polygons.

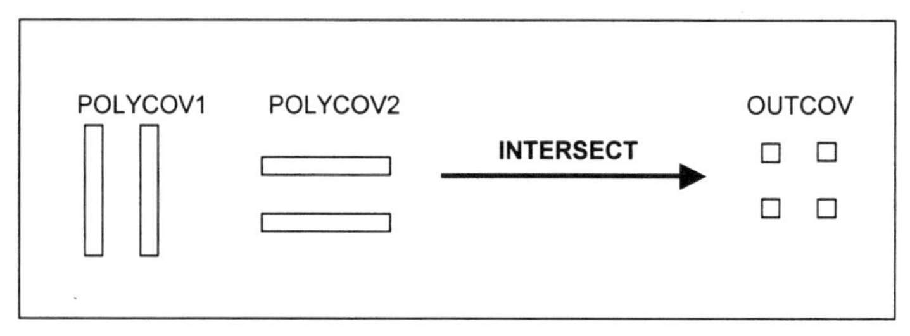

UNION creates a new theme by overlaying two polygon themes. The output theme contains the combined polygons and attributes of both themes. Note that ***only*** polygon themes can be combined using UNION.

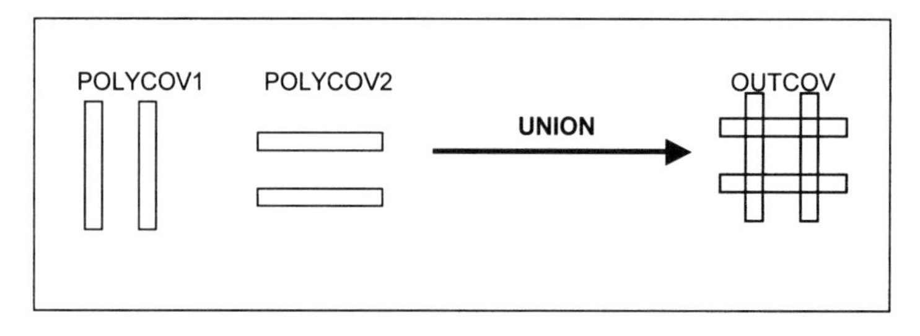

IDENTITY creates a new theme by overlaying a point, line, or polygon theme with an intersecting polygon theme. The output theme contains only the all original point, line, or polygons as well as the attributes transferred by the intersecting polygon theme.

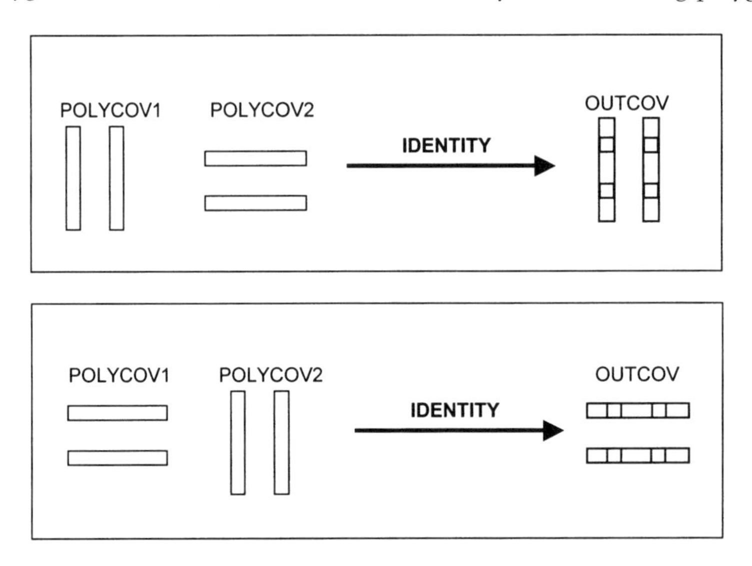

POLYGON ANALYSIS EXERCISES

1) Polygons are assumed to be pure in their content and discrete and absolute in their boundary locations. In other words, they are assumed to be homogeneous with sharp, distinct boundaries. This is rarely the case. Rank the following polygon themes in terms of their likely purity and boundary sharpness.

Theme	Purity Rank	Boundary Sharpness Rank
Wetland Polygons		
Tax Parcels		
Soil Polygons		

2) You have polygon themes of a wildfire burn from 1963 and a burn from the year 2000. Fill in the following flowchart to determine the total number of hectares from the 1963 burn that burned in 2000.

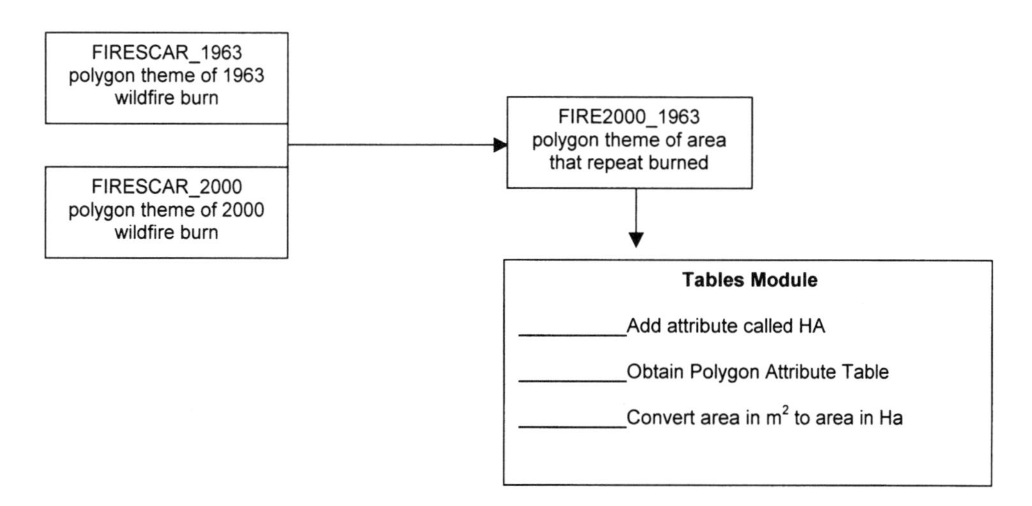

3) You have a line theme of rivers and streams and a polygon theme of three watershed areas (1= Rock Creek Watershed, 2=Clear Creek Watershed, 3=Willow Creek Watershed). For each watershed, you want to know the density of rivers and streams in meters per hectare of the watershed. Your GIS coordinate system is in meters.

Fill in the following flowchart to solve your GIS problem.

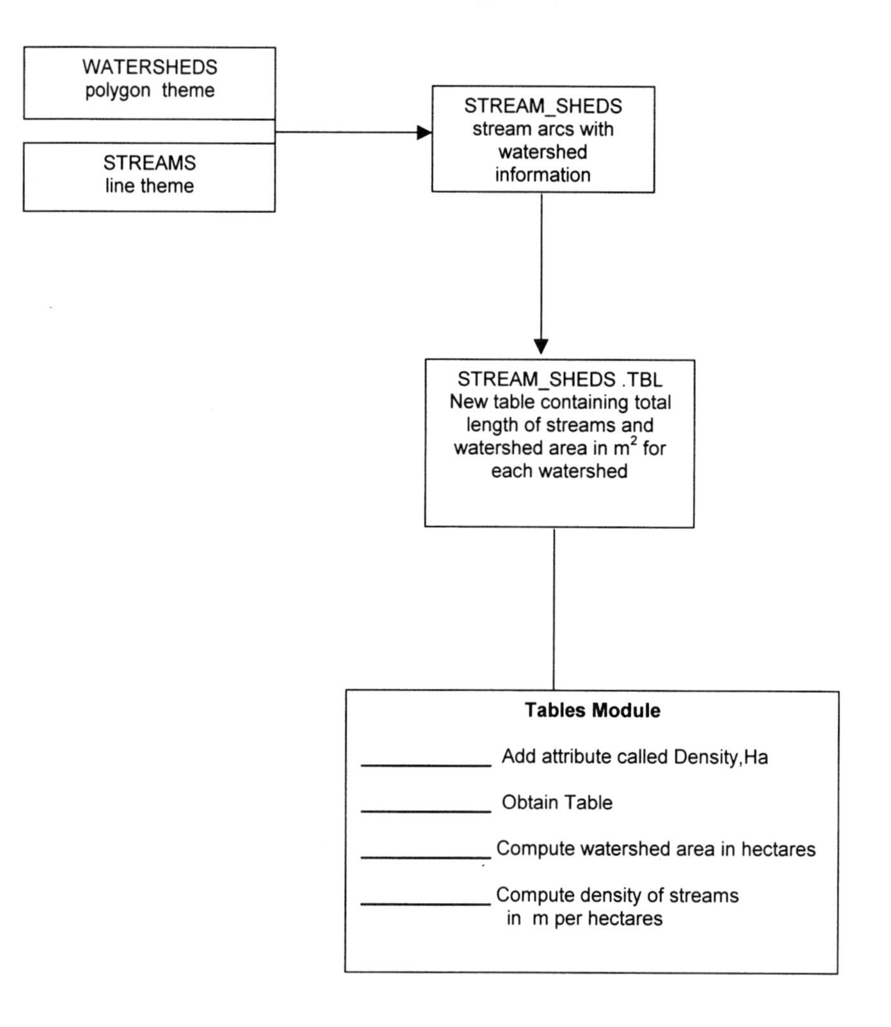

4)You have a point theme of mallard nest locations with three age classes of nesting mallards. You also have a polygon theme of land cover types (1= tussock, 2=low shrub, 3=high shrub, 4=barren). Your GIS coordinate system is in meters.

You want to estimate the density of nests in each land cover type by mallard age class as follows:

Tussock Type (hectares)	Low Shrub Type (hectares)	High Shrub Type (hectares)	Barren Type (hectares)

Mallard Age Class	Tussock Type (total nests)	Low Shrub Type (total nests)	High Shrub Type (total nests)	Barren Type (total nests)
1				
2				
3				

Fill in the following flowchart to solve your GIS problem.

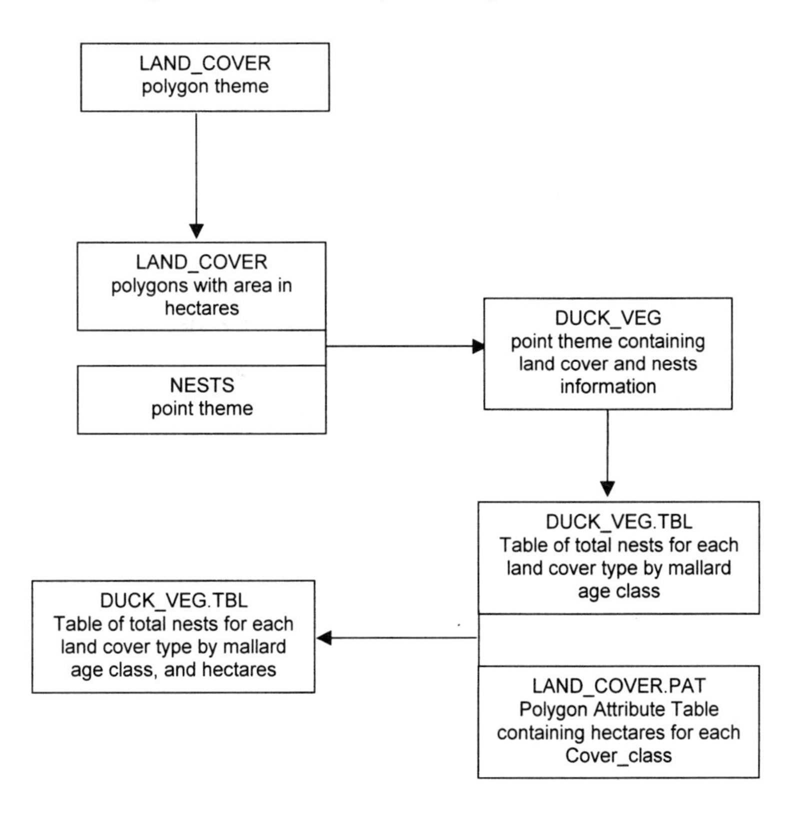

5) You have a polygon theme of vegetation types (*vegpolys)* and a polygon theme of watershed areas *(watersheds)*. You want a table showing the area of each watersed and the area of black spruce in each watershed.

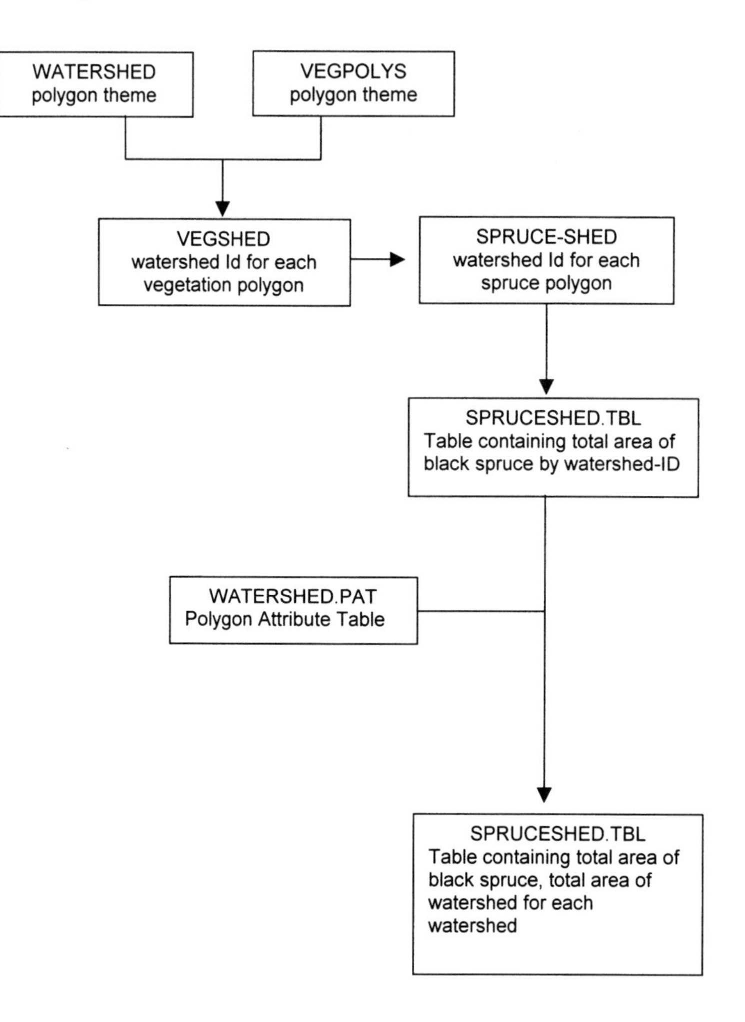

6) You have three polygon themes: SOIL, VEG, and TINPOLYS. SOIL contains soil information, VEG contains vegetation information, and TINPOLYS are triangles where the slope and aspect of each triangle has been estimated.

You are checking for possible errors in the vegetation theme. You want to find if there are any aspen polygons that occur over poorly drained soils or on northerly facing slopes that are over 10 percent in slope gradient. These would likely be errors in vegetation mapping because aspen typically grows on relatively dry, warm sites.

Fill in the following flowchart to solve your GIS problem.

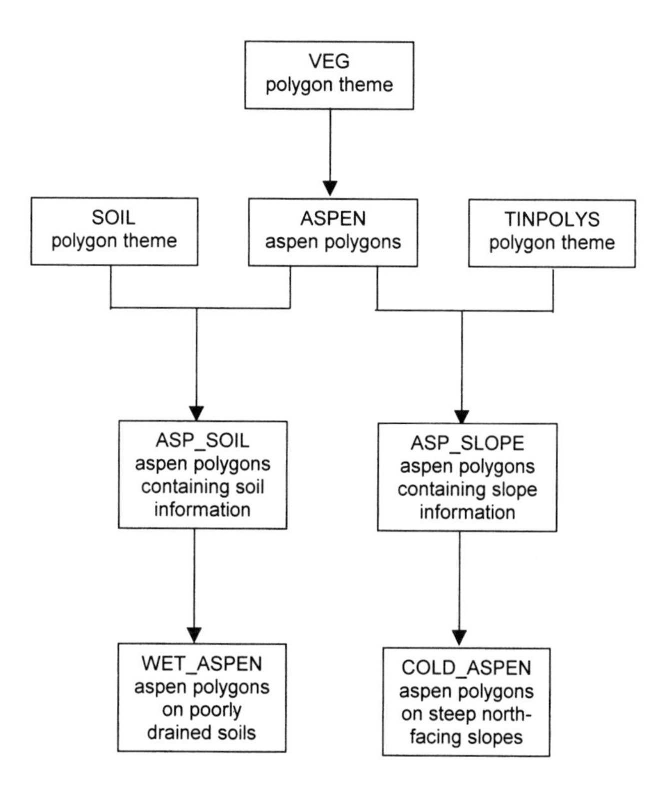

7) You have three candidate routes for a new power line. The cost of building the line varies by amount of land owned privately, by soil type to build on, and by amount of wetlands to cross and as well as total distance. You want to know this type of information for the three routes. Fill in the following flowchart with the most appropriate GIS tools:

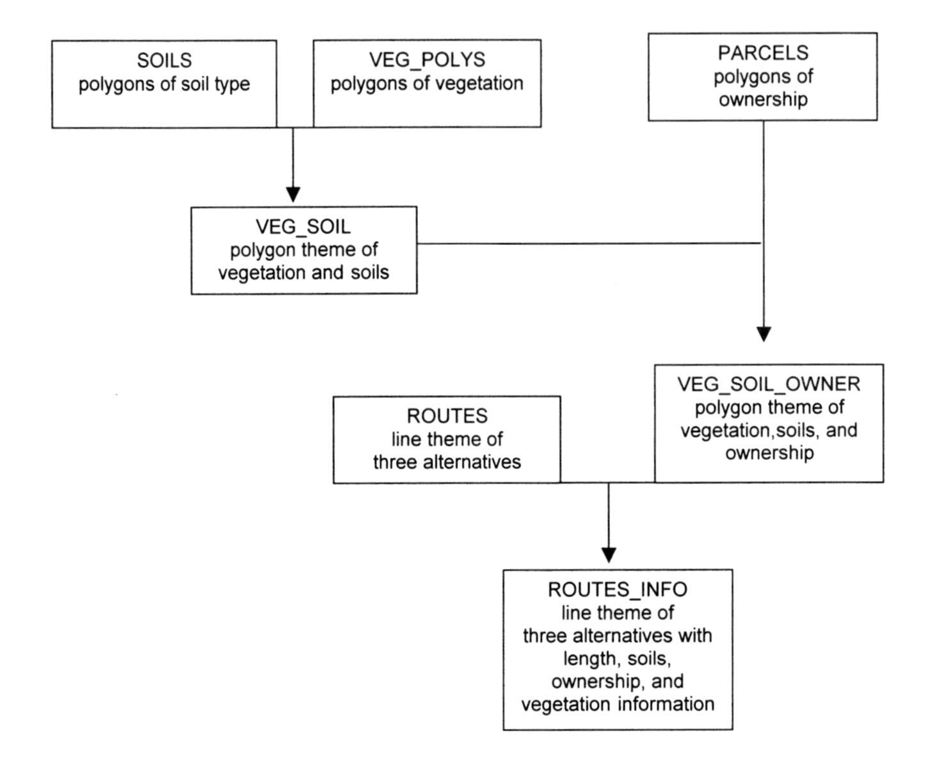

8) You have a pretimber harvest polygon theme of alder clumps and a post-timber harvest point theme of planted spruce seedlings. You want to find the mean height growth of the spruce seedlings in areas that were in or within 10 meters of pre-harvest alder versus areas that were at least 10 meters away from the pre-harvest alder clumps. Fill in the following flowchart with the most appropriate GIS tools:

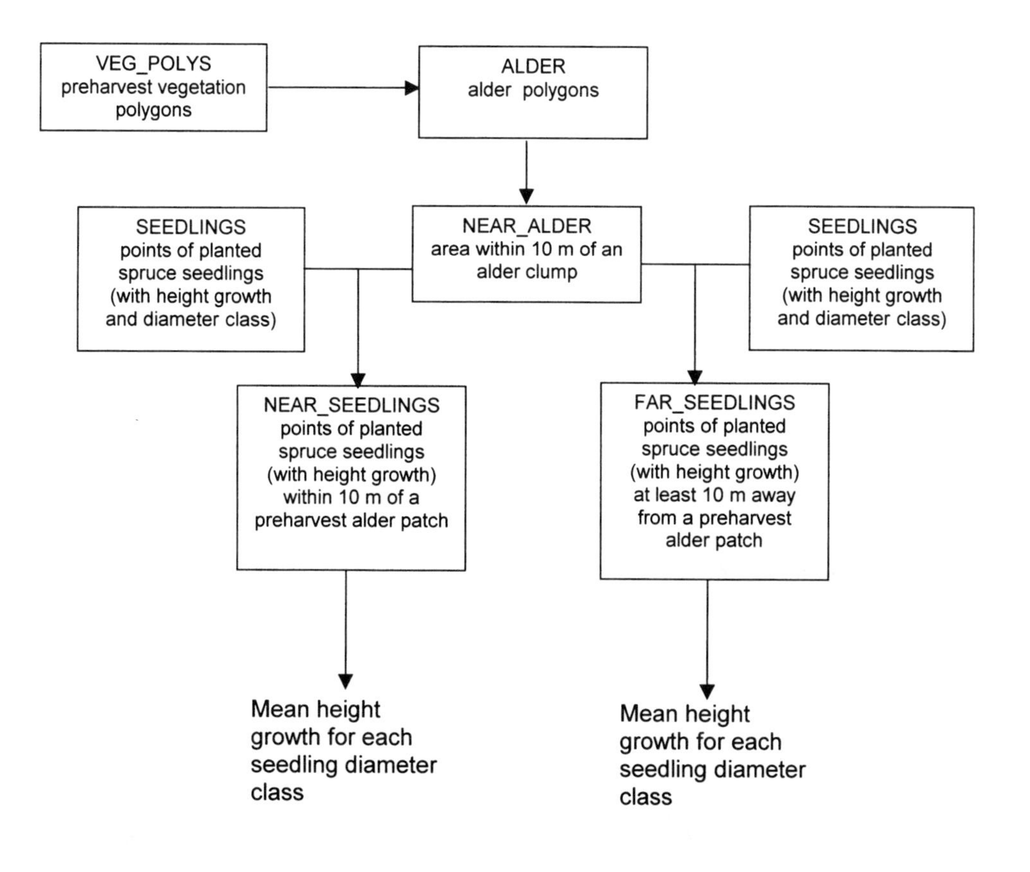

9) You have polygon themes of lakes. One theme is from early summer and the other theme is from late summer. You want to create a new theme of the best lakes for waterfowl habitat . . . lakes that had at least 1 hectare of exposed mudflats in late summer due to drawdown of the lake water level. Fill in the following flowchart to solve your GIS problem.

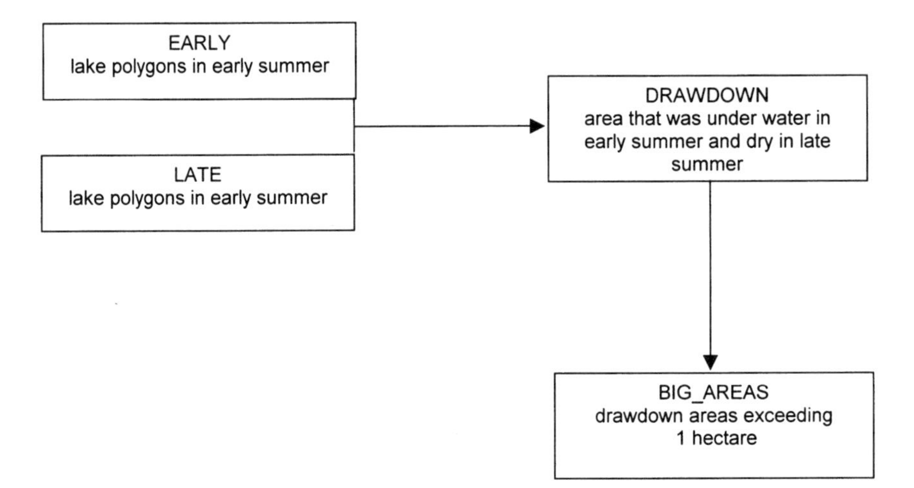

10) You have a point theme of well locations, a point theme of gas station locations, a soils polygon theme, and a parcels polygon theme containing the address of the owner of each parcel.

Outline what tools you would use to develop a text file of all well owners that have wells on sand, loamy sand, or sandy loam soils within a kilometer of a gas station.

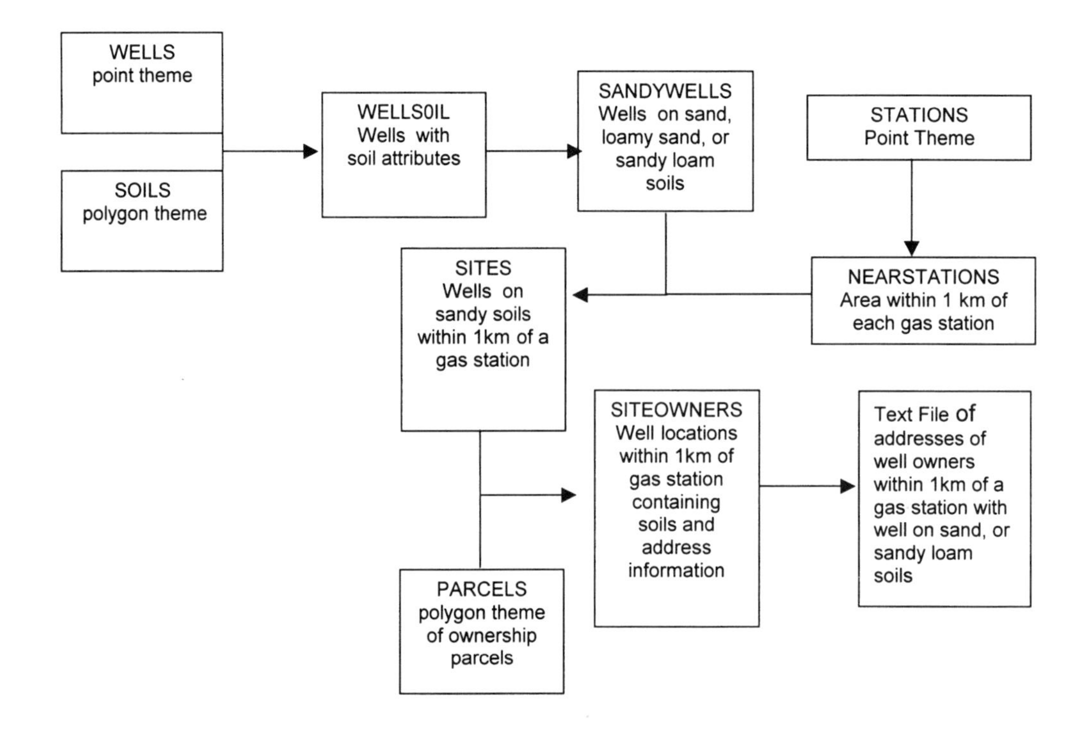

Chapter 8

Grid Analysis

INTRODUCTION

GRID AND VECTOR GIS WORLDS

Until now, you have worked mainly in the vector world of points, lines, and polygons. These are discrete features represented by X,Y coordinates, and each feature has a unique identification number. Grids are different because they may represent a continuous surface instead of discrete features. In general, the grid GIS data model has several advantages relative to the vector GIS data model of points, lines, and polygons:

- Grids allow for faster analysis, especially for any overlay type analysis
- Grids allow for modeling continuous surfaces such as wildfire or oil spill spreads, or determining the optimal path through a surface.
- Grid is the natural model for many data sources in natural resources such as digital elevation models, scanned maps, and land cover data derived from satellite data.
- Grids allow you to represent the world as a "fuzzier system." For example, in the vector world, vegetation polygons are often assigned a vegetation type from visual interpretation of aerial photographs. Typically, relatively large polygons are delineated because of an arbitrary minimum mapping unit criteria used during the air photo interpretation. With land cover grids, the minimum mapping unit is the size of the grid cell, and therefore heterogeneous areas can be better represented. In the vector world, polygons are assigned as one homogeneous land cover class. In the grid world, the same area may be several land cover classes, and the probability of correct classification of each grid cell can also be stored.

The vector model of points, lines, and polygons does have some advantages over the grid model:

- Since features are stored as X, Y coordinates, it usually takes less disk space to store points, lines, and polygons relative to grids that have the same precision of spatial detail.
- The vector model allows for modeling of linear networks and the incorporation of measured events via dynamic segmentation.
- The vector model is a natural model for many data sources derived from GPS surveying, or for utility, urban, and transportation data.

- The vector model allows you to represent the spatial location of features with high precision relative to grid representations.

Your choice of vector versus grid analysis will depend largely on your data sources and applications. In this chapter, you will learn that many of the grid operations are analogous to vector operations.

There are potentially hundreds of grid operations. In this chapter, you will learn about some common grid operations, especially applicable for natural resource applications. These operations include grid arithmetic operations, selection operations, grouping operations, distance operations, optimal path operations, and common topographic operations.

GRID ARITHMETIC OPERATORS

Grids can be added, subtracted, multiplied, and divided. If an output cell value is undefined, it is assigned a value of NODATA. For example, imagine that you have two grids that you want to divide with:

Grid1

2	2	1	3	1
1	1	4	4	1
1	1	4	4	1
3	3	3	4	3
3	3	4	3	3

Grid2

2	2	1	1	1
2	2	1	1	1
3	3	1	1	1
3	3	3	0	0
3	3	4	0	0

GRID3 = GRID1 / GRID2

The two input grids are integer type, therefore the output grid will also be integer type. For example, cell(row 2, column 1) is output as ½ or 0.5 and this is output as an integer value of 0. The cells that are undefined because they were divided by zero are assigned to NODATA. Notice that zero is a valid grid cell value while NODATA indicates no valid data for that cell.

Grid3

1	1	1	3	1
0	0	4	4	1
0	0	4	4	1
1	1	1		
1	1	1		

In order to have an arithmetic expression return fractions, at least one of the input grids must be floating point type. You can use the FLOAT function to convert integer grids to floating-point grids. For example:

GRID4 = FLOAT(GRID1) / FLOAT(GRID2)

Grid4

1.0	1.0	1.0	3.0	1.0
0.5	0.5	4.0	4.0	1.0
0.33	0.33	4.0	4.0	1.0
1.0	1.0	1.0		
1.0	1.0	1.0		

Keep in mind that NODATA cells always remain NODATA cells in arithmetic operations.

Grid5 = Grid1 + Grid3

Grid5

3	3	2	6	2
1	1	8	8	2
1	1	8	8	2
4	4	4		
4	4	5		

Grid1

2	2	1	3	1
1	1	4	4	1
1	1	4	4	1
3	3	3	4	3
3	3	4	3	3

Grid3

1	1	1	3	1
0	0	4	4	1
0	0	4	4	1
1	1	1		
1	1	1		

You can change NODATA to any valid value by using the CON function which you will learn about soon.

SELECTION OPERATIONS: TEST, SELECT, CON

TEST

- **Returns a 1 if the cell meets some logical expression criteria, otherwise the output cell is assigned a zero.**

As an example, imagine that you have a grid of land cover types with 1= water, 2= coniferous forest, 3= broadleaf forest, 4= mixed forest, and 5=marsh. You want to create a grid where 0 represents nonforest and 1 represents forest.

Land Cover Grid

1	1	5	3	4
1	1	5	3	4
5	5	3	4	2
3	3	4	2	2
3	4	2	2	2

Forest = TEST(Land_Cover, 'Value NE 1 and Value NE 5')

Forest Grid

0	0	0	1	1
0	0	0	1	1
0	0	1	1	1
1	1	1	1	1
1	1	1	1	1

Forest Value Attribute Table

Value	Count
0	8
1	17

SELECT

- **Returns the original cell value if the cell meets some logical expression criteria, otherwise the output cell is assigned as NODATA.**

As an example, imagine that you have a grid of elevation values with 255 representing fill pixels along the edge of the grid. If you asked for the mean elevation value of the grid, zeros would be included in the calculation. You want to use the **SELECT** function to convert the 255s to NODATA.

Elevation Grid

255	101	103	105	255
255	102	104	105	255
255	104	106	109	255
255	105	108	112	255
255	110	112	115	255

Elev2 = SELECT (ELEVATION, 'Value NE 255')

Elev2 Grid

	101	103	105	
	102	104	105	
	104	106	109	
	105	108	112	
	110	112	115	

CON

- **Tests for a user-specified logical expression and returns user-specified values.**

The CON function stands for conditional test. The user specifies what happens if the test is true or false. For example you could create a HIGH_ELEV grid as follows:

HIGH_ELEV = CON (ELEV > 1000, 1, 0)

If the elevation value is greater than 1000, the output cell receives a 1, and if it is less than or equal to 100, an output cell receives a zero.

ELEV Grid:

980	990	1003	1004	990
997	995	1010	1007	993
998	998	1004	1005	995
999	999	1002	1001	997
997	999	1000	1000	998

HIGH_ELEV Grid:

0	0	1	1	0
0	0	1	1	0
0	0	1	1	0
0	0	1	1	0
0	0	0	0	0

If you do not indicate the output value for a false test, NODATA is assigned automatically. For example, you could create a HIGH_ELEV grid as follows:

HIGH_ELEV = CON (ELEV > 1000, ELEV)

If the elevation value is greater than 1000, the output cell receives the elevation value, and if it is less than or equal to 100, an output cell receives a NODATA .

ELEV Grid:

980	990	1003	1004	990
997	995	1010	1007	993
998	998	1004	1005	995
999	999	1002	1001	997
997	999	1000	1000	998

HIGH_ELEV

		1003	1004	
		1010	1007	
		1004	1005	
		1002	1001	

You can convert NODATA cells to some other value by using the CON(ISNULL . . . as follows:

HIGH_ELEV

		1003	1004	
		1010	1007	
		1004	1005	
		1002	1001	

FILLZEROS = CON(ISNULL(HIGH_ELEV), 0, HIGH_ELEV)

The condition tested asks: Is the grid cell value of HIGH_ELEV NODATA or null? If it is, a zero value is returned, and if it is not, the original cell value of HIGH_ELEV is returned:

FILLZEROS

0	0	1003	1004	0
0	0	1010	1007	0
0	0	1004	1005	0
0	0	1002	1001	0
0	0	0	0	0

GROUPING OPERATIONS: RECLASS, REGIONGROUP

RECLASS

- **Changes cell values based on the rules in a remap table.**

Let's start with a simple example. You have a single line remap table as follows: **2 3 : 1**

The remap table specifies a rule that cells with values of 2 or 3 are reclassed to a value of 1 in the output grid. A value of 1 or 4 would be assigned NODATA if you have a NODATA flag set on. If you have a NODATA flag set to off, the original values of 1 or 4 would be output, while 2 or 3s would be output as 1 values.

The RECLASS function is often used to classify quantitative data into categories. As an example, you could use the RECLASS function to create elevation classes from a grid of elevation values. Elevations of 100 to 300, be reclassed to cell values of 1. Elevations above 300 and less than or equal to 400 will be reclassed to cell values of 2. And elevations above 400 and less than or equal to 500 will be reclassed to cell values of 3.

ELEV Grid:

0	150	370	390	360
150	200	360	400	370
220	280	340	410	380
330	300	400	420	380
350	360	390	390	0

ELEV.txt remap table

```
100 300 : 1
300 400 : 2
400 500 : 3
```

ELEV_CLASS = RECLASS (ELEV, ELEV.TXT, NODATA)

ELEV_CLASS Grid:

	1	2	2	2
1	1	2	2	2
1	1	2	3	2
2	1	2	3	2
2	2	2	2	

REGIONGROUP

- **Groups cells of the same value that touch each other.**

REGIONGROUP allows you to aggregate cells that have the same value into groups. As an example, you might have a vegetation grid and you are interested in small patches of a particular class. REGIONGROUP starts at the upper left corner cell and proceeds left to right assigning group numbers based on cells that touch and have the same cell values.

VEG Grid:

4	3	4	3	4
3	3	4	3	4
4	4	4	3	4
4	4	3	3	4
4	4	3	4	4

GROUPS = REGIONGROUP (VEG)

GROUPS Grid:

1	2	3	4	5
2	2	3	4	5
3	3	3	4	5
3	3	4	4	5
3	3	4	5	5

GROUPS Value Attribute Table:

Value	Count	Link
1	1	4
2	3	3
3	9	4
4	6	3
5	6	4

Where the Value is the group number, and the Link is the original value before the grouping occurred. For example, group 5 originally had values of 4 and were cells that touched.

DISTANCE OPERATIONS: EXPAND, EUCDISTANCE

EXPAND

- **Expands user-specified cell values (analogous to BUFFER).**

The user specifies the number of cells to expand and the cell values that are candidates for expansion. For example, EXPAND(Grid, 4, list, 0,1) would select any cells that have values of 0 or 1 and expand them out by 4 cells. When two values compete to expand into the same area, the conflict is resolved based on the value of the majority of surrounding cells. Cells can also expand into NODATA cells.

As an example, imagine that you have a grid of lakes and you want to buffer the lakes by one grid cell by using the EXPAND function.

LAKES Grid:

1	1	0	0	0	0	0	0	0	0
1	1	0	0	0	0	0	0	0	0
0	0	0	0	0	0	0	0	0	0
0	0	0	3	3	3	0	0	0	0
0	0	0	3	3	3	0	0	0	0
0	0	0	3	3	3	0	0	0	0
0	0	0	3	3	3	0	0	0	0
0	0	0	0	0	0	0	0	0	0
0	0	0	0	0	0	0	0		
0	0	0	0	0	0	0	0		5

BUFLAKES = EXPAND(LAKES, 1, LIST, 1,3,5)

BUFLAKES Grid:

1	1	1	0	0	0	0	0	0	0
1	1	1	0	0	0	0	0	0	0
1	1	1	3	3	3	3	0	0	0
0	0	3	3	3	3	3	0	0	0
0	0	3	3	3	3	3	0	0	0
0	0	3	3	3	3	3	0	0	0
0	0	3	3	3	3	3	0	0	0
0	0	3	3	3	3	3	0	0	0
0	0	0	0	0	0	0	0	5	5
0	0	0	0	0	0	0	0	5	5

EUCDISTANCE

- **Calculates the distance to the closest non-NODATA cell for each cell.**

The distance calculation is from the center of one cell to the center of the closest non-NODATA cell. The distance is in grid cell units, rather than the GIS coordinate system units. For example,

Input Grid:

	100	

Distances

1.41	1.00	1.41
1.00	0	1.00
1.41	1.00	1.41

As an example, imagine that you have a grid of lakes and you compute the closest distance to any lake for each grid cell.

LAKES Grid:

1	1								
1	1								
			3	3	3				
			3	3	3				
			3	3	3				
			3	3	3				
									5

DISTLAKES = EUCDISTANCE(LAKES)

DISTLAKES Grid:

0.00	0.00	1.00	2.00	3.00	3.00	3.16	3.61	4.24	5.00
0.00	0.00	1.00	2.00	2.00	2.00	2.24	2.83	3.61	4.47
1.00	1.00	1.41	1.00	1.00	1.00	1.41	2.24	3.16	4.12
2.00	2.00	1.00	0.00	0.00	0.00	1.00	2.00	3.00	4.00
3.00	2.00	1.00	0.00	0.00	0.00	1.00	2.00	3.00	4.00
3.00	2.00	1.00	0.00	0.00	0.00	1.00	2.00	3.00	4.00
3.00	2.00	1.00	0.00	0.00	0.00	1.00	2.00	3.00	3.00
3.16	2.24	1.41	1.00	1.00	1.00	1.41	2.24	2.24	2.00
3.61	2.83	2.24	2.00	2.00	2.00	2.24	2.24	1.41	1.00
4.24	3.61	3.16	3.00	3.00	3.00	3.00	2.00	1.00	0.00

OPTIMAL PATH OPERATIONS: COSTDISTANCE, COSTPATH

You learned about optimal paths in the Network Analysis chapter. With grids, you can estimate the optimal path from source cells to destination cells in an analogous manner.

COSTDISTANCE

- **Estimates the minimum accumulative cost to resource cells.**

The COSTDISTANCE function computes the minimum accumulative cost for a grid and outputs this cost grid, a grid of directions associated with the minimum cost, and a grid allocating the source cells. This information is required to compute the optimal path using the COSTPATH function.

Imagine that you have two resources, coded with values 1 and 2. You also have a grid representing the cost of crossing each grid cell. This cost might be travel time, road construction cost, fuel consumption cost, or cost in terms of aesthetics. You want to compute the minimum accumulative cost for all cells in your area.

COST Grid:

1	3	4	4	3	2
4	6	2	3	7	6
5	8	7	5	6	6
1	4	5		5	1
4	7	5		2	6
1	2	2	1	3	4

RESOURCE Grid:

	1	1			
		1			
2					

What is the least costly path to reach a 1 or a 2 in the source grid? The cost is zero if you are already sitting in a source grid cell . . . assign these to the output grid.

Minimum Accumulative Costs

	0	0			
		0			
0					

The calculation of accumulative costs is modeled based on a node-link model of graph theory. The center of each cell is represented by nodes and all possible costs of movement from one cell to another are represented by links. For example, our grid is represented as follows.

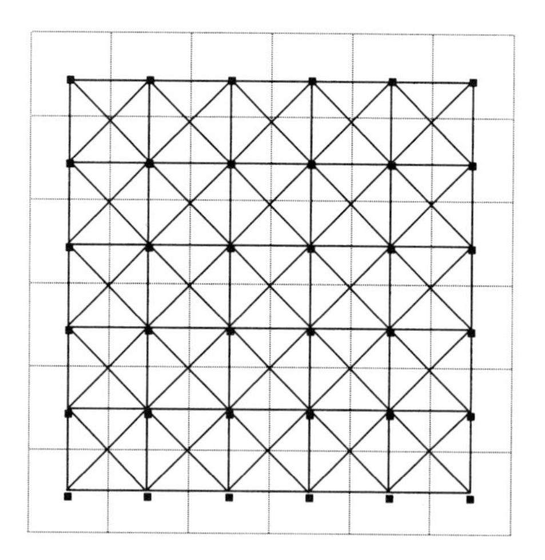

The cost to move horizontally or vertically from one cell to another is the total cost of the two cells divided by 2 cells. For example, to move from cell (1,1) to cell (2,1) would cost you (1+3) / 2 cells = 2.0.

The cost to move diagonally from one cell to another is the total cost of the two cells divided by 2 cells times the diagonal distance (square root of 2 = 1.4142). For example, to move from cell (4,3) to cell (3,2) would cost (2+5)/2 * 1.4142 = 4.9497

The algorithm computes the lowest possible accumulative cost for all cells neighboring the most recently assigned output cells. For example, to get to a resource from cell (4,3) you could move diagonally to cell (3,2) and it would cost you 4.9497. Or you could move from cell (4,3) up to cell (4,2) and then left to cell (3,2) and it would cost you (5+3)/2 + (3+2)/2 or 6.5. Or you could move from cell (4,3) left to cell (3,3) and then up to cell (3,2) and it would cost you (5+7)/2 + (7+2)/2 = 10.5. The lowest of the possible moves is the diagonal move which costs you 4.9.

Minimum Accumulative Costs

2.0	0	0	4.0		
4.5	4.0	0	2.5		
	7.1	4.5	4.9		
2.5	5.7				
0	1.5				

COST Grid:

1	3	4	4	3	2
4	6	2	3	7	6
5	8	7	5	6	6
1	4	5		5	1
4	7	5		2	6
1	2	2	1	3	4

The minimum accumulative costs of all neighboring cells is assigned to the output grid. In this example, it is 1.5 at cell (2,6). Then the accumulative cost of the new neighboring cells is computed. For cell (3,6) the accumulative cost is the cost of moving from cell (3,6) to cell (2,6) to cell (1,6) = (2+2) / 2 + 1.5 = 3.5. For cell (3,5) the accumulative cost is the cost of moving from to cell (2,6) to cell (1,6) = 1.4142(5+2)/2 + 1.5 = 4.9497 + 1.5 = 6.4497.

Minimum Accumulative Costs

2.0	0	0	4.0		
4.5	4.0	0	2.5		
	7.1	4.5	4.9		
2.5	5.7	6.4			
0	1.5	3.5			

COST Grid:

1	3	4	4	3	2
4	6	2	3	7	6
5	8	7	5	6	6
1	4	5		5	1
4	7	5		2	6
1	2	2	1	3	4

The minimum accumulative costs of all neighboring cells is assigned to the output grid. In this example, it is 2.0 at cell (1,1). Then the accumulative cost of the new neighboring cells is computed. In this case there are no new costs to compute.

Minimum Accumulative Costs

2.0	0	0	4.0		
4.5	4.0	0	2.5		
	7.1	4.5	4.9		
2.5	5.7	6.4			
0	1.5	3.5			

COST Grid:

1	3	4	4	3	2
4	6	2	3	7	6
5	8	7	5	6	6
1	4	5		5	1
4	7	5		2	6
1	2	2	1	3	4

The minimum accumulative costs of all neighboring cells is assigned to the output grid. In this example, it is 2.5 at cell (4,2). Then the accumulative cost of the new neighboring cells is computed. For example, for cell (5,1) the accumulative cost is the cost of moving from cell (5,1) to cell (4,2) to cell (3,2), which is 1.4142(3+3/2) + 2.5 = 6.7426. For cell (5,2) the accumulative cost is the cost of moving from cell (5,2) to cell (4,2) to cell (3,2), which is (7+3)/2 + 2.5 = 7.5. For cell (5,3) the accumulative cost is the cost of moving from cell (5,3) to cell (4,2) to cell (3,2), which is 1.4142 (6+3)/2 + 2.5 = 8.8639.

Minimum Accumulative Costs

2.0	0	0	4.0	6.7	
4.5	4.0	0	2.5	7.5	
	7.1	4.5	4.9	8.9	
2.5	5.7	6.4			
0	1.5	3.5			

COST Grid:

1	3	4	4	3	2
4	6	2	3	7	6
5	8	7	5	6	6
1	4	5		5	1
4	7	5		2	6
1	2	2	1	3	4

The minimum accumulative cost of all neighboring cells is assigned to the output grid. In this example, it is 2.5 at cell (1,5). Then the accumulative cost of the new neighboring cells is computed. For cell (1,4), the accumulative cost is (1+4)/2 + 2.5 = 5.0. And for cell (2,4), the accumulative cost is (4+1)/2 + (1+4)/2 + (4+1)/2 = 7.5.

Minimum Accumulative Costs

2.0	0	0	4.0	6.7	
4.5	4.0	0	2.5	7.5	
	7.1	4.5	4.9	8.9	
5.0	7.5				
2.5	5.7	6.4			
0	1.5	3.5			

COST Grid:

1	3	4	4	3	2
4	6	2	3	7	6
5	8	7	5	6	6
1	4	5		5	1
4	7	5		2	6
1	2	2	1	3	4

The minimum accumulative cost of all neighboring cells is assigned to the output grid. In this example, it is 3.5 at cell (3,6). Then the accumulative cost of the new neighboring cells is computed. For cell (4,5), the accumulative cost is NODATA because that is the COST Grid cell value. And for cell (4,6), the accumulative cost is (1+2)/2 + 3.5 = 5.0

Minimum Accumulative Costs

2.0	0	0	4.0	6.7	
4.5	4.0	0	2.5	7.5	
	7.1	4.5	4.9	8.9	
5.0	7.5				
2.5	5.7	6.4	nodata		
0	1.5	3.5	5.0		

COST Grid:

1	3	4	4	3	2
4	6	2	3	7	6
5	8	7	5	6	6
1	4	5		5	1
4	7	5		2	6
1	2	2	1	3	4

The minimum accumulative cost of all neighboring cells is assigned to the output grid. In this example, it is 4.0 at cells (4,1) and (2,2). Then the accumulative cost of the new neighboring cells is computed.

Minimum Accumulative Costs

2.0	0	0	4.0	6.7	
4.5	4.0	0	2.5	7.5	
8.0	7.1	4.5	4.9	8.9	
5.0	7.5				
2.5	5.7	6.4	nodata		
0	1.5	3.5	5.0		

COST Grid:

1	3	4	4	3	2
4	6	2	3	7	6
5	8	7	5	6	6
1	4	5		5	1
4	7	5		2	6
1	2	2	1	3	4

The minimum accumulative cost of all neighboring cells is assigned to the output grid. In this example, it is 4.5 at cells (1,2) and (3,3). Then the accumulative cost of the new neighboring cells is computed.

Minimum Accumulative Costs

2.0	0	0	4.0	6.7	
4.5	4.0	0	2.5	7.5	
8.0	7.1	4.5	4.9	8.9	
5.0	7.5	10.5	nodata		
2.5	5.7	6.4	nodata		
0	1.5	3.5	5.0		

COST Grid:

1	3	4	4	3	2
4	6	2	3	7	6
5	8	7	5	6	6
1	4	5		5	1
4	7	5		2	6
1	2	2	1	3	4

The iterative process continues until the minimum accumulative cost has been assigned to all cells. Then the grid is output as the result of the COSTDISTANCE function. And a companion Backlink grid is output as a grid of the direction of minimum accumulative costs. The following are the codes used for these directions.

6	7	8
5	0 Source Cell	1
4	3	2

Minimum Accumulative Costs Grid

2.0	0	0	4.0	6.7	9.2
4.5	4.0	0	2.5	7.5	13.1
8.0	7.1	4.5	4.9	8.9	12.7
5.0	7.5	10.5	nodata	10.6	9.2
2.5	5.7	6.4	nodata	7.1	11.1
0	1.5	3.5	5.0	7.0	10.5

Backlink Grid

1	0	0	5	4	5
7	1	0	5	5	6
3	8	7	6	6	3
3	5	7		3	4
3	4	4		4	5
0	5	5	5	5	5

Another companion grid that is output is a minimum cost allocation grid. This grid indicates which resource is the least costly to get to from each cell.

Resource Allocation Grid

1	1	1	1	1	1
1	1	1	1	1	1
2	1	1	1	1	2
2	2	1		2	2
2	2	2		2	2
2	2	2	2	2	2

COSTPATH

- **Estimates the least cost path from source cells to destination cells.**

In order to estimate the optimal path, information about the minimum accumulative cost and associated backlink direction is needed. Typically this cost information is supplied by first running the COSTDISTANCE function.

As an example, imagine that we have the following cost information from our previous COSTDISTANCE example. You want to estimate the optimal path to get from one cell location to a resource cell as "cheaply" as possible. The minimum accumulative costs grid might be expressing the cost in terms of dollars, time, distance, or any other cost measurement. The output optimal path grid has special codes as follows: 1 = a resource grid cell, 2= shared paths, 3= first optimal path, 4= second optimal path, 5= third optimal path, and so on.

In the following example, there is one optimal path (the first which is coded with 3s) and the total cost of reaching a resource is 13.1.

Minimum Accumulative Costs Grid

2.0	0	0	4.0	6.7	9.2
4.5	4.0	0	2.5	7.5	**13.1**
8.0	7.1	4.5	4.9	8.9	12.7
5.0	7.5	10.5	nodata	10.6	9.2
2.5	5.7	6.4	nodata	7.1	11.1
0	1.5	3.5	5.0	7.0	10.5

Backlink Grid

1	0	0	5	4	5
7	1	0	5	5	6
3	8	7	6	6	3
3	5	7		3	4
3	4	4		4	5
0	5	5	5	5	5

From-Cell Grid —— **COSTPATH** ——▶ Path Grid

From-Cell Grid

					100

Path Grid

				3	
		1	3		3

In the next example, there are two optimal paths for the two from-cells. The first optimal path is from the 200 cell and is straight left and costs 7.5. The second optimal path (coded with 4s) is the same as the path from our first example and costs 13.1. The Path Grid cell coded with a 2 is a cell that contains both paths going through it.

Minimum Accumulative Costs Grid

2.0	0	0	4.0	6.7	9.2
4.5	4.0	0	2.5	**7.5**	**13.1**
8.0	7.1	4.5	4.9	8.9	12.7
5.0	7.5	10.5	nodata	10.6	9.2
2.5	5.7	6.4	nodata	7.1	11.1
0	1.5	3.5	5.0	7.0	10.5

Backlink Grid

1	0	0	5	4	5
7	1	0	5	5	6
3	8	7	6	6	3
3	5	7		3	4
3	4	4		4	5
0	5	5	5	5	5

From-Cell Grid —— **COSTPATH** ——▶ Path Grid

From-Cell Grid

				200	100

Path Grid

				4	
		1	2	3	4

In some applications, you may be interested in the optimal path from any cell instead of from each individual cell. You can specify this by using the BYLAYER flag with the COSTPATH function. As an example, imagine that you have an existing road and want to estimate the optimal path from the existing road to any resource cell:

Minimum Accumulative Costs Grid

2.0	0	0	4.0	6.7	9.2
4.5	4.0	0	2.5	7.5	13.1
8.0	7.1	4.5	**4.9**	8.9	12.7
5.0	7.5	10.5	nodata	10.6	9.2
2.5	5.7	6.4	nodata	7.1	11.1
0	1.5	3.5	5.0	7.0	10.5

Backlink Grid

1	0	0	5	4	5
7	1	0	5	5	6
3	8	7	6	6	3
3	5	7		3	4
3	4	4		4	5
0	5	5	5	5	5

From-Cell Grid —— **COSTPATH** BYLAYER ——▶ Path Grid

From-Cell Grid

				1	1
1			1		
	1	1			

Path Grid

		1			
			3		

The COSTPATH function finds the minimum accumulative costs grid and looks at the values for all road cells. The lowest minimum accumulative cost is 4.9 and the COSTPATH function codes that cell location with a 3 as the optimal path from the road to any resource.

TOPOGRAPHIC OPERATIONS: SLOPE, ASPECT

SLOPE

- **Estimates rate of maximum slope change for each cell.**

The slope is computed from the 3 row by 3 column neighborhood of each cell. Basically, a plane is computed for this neighborhood and the maximum slope of the plane is returned to the cell being processed. At the edge of the grid, at least three cells (outside the grid's extent) will contain NODATA as their z values. These cells will be assigned the center cell's value. The result is a flattening of the 3x3 plane that is fit to these edge cells, which thus usually leads to a reduction in the slope.

As an example, imagine that you have a grid of elevation values and you want to estimate the percent slope of each cell. In this example, the cell size is 100 meters wide and long and the elevation values are in meters above sea level.

ELEVATION Grid:

100	100	100	100	200	300	300	300	200	100
100	100	100	100	200	300	300	300	200	100
100	100	100	100	200	300	300	300	200	100
100	100	100	100	200	300	300	300	200	100
100	100	100	100	200	300	300	300	200	100
500	500	500	500	200	300	300	300	200	100
1000	1000	1000	1000	200	300	300	300	200	100
1500	1500	1500	1500	500	500	500	500	500	100
2000	2000	2000	2000	500	500	500	500	500	100
3000	3000	3000	3000	500	500	500	500	500	100

SLOPE = SLOPE(ELEVATION, PERCENTRISE)

SLOPE Grid:

0.00	0.00	0.00	39.53	75.00	39.53	0.00	39.53	75.00	39.53
0.00	0.00	0.00	50.00	100.00	50.00	0.00	50.00	100.00	50.00
0.00	0.00	0.00	50.00	100.00	50.00	0.00	50.00	100.00	50.00
0.00	0.00	0.00	50.00	100.00	50.00	0.00	50.00	100.00	50.00
158.11	200.00	200.00	150.00	70.71	50.00	0.00	50.00	100.00	50.00
337.73	450.00	450.00	374.58	159.10	50.00	0.00	50.00	100.00	50.00
375.00	500.00	500.00	549.15	395.28	118.59	100.00	118.59	160.08	95.20
375.00	500.00	500.00	677.54	571.18	113.19	100.00	113.19	201.56	166.77
565.96	750.00	750.00	988.21	833.85	0.00	0.00	0.00	200.00	200.00
395.28	500.00	500.00	1064.34	833.85	0.00	0.00	0.00	158.11	158.11

ASPECT

- **Estimates the slope direction of each cell.**

Aspect estimates the maximum slope direction of each cell. The output value is the compass direction expressed from 0 to 360 (for example, 0=north, 45=northeast, 90 = east). If the estimate slope is zero, there is no slope direction and a value of –1 is returned for aspect of flat cells.

As an example, imagine that you have a grid of elevation values and you want to estimate the slope direction of each cell. In this example, the cell size is 100 meters wide and long and the elevation values are in meters above sea level.

ELEVATION Grid:

100	100	100	100	200	300	300	300	200	100
100	100	100	100	200	300	300	300	200	100
100	100	100	100	200	300	300	300	200	100
100	100	100	100	200	300	300	300	200	100
100	100	100	100	200	300	300	300	200	100
500	500	500	500	200	300	300	300	200	100
1000	1000	1000	1000	200	300	300	300	200	100
1500	1500	1500	1500	500	500	500	500	500	100
2000	2000	2000	2000	500	500	500	500	500	100
3000	3000	3000	3000	500	500	500	500	500	100

ASP_GRID = ASPECT(ELEVATION)

ASP_GRID Grid:

-1	-1	-1	288	270	252	-1	108	90	72
-1	-1	-1	270	270	270	-1	90	90	90
-1	-1	-1	270	270	270	-1	90	90	90
-1	-1	-1	270	270	270	-1	90	90	90
342	0	0	0	315	270	-1	90	90	90
358	0	0	26	45	270	-1	90	90	90
0	0	0	41	55	342	0	18	51	67
0	0	0	52	66	354	0	6	60	77
354	0	0	55	77	-1	-1	-1	90	90
18	0	0	50	103	-1	-1	-1	72	108

GRID ANALYSIS EXERCISES

1) You run **REGIONGROUP** on the following grid. Fill in the output grid:

Input Grid

0	90	90	270	270	90
0	90	90	270	270	90
90	90	90	270	270	90
90	90	90	270	270	90
180	180	180	180	180	180
180	180	180	180	180	180

Output Grid

Value	Count

2) What would the output grid contain if you use **TEST** with the following logical expression: **Value in {1,3,8,9}**

Input Grid

0	1	2	2	3	1
0	1	2	2	2	8
1	2	3	8	2	1
2	2	3	7	7	3
4	4	3	6	8	9
4	5	5	6	8	0

Output Grid

Output Grid Value Attribute Table

Value	Count

3) What would the output grid contain if you use **SELECT** with the following logical expression: **Value in {1,3,8,9}**

Input Grid

0	1	2	2	3	1
0	1	2	2	2	8
1	2	3	8	2	1
2	2	3	7	7	3
4	4	3	6	8	9
4	5	5	6	8	0

Output Grid

Output Grid Value Attribute Table

Value	Count

4) You have a grid of willow cells. You want to buffer all willow areas by 1 cell. Fill the output grid with the appropriate values after the **EXPAND** function is executed.

BUFWILLOW = EXPAND (WILLOW, 1, LIST, 4, 5, 6,)

WILLOW Grid

0					5				
0	4					5			
	4						5		
							5		
			6	6					
			6	6					
					6				
								19	
							19	19	

BUFWILLOW Grid

5) You have the following grids output from running the **COSTDISTANCE** function.

Accumulative Cost Grid

0	200	350	650	1183	1045
150	283	483	633	833	1033
350	574	566	866	1216	1045
750	845	786	612	716	750
1300	1257	745	362	450	300
1200	950	550	250	150	0

Backlink Grid

0	5	5	5	3	4
7	6	6	6	5	5
7	6	6	5	3	6
7	6	2	3	4	3
7	2	2	2	3	3
1	1	1	1	1	0

Fill in the output grid resulting from estimating an optimal path using the **COST-PATH** function:

From Cell Grid

		1			

Optimal Path Grid

6) You have a grid of forest types and a grid of three watersheds. You want to produce the following information.

	Number of Grid Cells in Watershed by Vegetation Type				
Watershed	**Black/White Spruce**	**Aspen-Birch**	**Riparian Shrub**	**Other**	**Total**
1					
2					
3					

Fill in the following flowchart with the appropriate grid tools to solve your problem.

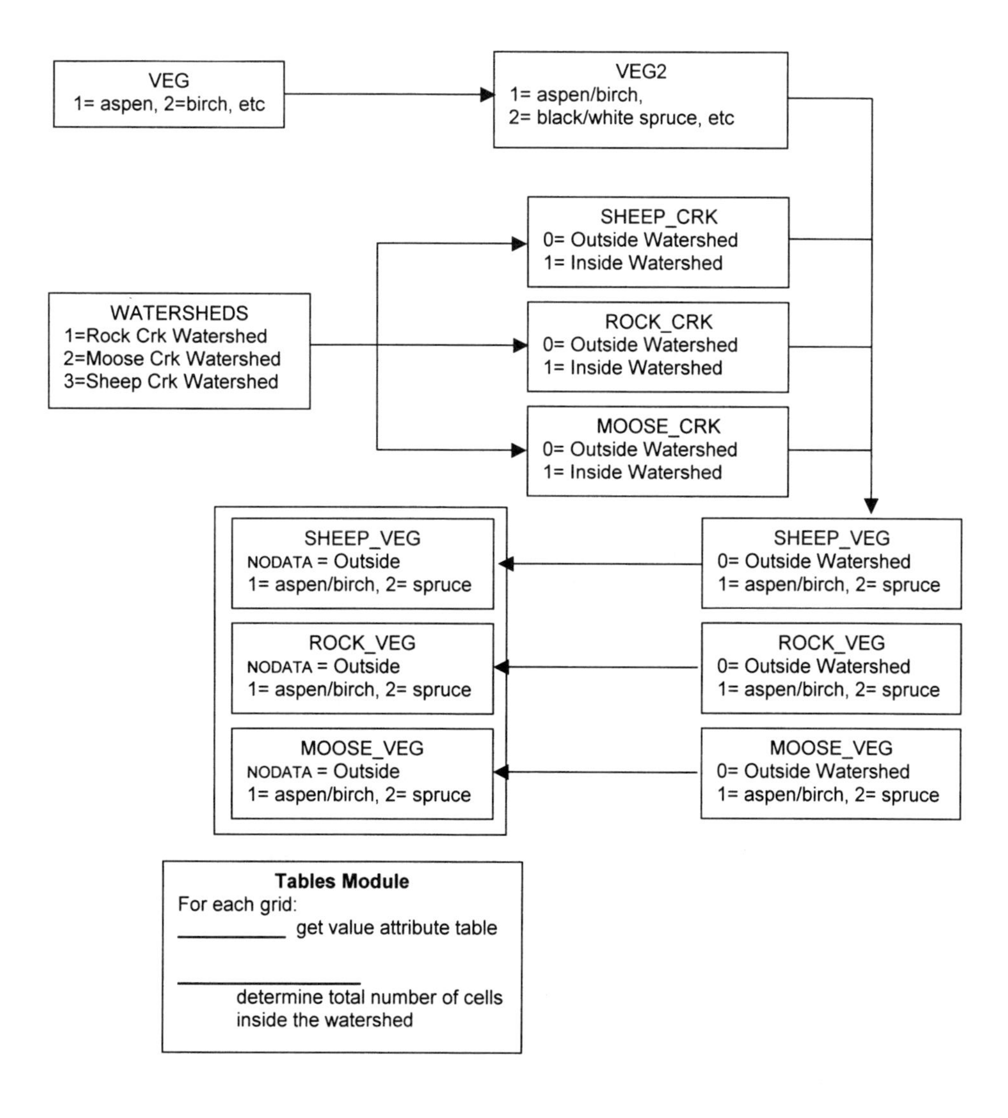

7) You have a 1/0 grid of all lightning strike locations in interior Alaska during last summer. You also have a grid of elevation values in meters for the same area. Each grid cell is 1000 by 1000 meters. You want to determine the following information:

Elevation Class	Total Number of Strikes	Total Area (ha)
Less than 500 m		
500 – 1,000 m		
> 1000 m		

Fill in the following flowchart with appropriate grid operations to solve your GIS problem:

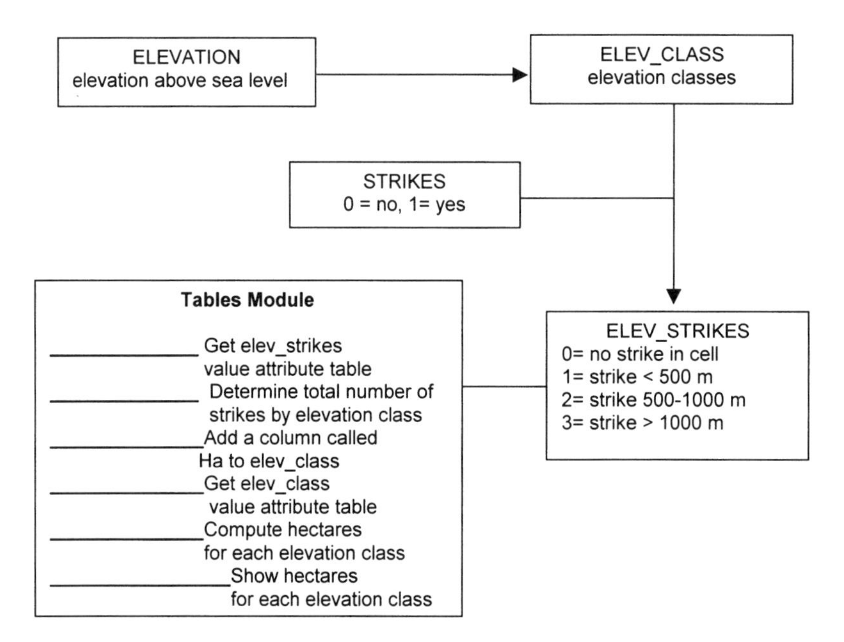

8) You have one grid (*veg*) that contains values of 71,72,73,74,75 for open water and values of 63,64,65,67 for wetland. The grids have a cell size of 25 meters. Draw a flow chart showing how you would produce a grid of wetlands cells that are on the shorelines of open water.

Fill in the following flowchart with appropriate grid operations to solve your GIS problem:

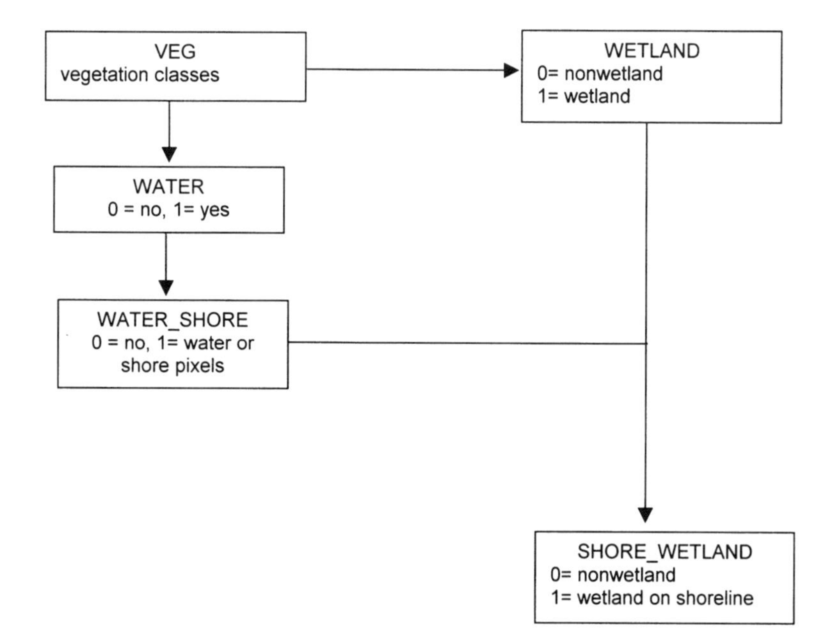

9) A forester is planning potential harvest units. She wants to find large (>100 ha) stands of closed aspen, birch, or aspen/birch forest (VEGCLASS 16-19) The SIZECLASS of the stands should be sawtimber or pole size (3 or 2) and the stands should be at least 100 meters from open water (VEGCLASS 80-89) and at least 50 meters from marsh areas (VEGCLASS 77-79). All grids have a cell size of 25 by 25 meters.

Fill in the following flowchart with the appropriate Grid tools to solve your GIS problem:

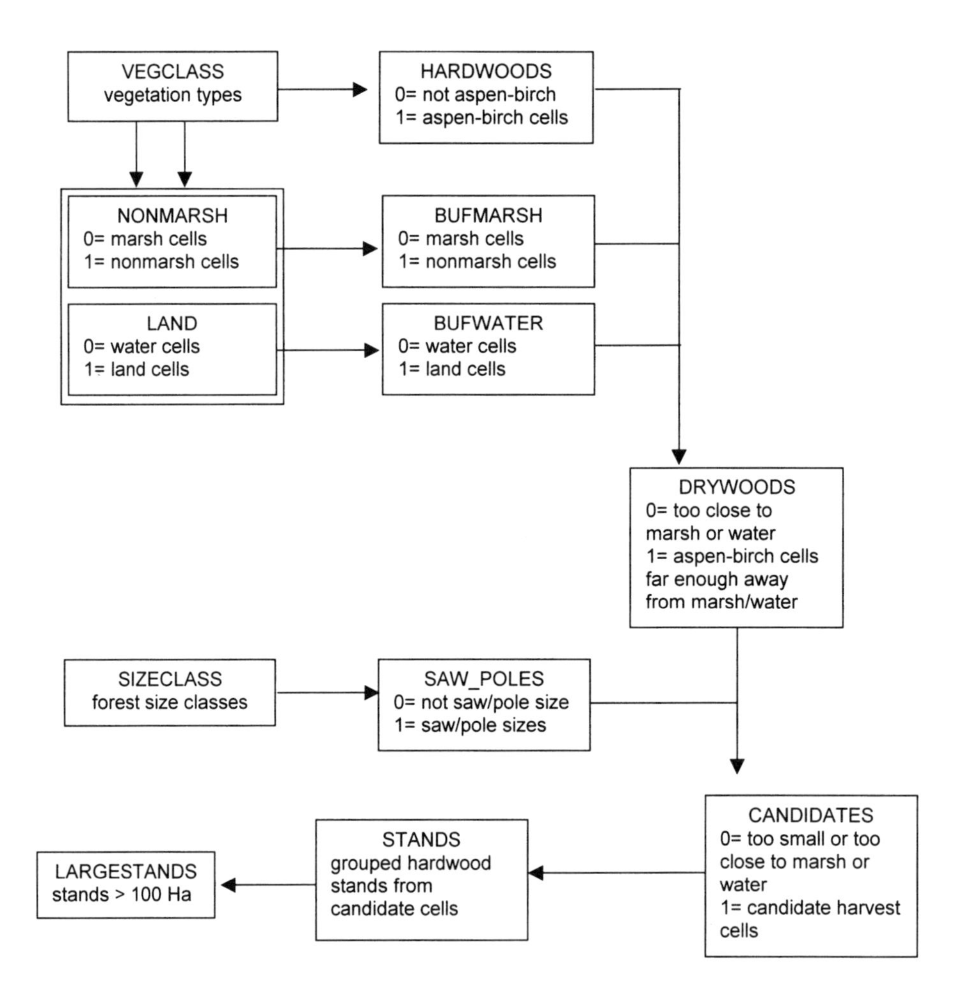

10) You have a grid of vegetation types (VEG), soil drainage class (DRAINAGE) and a grid of elevation values (ELEV). You are checking for possible errors in the VEG grid. Aspen typically grows on warm sites such as coarse, well-drained soils or warm, southerly-facing slopes. You want to find any cells that have an aspen vegetation class occurring on either poorly drained soils or cold slopes (slopes with an aspect from north to east, northwest to west and a gradient greater than 10%).

Fill in the following flowchart with the appropriate Grid tools to solve your GIS problem:

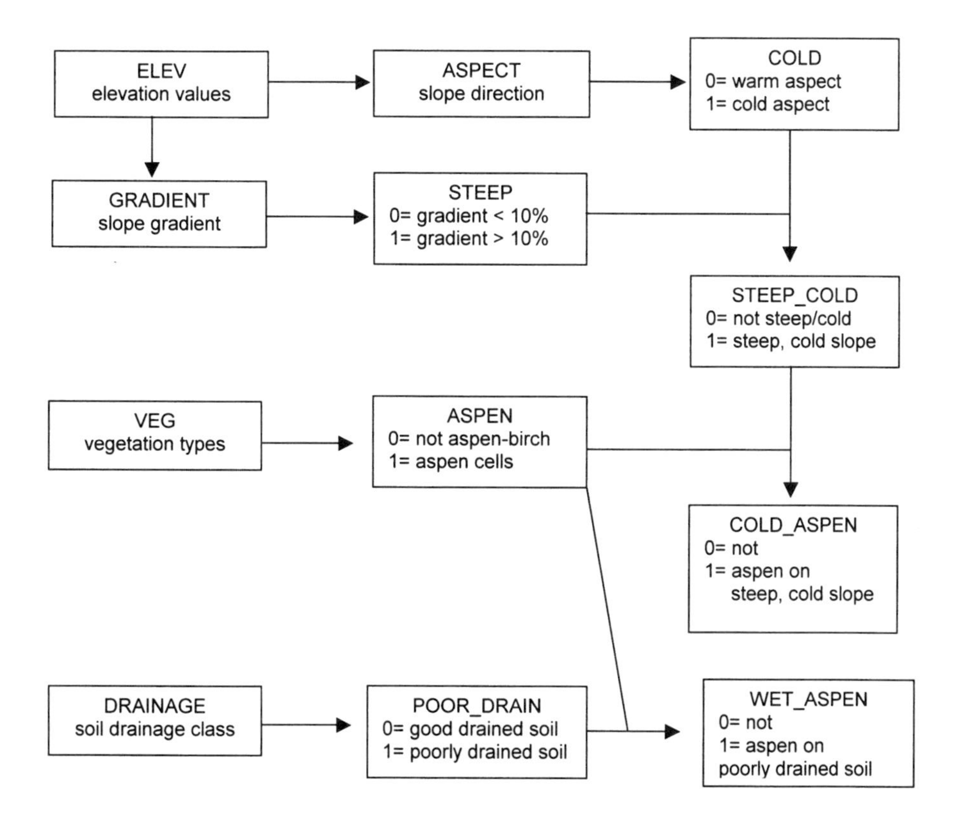

Chapter 9

Image Analysis

INTRODUCTION

An image is a special type of grid taken from some remote sensing device like a camera or digital scanner. Common image sources in GIS are digital orthophotos, satellite images, aerial video images, and scanned documents. Image processing techniques can be very sophisticated and entire software packages for image processing of remotely sensed images exist. Basic image processing operations are available in most GIS systems and are therefore covered in this chapter. These operations include image display and enhancement, image rectification, unsupervised and supervised image classification, and classification accuracy assessment.

IMAGE DISPLAY

A single band image is a grid with one value in each grid cell. These values are sometimes called digital numbers or DNs in remote sensing. To display a single band image, you control your output display intensity with image DNs.

Let's start with a simple example. Imagine that you have a display device that can only display in four patterns.

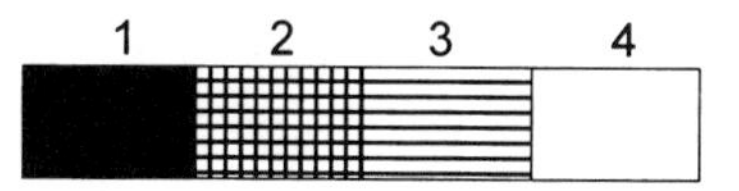

The range of digital numbers for your particular sensor is 0 to 255. Therefore one possible display rule is to use this range to assign pixel display as follows:

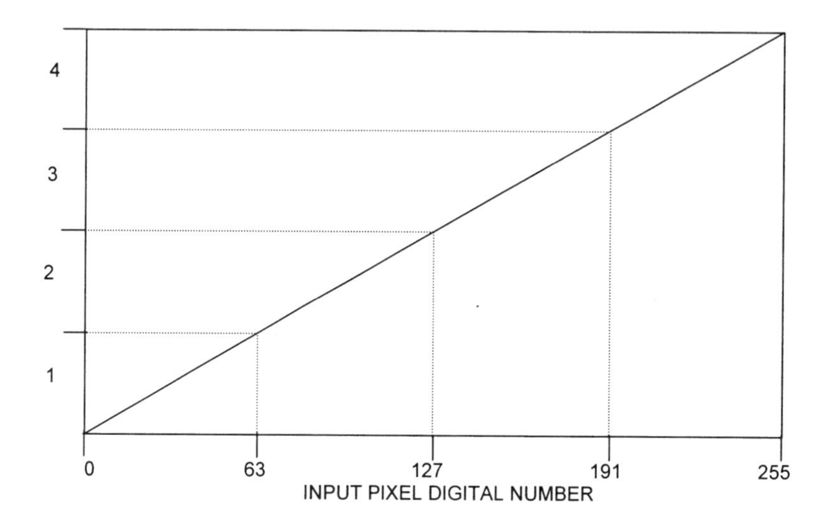

The rule from the previous linear function means the following:

Input Pixel DN	Output Shade
0 - 63	
64- 127	
128-191	
192-255	

Using the above linear function we could display your digital image as follows:

Single Band Image

98	96	95	90	13	12	11	85	85	86
98	95	90	11	10	10	85	84	84	85
97	96	11	11	10	14	89	83	86	83
97	95	10	10	11	85	84	86	85	86
96	91	10	10	11	85	85	85	84	85

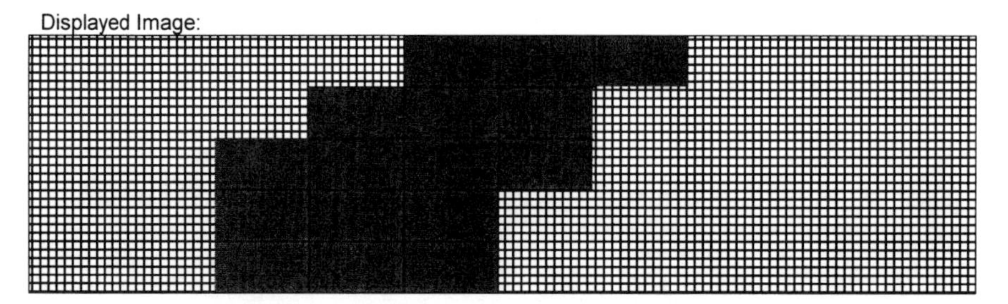

Imagine that coniferous vegetation has DNs ranging from 80 to 90, and broadleaf vegetation has DNs ranging from 90 to 100. The conifer pixels do not appear different than broadleaf pixels on your displayed image.

You could enhance the visual contrast between broadleaf and conifer vegetation by changing the linear contrast function as follows:

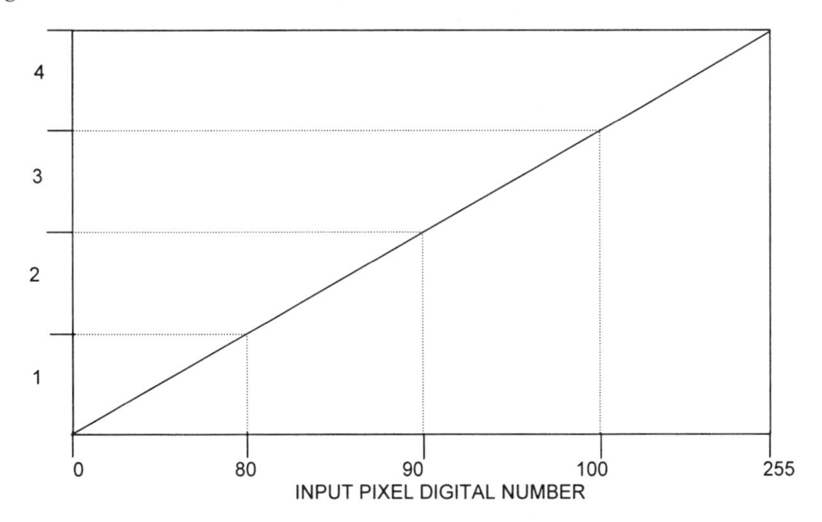

The rule from the previous linear function means the following:

Input Pixel DN	**Output Shade**
0 - 80	
80-90	
90-100	
100-255	

And apply your linear function to display your digital image in the four output shades:

Single Band Image:

98	96	95	90	13	12	11	85	85	86
98	95	90	11	10	10	85	84	84	85
97	96	11	11	10	14	89	83	86	83
97	95	10	10	11	85	84	86	85	86
96	91	10	10	11	85	85	85	84	85

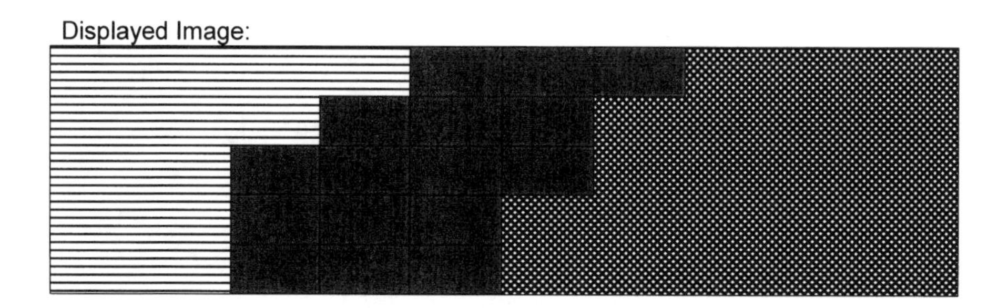

Notice that with the increased image contrast, the conifer and broadleaf pixels are displayed with different patterns.

LINEAR CONTRAST STRETCH

The linear function that you have been using is called a linear contrast stretch. Typically it is used with an image and a computer display that displays in video intensity values from 0 to 255. The GIS can compute the output video intensity for any pixel by using the following formula:

Video Intensity = [(Pixel Value – Minimum Pixel Value) / (Max – Min)] * 255

Here is a picture of the same linear contrast function:

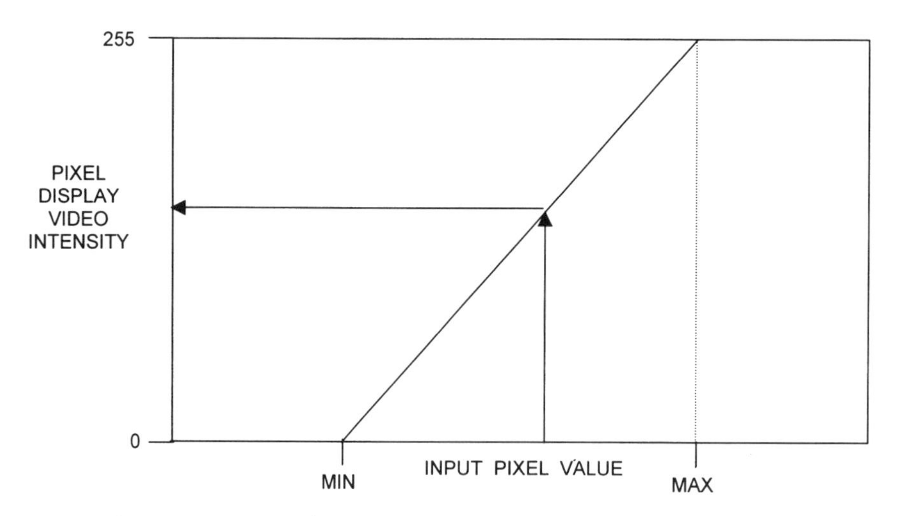

Let's say that you want to display your image to your computer screen using the linear contrast stretch. The minimum pixel value of your image is 10 and the maximum pixel value is 98. You could calculate the linear stretch output values and then fill in the appropriate video intensity values. As an example, the first row video intensity values are:

Input Pixel	Linear Contrast Stretch	Output Video Intensity
98	(98 – 10) / (98 – 10) * 255	255
96	(96 – 10) / (98 – 10) * 255	249
95	(95 – 10) / (98 – 10) * 255	246
90	(90 – 10) / (98 – 10) * 255	232
13	(13 – 10) / (98 – 10) * 255	9
12	(12 – 10) / (98 – 10) * 255	6
11	(11 – 10) / (98 – 10) * 255	3
85	(85 – 10) / (98 – 10) * 255	217
86	(86 – 10) / (98 – 10) * 255	220

Single Band Image:

98	96	95	90	13	12	11	85	85	86
98	95	90	11	10	10	85	84	84	85
97	96	11	11	10	14	89	83	86	83
97	95	10	10	11	95	84	86	85	86
96	91	10	10	11	85	85	85	84	85

Displayed Image Video Intensity Values:

255	249	246	232	9	6	3	217	217	220
255	246	232	3	0	0	217	214	214	217
252	249	3	3	0	12	229	212	220	212
252	246	0	0	3	246	214	220	217	220
249	235	0	0	3	217	217	217	214	217

By applying a linear contrast stretch, you can dramatically increase the range or contrast of your image display. For example, the original range of pixel values from your image was : 98 –10 = 88 and yet by using the linear contrast stretch, you can stretch the pixel display intensity values to 0 to 255.

The linear contrast stretch is analogous to film exposure. As an example, here is the same digital image of the Baltimore area displayed using different linear contrast stretches:

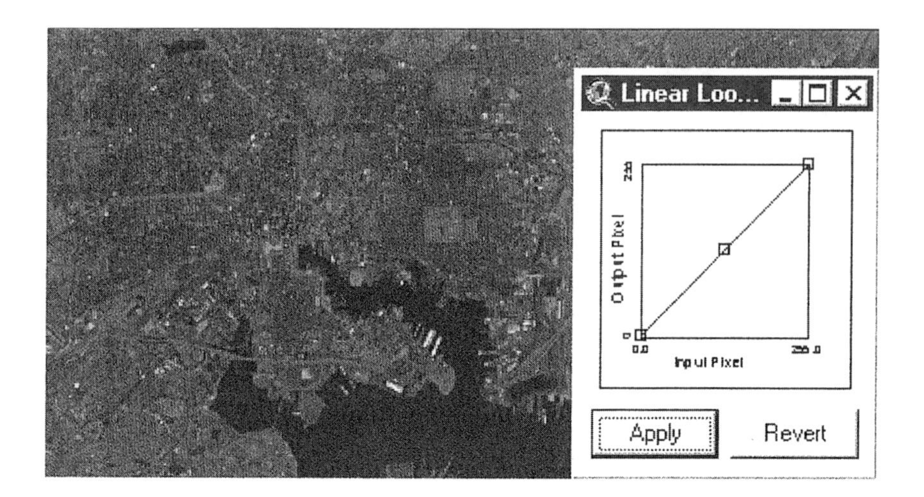

In the above example, each pixel value is controlling the video intensity value with no stretch. But since there are few pixels with values greater than 150, essentially the video intensity of 150 to 255 is being wasted on those few pixels.

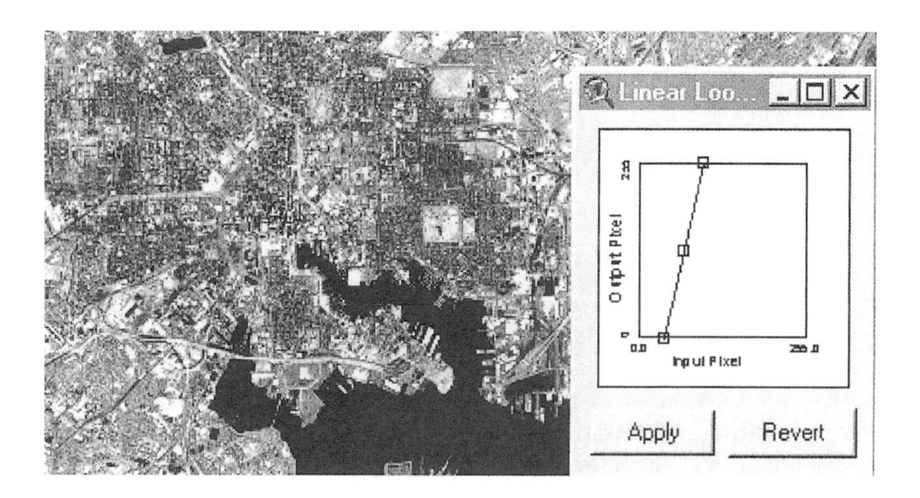

Most of the land pixels have values between 25 and 100. The linear contrast function above enhances the contrast of these pixels. Any pixels with values less than 25 will be black (the water pixels for example). And any pixels with values above 100 will be displayed with a video intensity of 255. The pixels between 25 and 100 are displayed with stretched video intensities ranging from 0 to 255.

IMAGE RECTIFICATION

Image rectification is the process of transforming a digital image from pixel (column, row) coordinates into your GIS coordinate system. It is a relatively simple process with satellite imagery that has little relief displacement. Typically a transformation called the *affine transformation* is used. This transformation is the same transformation that is used in GIS work when registering tics before digitizing new theme features.

THE AFFINE TRANSFORMATION

The affine transformation is a model of 2 equations that describe or fit a flat plane as follows:

Xmap = a + b1(image col) + b2(image row)

Ymap = a + c1(image row) + c2(image col)

or

Xmap = a + b1(pixel X) + b2(pixel Y)

Ymap = a + c1(pixel Y) + c2(pixel X)

Imagine the following simple example. You identify four locations on your image and estimate their map coordinates in the field using a GPS receiver.

Image GCP Pixels	***Xpixel***	**Ypixel**
1	200	1 200
2	400	1 400
3	100	1 600
4	500	1 800

Field GCP Point	***X_MAP***	**Y_MAP**
1	429 000	7 186 000
2	431 000	7 184 000
3	428 000	7 182 000
4	432 000	7 180 000

From these four ground control points or GCPs, you can develop linear models to predict any map coordinate for any image pixel (col,row) coordinate as follows:

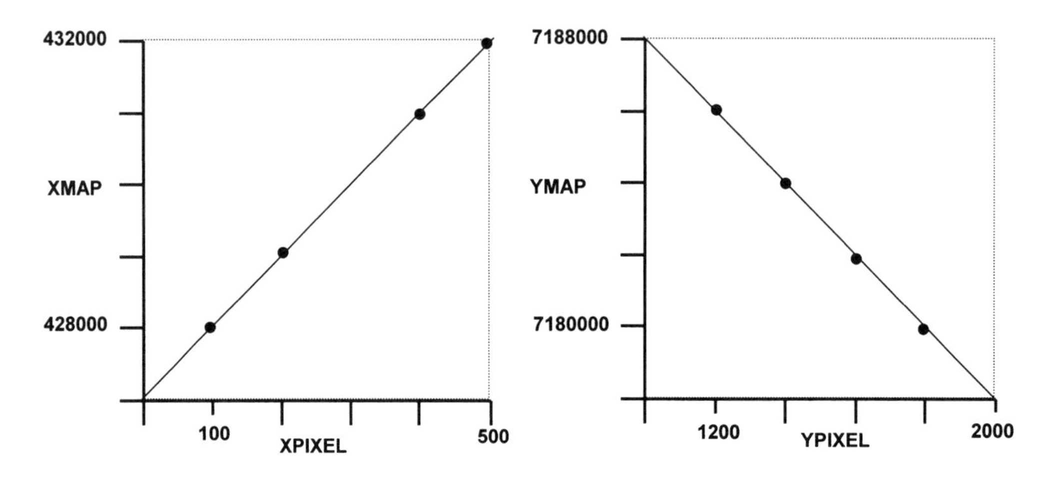

The affine transformation equations for these control points are:

XMAP = 427,000 + 10(XPIXEL) + 0(YPIXEL)

YMAP = 7,188,000 -10(YPIXEL) + 0(XPIXEL)

Notice that the slope of the line for the X linear model is positive while the slope of the line for the Y linear model is negative. This is because the image col and map X coordinate system increase in the same direction, while the image row and map Y coordinate system increase in opposite directions as follows:

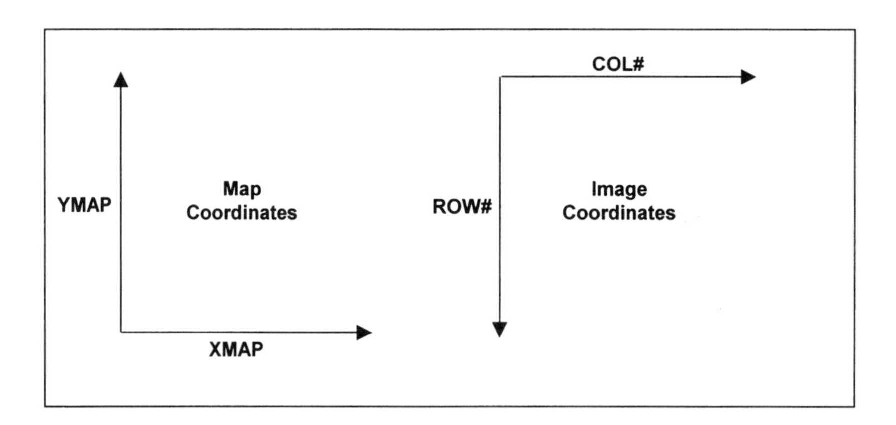

As a check, you could plug in any of the original image (col,row) coordinates and get your map coordinate estimates for that ground control location:

Image Pixel (col,row)		Xmap estimate	Ymap estimate
200	1 200	427,000 + 10(200) = 429,000	7,198,000 – 10(1200) = 7,186,000
400	1 400	427,000 + 10(400) = 431,000	7,198,000 – 10(1400) = 7,184,000
100	1 600	427,000 + 10(100) = 428,000	7,198,000 – 10(1600) = 7,182,000
500	1 800	427,000 + 10(500) = 432,000	7,198,000 – 10(1800) = 7,180,000

The affine transformation can "unwarp" an image by changing the following properties:

- Change in X, Y scale
- Translation in X,Y space
- Rotation of image in X,Y space

SCALING

The affine transformation can "stretch" the image in X and/or Y directions. For example, imagine that you located four image pixels and estimated their map coordinates.

GCP#	Image Column	Image Row	Map X	Map Y
1	125	500	2500	5000
2	250	500	5000	5000
3	125	1000	2500	2500
4	250	1000	5000	2500

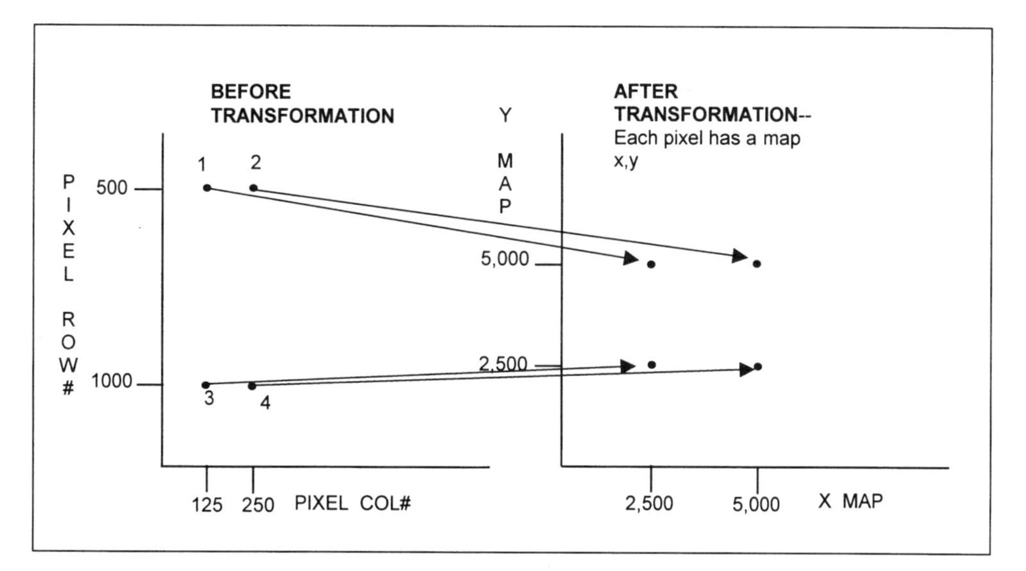

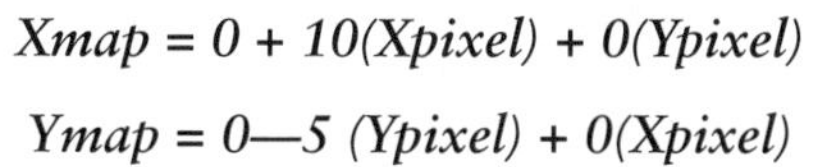

$$Xmap = 0 + 10(Xpixel) + 0(Ypixel)$$

$$Ymap = 0—5\ (Ypixel) + 0(Xpixel)$$

In this example, the X stretch factor is (5000 – 2500) / (250 – 125) = 2500 / 125 = 20. This means that the rectified pixels will have a width of 20 in the map coordinate space. The Y stretch factor is (5000 – 2500) / (500—1000) = 2500/ 500 = -5. This means that the rectified pixels will have a length of 5 in the map coordinate system.

TRANSLATION

The affine transformation can "translate" or shift the image in X and/or Y directions. For example, imagine that you located four image pixels and estimated their map coordinates.

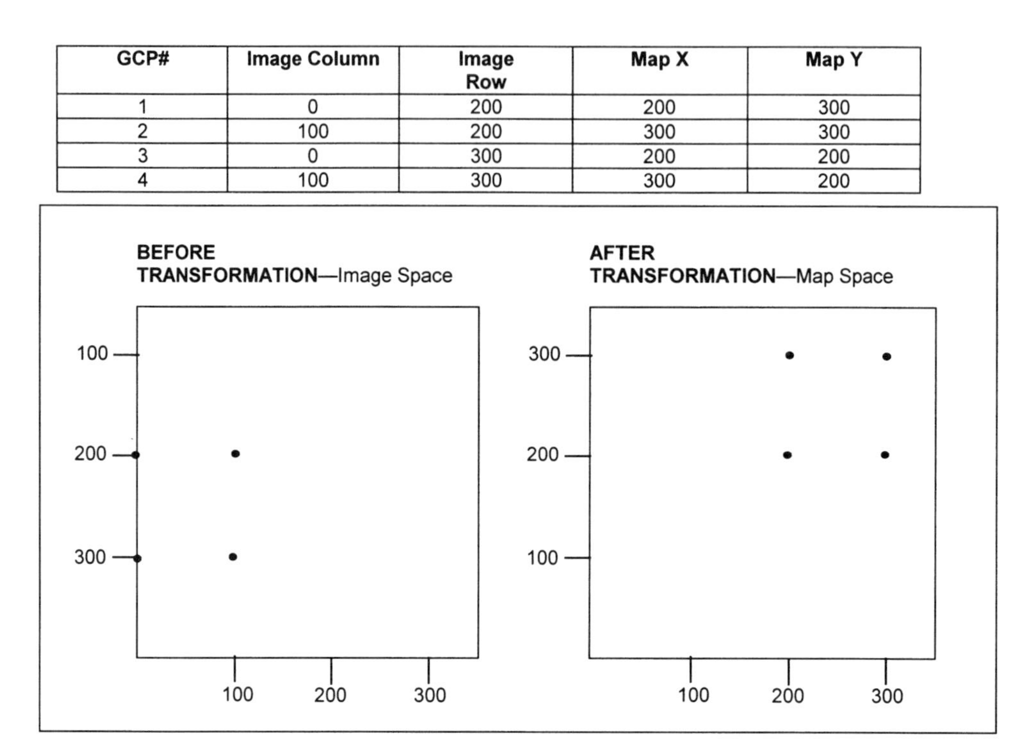

GCP#	Image Column	Image Row	Map X	Map Y
1	0	200	200	300
2	100	200	300	300
3	0	300	200	200
4	100	300	300	200

$$Xmap = 200 + 1(Xpixel) + 0(Ypixel)$$

$$Ymap = 500 - 1(Ypixel) + 0(Xpixel)$$

In this example, the affine transformation picks up the image and shifts it 200 units to the right and 100 units up into map space.

ROTATION

The affine transformation can "rotate" the image. For example, imagine that you located four image pixels and estimated their map coordinates.

GCP#	Image Column	Image Row	Map X	Map Y
1	0	625	250	500
2	250	750	500	500
3	125	375	250	250
4	375	500	500	250

$$Xmap = 0.8\ (Xpixel) + 0.4\ (Ypixel)$$

$$Ymap = -0.4(Xpixel) + 0.8(Ypixel)$$

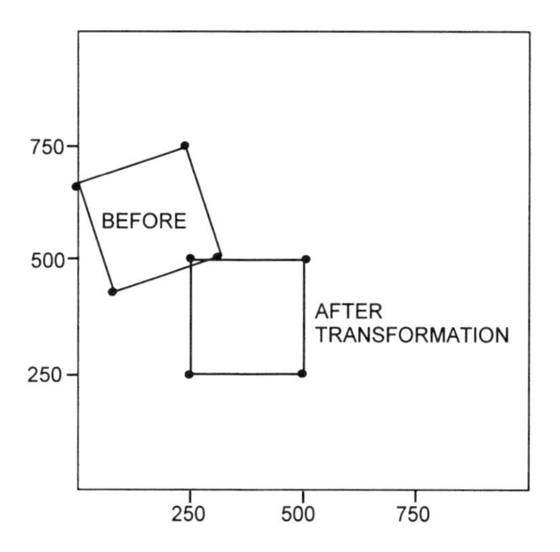

Rotation is especially important in rectifying polar-orbiting satellite imagery. Because the earth is rotating on its axis, a straight north to south orbit would result in images taken later in the day for southern regions relative to northern regions of the globe. To compensate for this, many satellites travel in a northeast to southwest orbit in an attempt to maintain a constant local time in images acquired during the orbit. These images need to be rotated in order for the north-south axis to be oriented up and down.

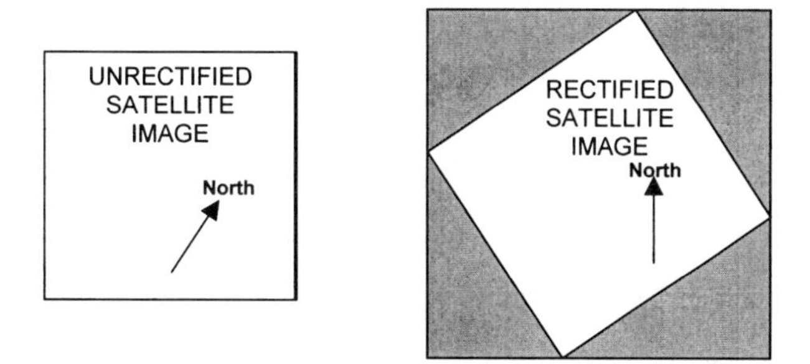

RECTIFICATION MODEL ERRORS

There will always be some amount of error in selection of ground control pixels and in estimating their associated map coordinates. For example, imagine that you selected eight ground control locations on an image as follows. The pixel coordinates in bold represented pixels that you selected that were not exactly the correct pixels for that ground control location.

GCP#	Pixel X (col)	Pixel Y (row)	Map X (GPS)	Map Y (GPS)
1	100	***98***	435000	7185000
2	**101**	***501***	435000	7175000
3	***499***	100	445000	7185000
4	500	***499***	445000	7175000
5	500	600	445000	7172500
6	***605***	***510***	447500	7175000
7	700	500	450000	7175000
8	700	***601***	450000	7172500

Using the eight GCP links, the GIS computes the best linear rectification model from your control point information . . .

Xmap = 432511.259 + 25.010 (Xpixel) + -0.074 (Ypixel)

Ymap = 7187448.139 – 24.906 (Ypixel) + 0.087 (Xpixel)

And the GIS computes the model X,Y and total error for each control point. For example, the predicted X,Y map coordinate for GCP#1 is:

Xmap Predicted = 432511.259 + 25.010 (100) + -0.074 (98) = 435005.007

Ymap Predicted = 7187448.139—24.906 (98) + 0.087 (100) = 7185016.051

And the X or Y errors are the GPS map coordinate minus the predicted map coordinate.

GCP #	X Pixel	Y Pixel	Xmap (GPS)	Xmap Predicted	X Error	Ymap (GPS)	Ymap predicted	Y Error
1	100	100	435000	435005.007	-5.007	7185000	7185016.051	-16.051
2	100	500	435000	435000.069	-0.069	7175000	7174978.791	21.209
3	500	100	445000	444983.982	16.018	7185000	7185000.695	-0.695
4	500	500	445000	444979.340	20.660	7175000	7175063.140	-63.140
5	500	600	445000	444971.834	28.166	7172500	7172547.597	-47.597
6	600	500	447500	447604.607	-104.607	7175000	7174798.259	201.741
7	700	500	450000	449981.332	18.668	7175000	7175055.545	-55.545
8	700	600	450000	449973.826	26.174	7172500	7172540.002	-40.002

Once the GIS has computed the X error and Y error for each ground control location, it can compute the total error for each GCP link as follows:

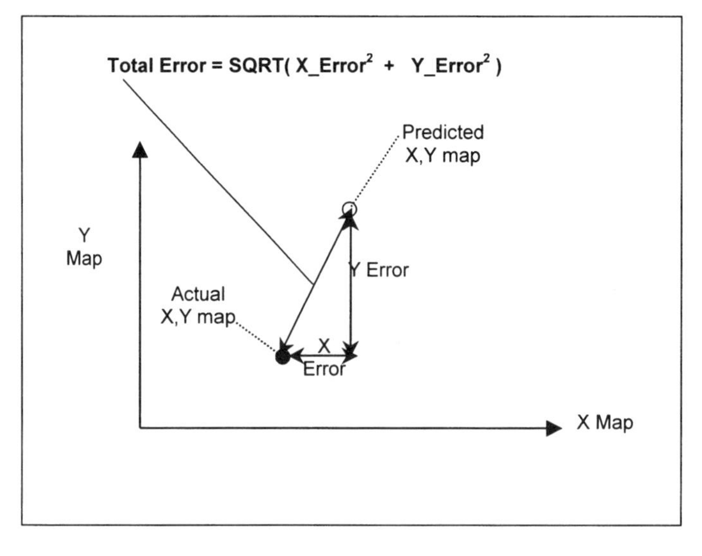

GCP#	Total Model Error
1	16.81
2	21.21
3	16.03
4	66.43
5	55.31
6	227.25
7	58.60
8	47.80

And the overall model error, called the *Root Mean Squared Error* is computed as:

RMS error = Square Root [(Sum of Total Errors Squared)/ Number of Links]
= SQRT [$(16.81^2 + 21.21^2 + 16.03^2 + 66.43^2 + 55.31^2 + 227.25^2 + 58.60^2 + 47.80^2) / 8$]
= 90.7 meters

IMAGE RESAMPLING OPTIONS

Once you have developed a rectification model with a satisfactory RMS error, the GIS applies the model to the image to translate the image from image (col, row) space to map (X,Y) space. The original image must be resampled because there is typically not a perfect match between image grid cells and the output map grid cells. With most GISs you can choose from three resampling options: nearest neighbor, bilinear interpolation, or cubic convolution.

Nearest Neighbor resampling selects the image pixel value that is closest to each output grid cell. It has the advantages of being the fastest resampling method and of preserving original pixel values. The disadvantage of nearest neighbor resampling is that the rectified image tends to have stair-step shaped linear features.

Bilinear Interpolation resampling selects the four of the image pixels that are closest to each output grid cell and outputs a weighted average based on these four pixels. The advantage of this resampling option is that it produces a smoother image and an image with better positional accuracy relative to the nearest neighbor option. The disadvantage of bilinear interpolation is that it is slower relative to nearest neighbor resampling.

Cubic Convolution resampling selects the 16 of the image pixels that are closest to each output grid cell and outputs a weighted average based on these 16 pixels. The advantage of this resampling option is that it produces a smoothest image relative to bilinear interpolation or nearest neighbor resampling. The disadvantage of bilinear interpolation is that it is very slow compared to bilinear interpolation or nearest neighbor resampling.

UNSUPERVISED CLASSIFICATION

Once an image has been rectified, you can classify it into a grid of land cover classes. There are two basic classification strategies: unsupervised and supervised classification. In unsupervised classification, you ask the GIS to group pixels that have similar spectral values into spectral classes. Then you decide what each spectral class represents in terms of land cover.

The first step in unsupervised classification is to ask the GIS to create spectral classes by grouping pixels that have similar values. You specify the parameters that the classifier should use such as the number of spectral classes, the number of computational iterations, the mininum number of pixels allowed in a spectral class, and the pixel sampling interval.

There are many clustering algorithms for unsupervised classification. One commonly used algorithm is the ISODATA or Iterative Self-Organizing Data Algorithm. As an example, imagine that you want to produce a land cover grid from a two-band image. You would first specify the clustering parameters as follows. A sample interval of 2 means every second pixel will be sampled starting with the upper left pixel and proceeding left to right, row by row. This speeds up the processing time for initial spectral clustering. The shaded pixels will be skipped in this step.

User-specified clustering parameters:
4 maximum number of classes, 2 iterations, sample_interval= 2

Input Image:

0 0	100 100	100 100	120 110	40 120	30 120
50 30	110 150	120 100	100 120	50 120	50 130
30 40	40 60	120 130	120 120	60 100	60 110
40 40	50 50	20 10	120 120	110 110	120 110

Start with 4 means at arbitrary intervals, then assign each sample to the closest class mean. The range of pixel values is 0 to 120. 120/4 classes = 30, so the starting means are (30,30), (60,60) (90,90), (120,120). Each sample pixel is assigned to the nearest mean.

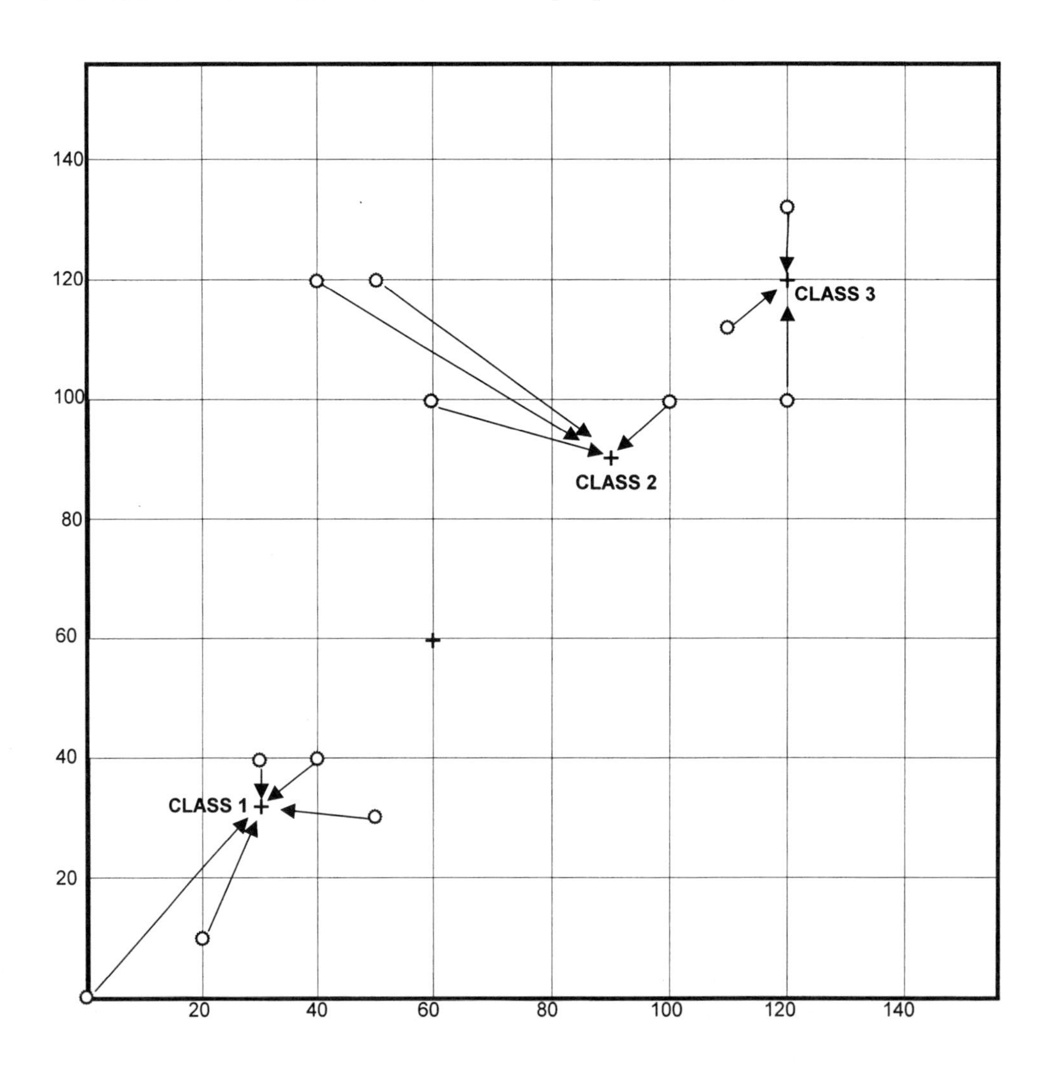

Interation #1:

Class	X-mean	Y-mean	Members
1	30	30	(0,0) (20,10) (50,30) (30,40) (40,40)
	60	60	NONE
2	90	90	(40,120) (50,120) (60,100) (100,100)
3	120	120	(110,100) (120,100) (120,130)

Interation#2: Compute new class means and plot new class means, then assign class members closest to each new class mean.

Iteration#2:

Class	X-mean	Y-mean	New Members
1	28	24	(0,0) (20,10) (50,30) (30,40) (40,40)
2	62.5	110	(40,120) (50,120) (60,100)
3	116.67	110	(100,100) (110,110) (120,100) (120,130)

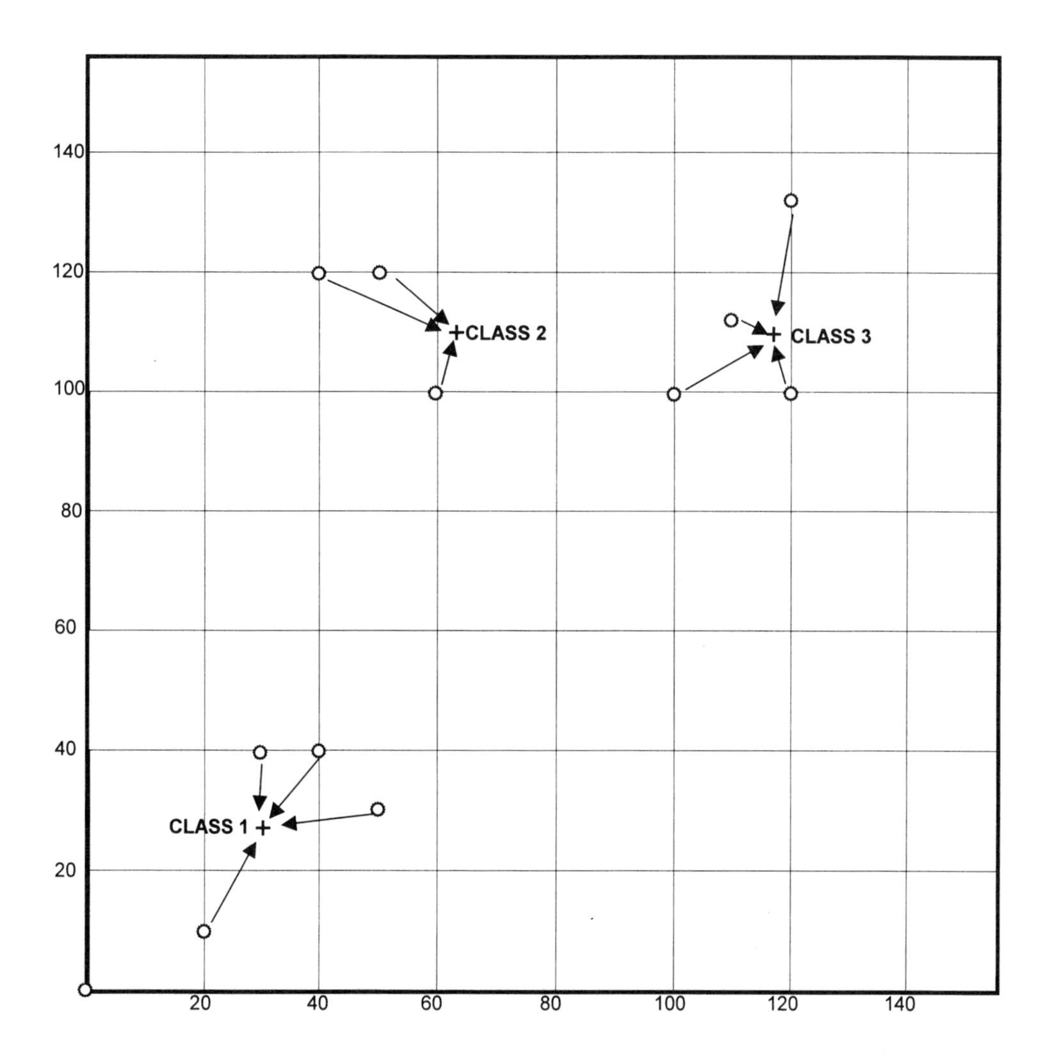

The next step is to use some rule to predict the spectral class of every pixel in the image. In this example, each pixel is assigned based on the closest spectral class means:

Spectral Class	X-mean	Y-mean
1	30	27.5
2	62.5	110
3	116.67	110

Input Image:

0	100	110	120	40	30
0	100	100	110	120	120
50	110	100	100	50	50
30	150	110	120	120	130
30	40	120	120	60	60
40	60	130	120	100	110
40	50	20	120	110	120
40	50	10	120	110	110

Classified Spectral Class Grid:

1	3	3	3	2	2
1	3	3	3	2	2
1	1	3	3	2	2
1	1	1	3	3	3

The final step would be for you to decide what each spectral class represents in terms of cover type and assign symbology for each class. For example, imagine that you determined that spectral class 1 represents water, 2 represents vegetation, and 3 represents sand. You could then assign symbols to each spectral class.

Classified Spectral Class Grid:

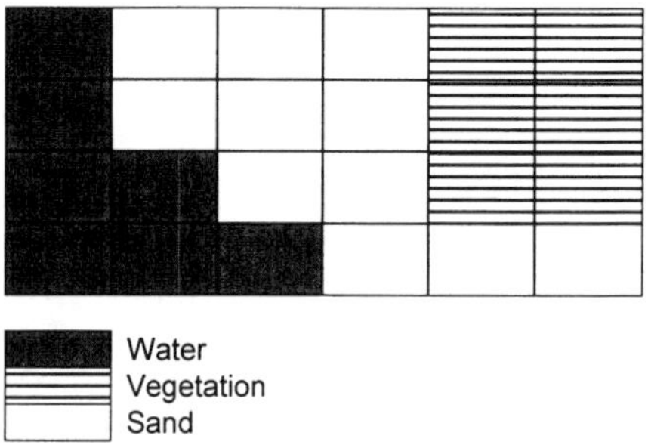

Water
Vegetation
Sand

SUPERVISED CLASSIFICATION

In supervised classification, you sample known cover class areas and the GIS uses the sample statistics from these areas to predict the cover class of each pixel in the image. The sample areas are typically called training fields or training areas and you need at least one training field for each cover class in your image.

Supervised classification is not necessarily better than unsupervised classification, it is simply a different strategy in classification. Supervised classification has the following advantages relative to unsupervised classification:

1) The user can delineate training fields from cover classes that could be mixed spectral classes in unsupervised classification, such as shadow versus water versus a burned area.
2) With maximum likelihood classifier, the user can produce probability estimates for each classified pixel and thus have an estimate of the "confidence" in the cover class predictions.

Supervised classification has the following disadvantages relative to unsupervised classification.

1) Rare cover types that exist in the image area may not be known by the user and therefore are not included in the classification.
2) An error in training field delineation can significantly alter the classification results. Each training field should be from large, homogeneous cover class areas and this is sometimes difficult to achieve.

SEED PIXEL TRAINING FIELDS

You have several options in delineating training fields. You could use existing land cover polygons to delineate training fields or you could digitize training polygons directly from your displayed image. One problem with these two approaches is that sometimes it is difficult to get a large training field without including some unrepresentative pixels.

Another approach, called the seed pixel approach, was invented to minimize this problem. As an example, imagine that you want to delineate a training field of a burned area from the following area of a large image. The first step is to point to a "seed pixel" that most represents the burned area. In this example, the seed pixel is shaded in gray. The seed pixel algorithm then searches neighboring pixels and develops a parallelepiped or rectangle based on the minimum and maximum values within the neighboring pixels. In this example, the parallelepiped would be as follows:

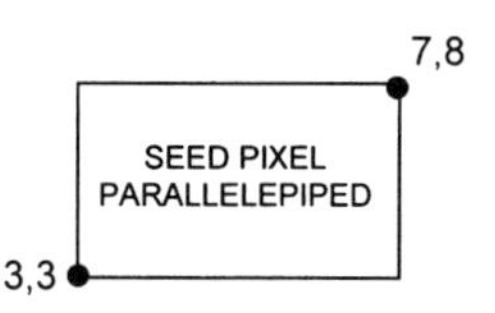

0 0	64 161	6 6	6 6	6 6	50 110	60 120	6 6	6 6	6 6
0 0	5 6	54 163	5 6	5 6	5 6	5 6	5 6	5 6	5 6
4 5	0 0	4 5	4 5	4 5	4 5	4 5	4 5	134 124	4 5
6 6	0 0	6 6	3 6	6 6	6 6	6 6	6 6	6 6	0 0
5 6	0 0	5 6	5 6	5 6	5 6	5 6	5 6	0 0	5 6
6 6	0 0	6 6	6 8	6 3	7 6	6 6	0 0	6 6	6 6
0 0	5 6	30 100	5 6	5 6	5 6	0 0	42 107	42 107	50 110
0 0	24 95	40 105	6 6	4 5	4 5	0 0	0 0	0 0	45 106
26 106	46 106	6 6	5 6	6 6	6 6	40 105	0 0	0 0	0 0
55 116	5 6	5 6	5 6	5 6	5 6	42 107	44 108	45 106	50 110

The GIS would then search all surrounding pixels and include any touching pixels that fall inside the seed pixel parallelepiped as a training field pixel representative of a burned pixel. These pixels are shaded in gray.

0 0	64 161	6 6	6 6	6 6	50 110	60 120	6 6	6 6	6 6
0 0	5 6	54 163	5 6	5 6	5 6	5 6	5 6	5 6	5 6
4 5	0 0	4 5	4 5	4 5	4 5	4 5	4 5	134 124	4 5
6 6	0 0	6 6	3 6	6 6	6 6	6 6	6 6	6 6	0 0
5 6	0 0	5 6	5 6	5 6	5 6	5 6	5 6	0 0	5 6
6 6	0 0	6 6	6 8	6 3	7 6	6 6	0 0	6 6	6 6
0 0	5 6	30 100	5 6	5 6	5 6	0 0	42 107	42 107	50 110
0 0	24 95	40 105	6 6	4 5	4 5	0 0	0 0	0 0	45 106
26 106	46 106	6 6	5 6	6 6	6 6	40 105	0 0	0 0	0 0
55 116	5 6	5 6	5 6	5 6	5 6	42 107	44 108	45 106	50 110

IMAGE CLASSIFIERS

There are many different types of classifiers that have been invented over the years. Many are complicated and available in sophisticated image processing software. In GIS, the more common classifiers include the Parallelepiped Classifier, and the Maximum Likelihood Classifier. The parallelepiped classifier is a very simple, quick classifier that is used in GIS. Because it is a fast classifier, it is commonly used in the seed tool algorithm. However, it has several disadvantages in image classification:

1) The parallelepiped classifier uses minimum and maximums that can be rare extreme values.
2) Overlapping cover class parallelepipeds commonly occur and must be dealt with in classification decisions.
3) Pixel values that fall outside of all cover class parallelepipeds result in unclassified pixels.
4) The parallelepiped classifier does not represent spectral correlation or class variability, which is common in remotely sensed images. For example, the following parallelepiped was developed using two correlated spectral bands and two classes that had different spectral variability. Because the parallelepiped classifier ignores spectral correlation and variability, it would incorrectly classify the candidate pixels into Class A instead of Class B in the example below.

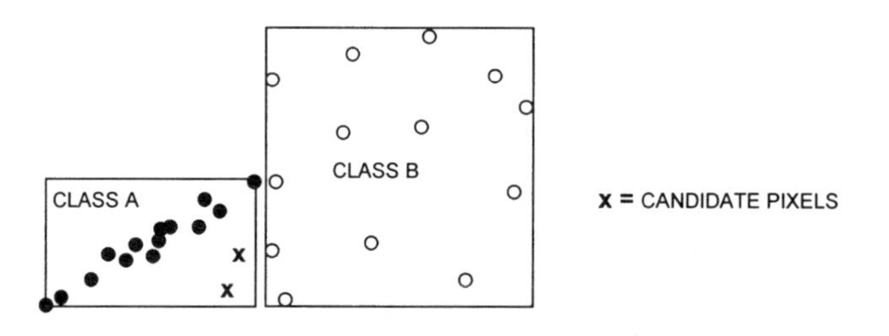

5) The parallelepiped classifier classifies each pixel with no estimate of "confidence" in its prediction.

MAXIMUM LIKELIHOOD CLASSIFIER

The maximum likelihood classifier assumes that pixel values from a given cover class are distributed as a bell-shaped curve centered around the mean for that cover class. For example, imagine that you are trying to classify water bodies on an alluvial floodplain. As your first step in the classification, you want to stratify pixels as either water or land pixels. You develop training fields for both water and land pixels as follows:

Training Polygon Statistics:

	Mean	Std Dev
Water	20	5
Land	30	5

The maximum likelihood classifier assumes that most of the land pixels will have a value of 30 and most of the water pixels will have a value of 20. The standard deviation is an estimate of variability and can be used with the formula for a bell-shaped curve to compute likelihood estimates for any pixel value as follows:

Likelihood(CoverClass) = [1/(sd * sqrt(2*pi))] exp –[(pixelvalue – mean)^2 / (2* sd ^{2})]

Likelihood Values for a given pixel being from water or land classes:

Pixel Value	Likelihood (Water)	Likelihood (Land)
5	0.00089	0.0000003
10	0.01079	0.00003
15	0.04840	0.00089
20	0.07979	0.01079
25	0.04840	0.04840
30	0.01079	0.07979
35	0.00089	0.04840
40	0.00003	0.01079
45	0.0000003	0.00089

The likelihood values can be plotted as a bell-shaped curve with the total area under each class curve summing to 1.0.

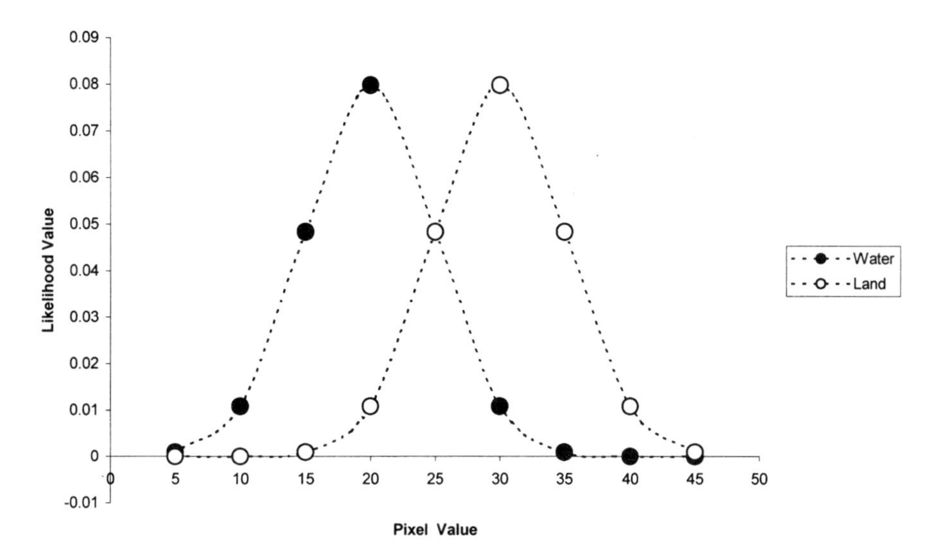

Once the likelihood values have been estimated, the GIS can look up the likelihood value for water versus land for any pixel value. The pixel is assigned the class that has the highest likelihood.

Original Image

19	32	33	34	19
30	20	35	18	27
32	29	19	27	28
32	30	26	21	26
31	32	30	26	22

Classified Image (1=water 2=land)

1	2	2	2	1
2	1	2	1	2
2	2	1	2	2
2	2	2	1	2
2	2	2	2	1

In this example, any pixel with a value less than 25 would have a higher likelihood for water relative to land. And any pixel with a value greater than 25 would have a higher likelihood for land relative to water. The likelihood values are equal at a pixel value of 25 . . . this type of tie rarely occurs in GIS applications since the likelihoods are computed as floating point values and the tie would rarely occur at a whole number value.

CLASSIFICATION ACCURACY ASSESSMENT

It is relatively easy to create a land cover grid by classifying remotely sensed images with a GIS. A thorough classification accuracy assessment is critical before accepting the classification as useful. In order to conduct a classification accuracy assessment, you need validation data to compare with your classification predictions. Such validatation data are often called "ground truth" or reference data.

REFERENCE DATA

How do you obtain reference data? One approach would be to randomly select image pixels and then locate these pixels in the field to determine their actual cover type. There are some problems with this approach:

1) It is often impossible to find single pixels in the field because of positional error in rectification, and errors in field navigation.
2) By randomly selecting pixels, most sampled pixels will be from dominant cover classes and yet the most important cover classes may be relatively rare on the landscape and not likely to be randomly selected.
3) Travel to randomly selected locations can be very expensive in remote areas and this could lead to an inadequate sample size of reference data, especially for the uncommon land cover classes.

Another approach is to use vector vegetation polygons as reference data in your accuracy assessment. There are some problems with this approach.

1) Vector polygons have a larger minimum mapping unit area relative to many classified images. For example, your vector polygons may have been mapped down to a minimum mapping unit area of 1 hectare while a 30-meter pixel is .09 hectares in area. So comparing polygons with grid cells can lead to a conservatively biased accuracy assessment.

2) Vector polygons may be from an older source like aerial photography or maps decades before the land cover source was taken. Changes in land cover due to the temporal differences between the polygons and the classified image may lead to a conservatively biased accuracy assessment.
3) There is positional error inherent in both vector polygons and in classified images. The positional error may lead to positional differences between correctly classified pixels and correctly classed polygons which may also lead to a conservative estimate of classification accuracy.

ERROR MATRIX

The error matrix is the foundation of accuracy assessment—it compares predicted classes with actual classes. The error matrix is typically structured as follows:

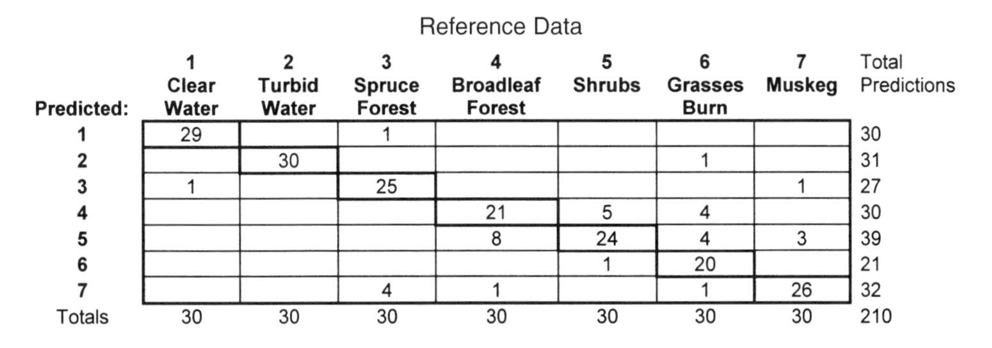

Reference Data

Predicted:	1 Clear Water	2 Turbid Water	3 Spruce Forest	4 Broadleaf Forest	5 Shrubs	6 Grasses Burn	7 Muskeg	Total Predictions
1	29		1					30
2		30				1		31
3	1		25				1	27
4				21	5	4		30
5				8	24	4	3	39
6					1	20		21
7			4	1		1	26	32
Totals	30	30	30	30	30	30	30	210

Each diagonal cell represents the total number of correctly predicted pixels for the cover class. For example, there were 29 correctly predicted clear water pixels. The sum of the diagonal cells divided by the total gives the overall classification accuracy.

Overall Classification Accuracy = [(29 + 30 + 25 + 21 + 24 + 20 + 26) / 210] * 100 = 83%

PRODUCER'S AND CONSUMER'S ACCURACY

For each class there are two types of accuracy: producer's and consumer's accuracy. Producer's accuracy is computed by looking at the predictions produced for a class and determining the percentage of correct predictions. For example, for the shrub class there were 39 pixels that were predicted to be shrub and the associated producer's accuracy for shrub is:

Shrub Producer's Accuracy = (24 correct predictions/ 39 total predictions) * 100 = 62%

Consumer's accuracy is computed by looking at the reference data for a class and determining the percentage of correct predictions for these samples. For example, for the shrub class there were 30 pixels that were actually shrub and the associated consumer's accuracy for shrub is:

Shrub
Producer's = (24 correct predictions/ 30 total shrub pixel) * 100 = 80%
Accuracy

IMAGE ANALYSIS EXERCISES

1) Match the images with their companion linear contrast stretch functions:

Image#1

Image#2

Image#3

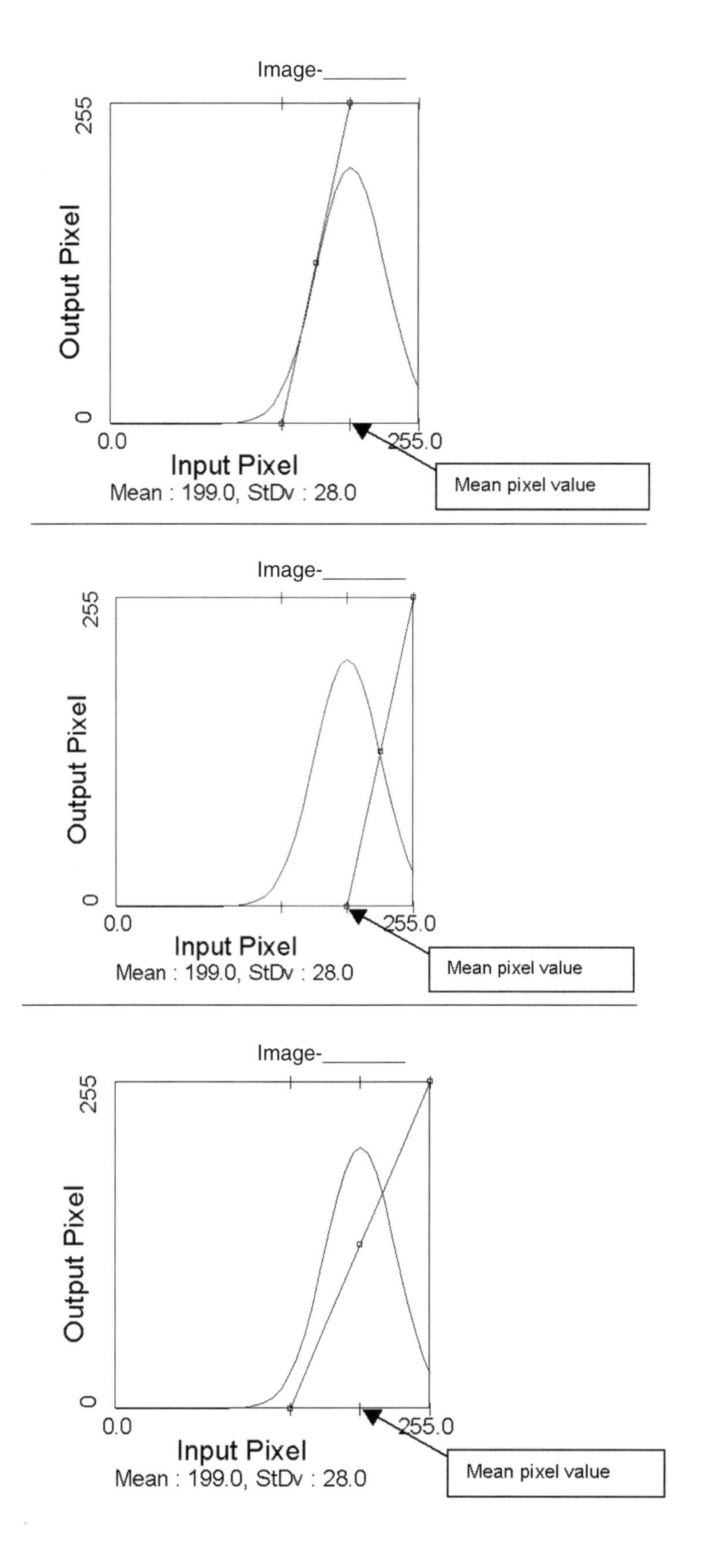
Image-________
255
Output Pixel
0
0.0
255.0
Input Pixel
Mean : 199.0, StDv : 28.0
Mean pixel value
Image-________
255
Output Pixel
0
0.0
255.0
Input Pixel
Mean : 199.0, StDv : 28.0
Mean pixel value
Image-________
255
Output Pixel
0
0.0
255.0
Input Pixel
Mean : 199.0, StDv : 28.0
Mean pixel value

2) Circle the appropriate resampling method (either nearest neighbor or cubic convolution) for the following two rectified images:

Nearest Neighbor or Cubic Convolution?

3) You are trying to co-register a 30-meter Landsat Thematic Mapper image with a geo-rectified digital raster graphics topographic map. Using five ground control pixels from your 30-meter TM image, you find the following:

Map Coordinates		Image Coordinates	
UTM X	UTM Y	Column	Row
518,000	5,190,940	1	3
518,030	5,190,880	2	5
518,060	5,190,970	3	2
518,090	5,190,910	4	4
518,120	5,191,000	5	1

Using these data, develop a linear transformation model for image rectification. Fill in the proper transformation coefficients.

UTM_X = _________ + ________(Image Column) + 0(Image Row)

UTM_Y = _________— ________(Image Row) + 0(Image Column)

4) You have the following table of ground control links for developing a linear rectification model. The output image grid cell size is 25 meters. What is the RMS error of the rectification model in image pixels?

GCP#	Total Model Error
1	26.81
2	11.21
3	26.03
4	66.43
5	45.31
6	37.25
7	58.60
8	47.80

5) A consulting group supplies you with rectified satellite imagery. The contract specifies the rectification RMS error to be less than 1 pixel. The rectified imagery has a pixel size of 25 meters. Circle any of the following statements if they are correct:

- The positional accuracy of each pixel in the rectified image is +/- 1 pixel.
- Every pixel in the rectified image is within 50 meters of its true map location.
- The positional accuracy of the rectified image is 25 meters.

6) You are delineating training fields using a parallelepiped seed-pixel approach. You pick the shaded pixel as your seed pixel and the parallelepiped is based on the eight neighboring pixels. Circle all pixels that would be included in your training field.

<table>
<tr><td>0
0</td><td>64
161</td><td>6
6</td><td>6
6</td><td>6
6</td><td>50
110</td><td>60
120</td><td>6
6</td><td>6
6</td><td>6
6</td></tr>
<tr><td>0
0</td><td>5
6</td><td>54
163</td><td>5
6</td><td>5
6</td><td>5
6</td><td>5
6</td><td>5
6</td><td>5
6</td><td>5
6</td></tr>
<tr><td>4
5</td><td>0
0</td><td>4
5</td><td>52
161</td><td>4
5</td><td>4
5</td><td>4
5</td><td>4
5</td><td>134
124</td><td>4
5</td></tr>
<tr><td>6
6</td><td>0
0</td><td>6
6</td><td>60
170</td><td>54
153</td><td>48
144</td><td>6
6</td><td>6
6</td><td>6
6</td><td>0
0</td></tr>
<tr><td>5
6</td><td>0
0</td><td>5
6</td><td>50
155</td><td>50
150</td><td>51
152</td><td>5
6</td><td>5
6</td><td>0
0</td><td>5
6</td></tr>
<tr><td>6
6</td><td>0
0</td><td>6
6</td><td>52
157</td><td>49
144</td><td>40
130</td><td>42
137</td><td>0
0</td><td>6
6</td><td>6
6</td></tr>
<tr><td>0
0</td><td>5
6</td><td>30
100</td><td>5
6</td><td>5
6</td><td>5
6</td><td>0
0</td><td>43
138</td><td>41
136</td><td>50
110</td></tr>
<tr><td>0
0</td><td>24
95</td><td>40
105</td><td>6
6</td><td>4
5</td><td>4
5</td><td>0
0</td><td>0
0</td><td>0
0</td><td>45
106</td></tr>
<tr><td>26
106</td><td>46
106</td><td>6
6</td><td>5
6</td><td>6
6</td><td>6
6</td><td>40
105</td><td>0
0</td><td>0
0</td><td>0
0</td></tr>
<tr><td>55
116</td><td>5
6</td><td>5
6</td><td>5
6</td><td>5
6</td><td>5
6</td><td>42
107</td><td>44
108</td><td>45
106</td><td>50
110</td></tr>
</table>

7) The following likelihood contours were developed using these training classes:

Training Class	Band 1 Mean	Band 2 Mean
Aspen	40	75
Spruce	20	65
Water	5	45

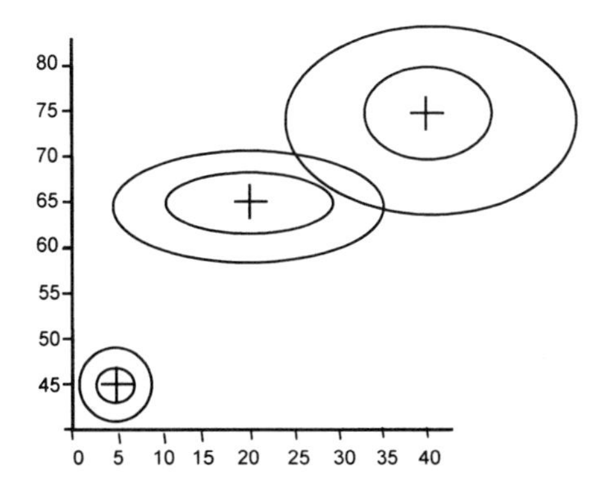

Using the maximum likelihood rule, classify the following image, by plotting the values:

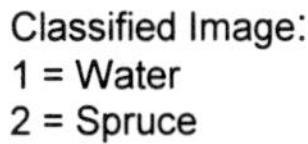

Original Image:

<table>
<tr><td>6
44</td><td>5
43</td><td>25
67</td><td>29
60</td></tr>
<tr><td>6
43</td><td>4
43</td><td>25
65</td><td>30
61</td></tr>
<tr><td>10
65</td><td>14
68</td><td>40
77</td><td>43
78</td></tr>
<tr><td>12
66</td><td>15
69</td><td>39
76</td><td>44
78</td></tr>
</table>

<table>
<tr><td></td><td></td><td></td><td></td></tr>
<tr><td></td><td></td><td></td><td></td></tr>
<tr><td></td><td></td><td></td><td></td></tr>
<tr><td></td><td></td><td></td><td></td></tr>
</table>

8) A consulting company conducts a 2001 image classification to locate all sawtimber white spruce stands. The image pixel size is 25 meters.

The accuracy of the classification is assessed by several groups. Match the following accuracy assessment scenarios with the most likely overall classification accuracies.

_____50% overall classification accuracy

_____90% overall classification accuracy

Group A: A guided approach was conducted in conjunction with the fieldwork for training data collection, thereby reducing the costs of data collection. Reference data were collected adjacent to training field areas. To minimize nonclassification errors (such as slight positional shifts) only homogeneous polygons of at least 5 hectares were used as reference data. The center pixel of these polygons was selected for each reference data sample location. The photos were taken the same month and year as the satellite image was acquired.

Group B: Using high altitude color infrared photographs (taken by NASA in 1979), we stereoscopically interpreted cover types to polygons with a minimum mapping unit of 2 hectares. The covertypes were transferred to the USGS 1:63,360 quadrangle series using a digitizing tablet and stored in a vector GIS. For accuracy assessment these GIS polygons were rasterized to correspond to the classified image. Classification accuracy assessment was conducted on a pixel by pixel basis between the reference data and the classified image.

9) You have an error matrix as follows. What is the overall classification accuracy?

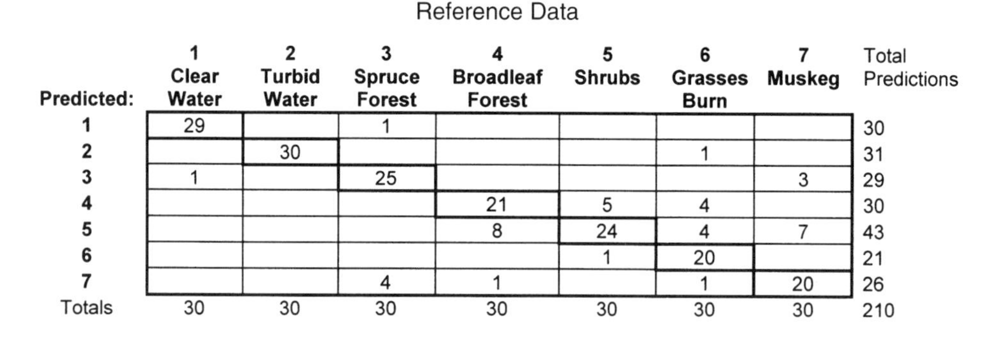

Reference Data

Predicted:	1 Clear Water	2 Turbid Water	3 Spruce Forest	4 Broadleaf Forest	5 Shrubs	6 Grasses Burn	7 Muskeg	Total Predictions
1	29		1					30
2		30				1		31
3	1		25				3	29
4				21	5	4		30
5				8	24	4	7	43
6					1	20		21
7			4	1		1	20	26
Totals	30	30	30	30	30	30	30	210

10) You have an error matrix as follows. What is the Producer's and User's accuracy for the muskeg class?

Reference Data

Predicted:	1 Clear Water	2 Turbid Water	3 Spruce Forest	4 Broadleaf Forest	5 Shrubs	6 Grasses Burn	7 Muskeg	Total Predictions
1	29		1					30
2		30				1		31
3	1		25				3	29
4				21	5	4		30
5				8	24	4	7	43
6					1	20		21
7			4	1		1	20	26
Totals	30	30	30	30	30	30	30	210

Chapter 10

Vector Exercises

INTRODUCTION

In this chapter, you can test your understanding of GIS tools for application of problem-solving in the vector GIS world. This world can include tools for tabular, point, line, and polygons analysis, as well as network analysis and dynamic segmentation. The following table summarizes these tools from the previous vector GIS chapters.

GENERAL VECTOR GIS TOOLS	
Tools for Managing GIS Features:	
LIST	List the contents of any GIS table.
COPY	Make a new copy theme from any point, line, or polygon theme.
APPEND	Append together two or more point, line, or polygon themes.
KILL	Delete a user-specified point, line, or polygon theme.
RENAME	Rename a user-specified point, line, or polygon theme.
DESCRIBE	Tell the user information about any point, line, or polygon theme.
BUILD	Build an attribute table for a point, line, or polygon theme.

TABULAR ANALYSIS TOOLS	
Record Selection:	
RESELECT	Select records using a logical expression.
NSELECT	Select all records not currently selected.
ASELECT	Add more records to currently selected record set.
Descriptive Statistics:	
STATISTICS	Computes mean, min, max, std. dev. by user-specified attribute class.
FREQUENCY	Sums attribute values by one or more attribute classes.
Viewing Tabular Information:	
DIR	List all tables available in your current workspace.
ITEMS	List item definitions (item names, type, input/output width) for your table.
LIST	List the information contained inside your table.

Modifying Tabular Information:	
ALTER	Allows you to alter item characteristics such as name or output width.
CALCULATE	Assigns new values to an item based on an arithmetic expression.
REDEFINE	Used to create new items that share columns space with existing items.
SORT	Allows you to sort selected records by specific table items.
UPDATE	Allows you to interactively type in new item values for selected table records.
Adding Tabular Information:	
ADD	Allows you to add records or rows of information to your table.
ADDITEM	Used to add new items or columns to your table.
Deleting Tabular Information:	
DROPITEM	Deletes any specified items from your table.
PURGE	Deletes the selected records from your table.
KILL	Deletes an entire table.

Exporting Tabular Information:	
UNLOAD	Writes selected tabular information to an ASCII text file.
COPY	Copies existing table to a new table.
SAVE	Writes selected tabular information to a binary file.
Linking or Merging Tables Together:	
JOINITEM	Permanently merges two tables based on a key item.
RELATE	Temporarily merges two or more tables based on a key item.
Indexing Tabular Informations:	
INDEXITEM	Creates an attribute index to increase query speed for that item.

POINT ANALYSIS TOOLS	
Tools for Managing Points:	
ADDXY	Add each point's X,Y coordinate as new items in the point attribute table.
NODEPOINT	Create a new point theme from nodes in a line or polygon theme.
ARCPOINT	Create a new point theme from arc vertices or polygon labels.
POINTNODE	Transfer attributes from points to the nearest node in a line or polygon theme.
Area Analysis Tools:	
THIESSEN	Creates polygons of proximity from points.
BUFFER	Create buffer areas of user-specified distance around points.
Distance Analysis Tools:	
NEAR	Compute the nearest distance from points to points, lines, or polygons.
POINTDISTANCE	Within a search radius limit, compute point distances between two point themes
Attribute Analysis Tools:	
RESELECT	Create a new point theme based on a user-specified logical query.
INTERSECT	Transfer polygon attributes to a point theme.

LINE ANALYSIS TOOLS	
Tools for Analyzing Lines:	
DISSOLVE	Merge adjacent arcs if they contain the same item value.
COUNTVERTICES	Count number of vertices in each arc, and add the count as a new attribute item.
BUFFER	Create buffer polygons of user-specified distance around lines.
RESELECT	Create new line theme based on a user-specified logical query.
INTERSECT	Transfer polygon attributes to a line theme.

POLYGON ANALYSIS TOOLS	
Polygon Generalization Tool:	
DISSOLVE	Merges adjacent polygons if they have the same attribute value.
ELIMINATE	Merges selected polygons with their neighbors and eliminates shared border.
Area Analysis Tool:	
BUFFER	Create buffer areas of user-specified distance around polygons.
Vertex/Node Analysis Tool:	
COUNTVERTICES	Count number of vertices in each polygon, and add the count as an attribute item.
Cutting Tools:	
CLIP	Cut and copy operation to produce one output theme.
SPLIT	Cut and copy operation to produce many output themes.
ERASE	Cut operation that eliminates polygons using eraser polygons.
UPDATE	Cut and paste operation to produce one output theme.
Spatial Join Tools:	
UNION	Combines all polygons by overlaying two input polygon themes.
INTERSECT	Transfers polygon attributes to intersecting point, line, or polygon theme.
IDENTITY	Combines an intersecting polygon with point, line, or another polygon theme.
Attribute Analysis Tools:	
RESELECT	Create a new polygon theme based on a user-specified logical query.

DYNAMIC SEGMENTATION TOOLS	
Tools for Analyzing Events:	
EVENTSOURCE	Links your event table to your route system.
ADDROUTEMEASURE	Creates an event table from your point theme.
POLYGONEVENTS	Creates an event table from your polygon theme.
OVERLAYEVENTS	Combines two or more event tables.
DISSOLVEEVENTS	Combines rows in linear event table if they contain the same attribute value.
EVENTPOINT	Converts point events into a new point theme.
EVENTARC	Converts events into a new line theme.
Tools for Analyzing Routes/Sections:	
RESELECT	Saves user-specified sections or routes to a new line theme.
ROUTEARC	Converts each route to an arc in the output theme.
SECTIONARC	Converts each section to an arc in the output theme.
ROUTESTATS	Computes descriptive statistics for routes and sections.

REAL ESTATE APPLICATION

You are interested in purchasing property where you can build a retreat cabin as a vacation home. You have the following polygon themes: parcels, elev_zone, vegetation. You also have a line theme of roads. You want to find land that is privately owned so that you can make an offer to purchase. You want to purchase an area of at least 1 hectare and is totally above 500 feet elevation (to avoid flooding problems). You want decent road access : at least partially within 100 meters of a road. And you want the area to contain some aspen vegetation (which typically grows on warm sites). Fill in the following flowchart with the appropriate vector GIS tools to solve your problem:

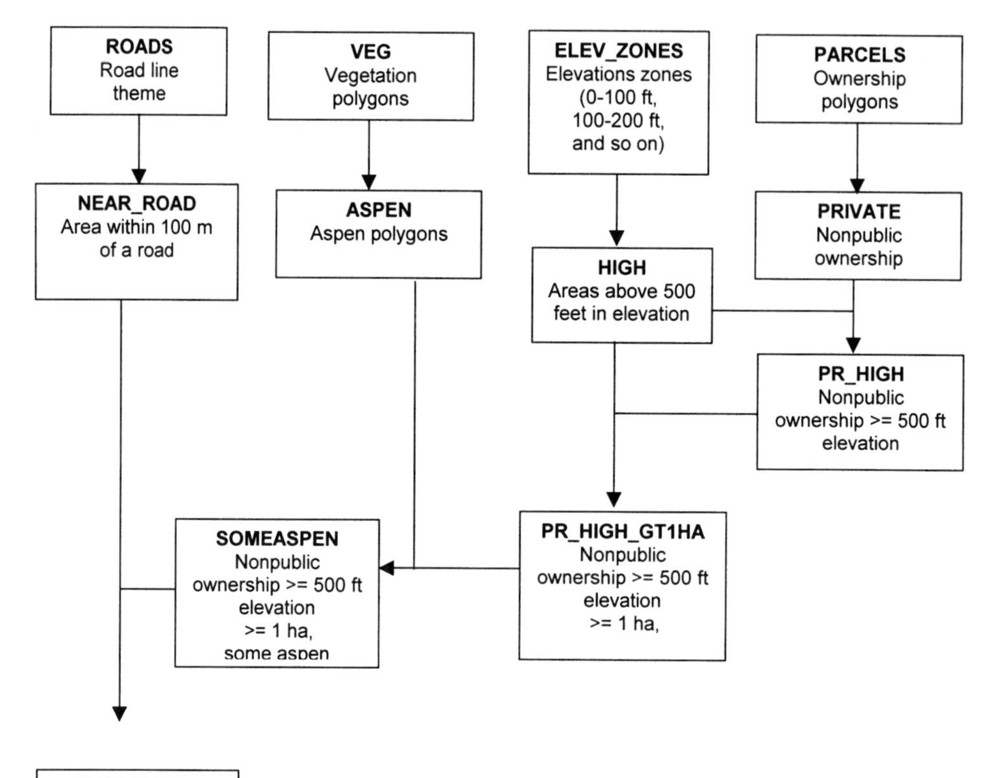
ROADS
Road line theme
VEG
Vegetation polygons
ELEV_ZONES
Elevations zones (0-100 ft, 100-200 ft, and so on)
PARCELS
Ownership polygons
NEAR_ROAD
Area within 100 m of a road
ASPEN
Aspen polygons
HIGH
Areas above 500 feet in elevation
PRIVATE
Nonpublic ownership
PR_HIGH
Nonpublic ownership >= 500 ft elevation
SOMEASPEN
Nonpublic ownership >= 500 ft elevation >= 1 ha, some aspen
PR_HIGH_GT1HA
Nonpublic ownership >= 500 ft elevation >= 1 ha,
CANDIDATES
Nonpublic ownership >= 500 ft elevation >= 1 ha, some aspen some of parcel within 100 m of road

MOOSE HABITAT ANALYSIS

You are hired as the GIS guru in a moose habitat study. You have the following polygon themes: rivers, vegetation. You also have a point theme of radio-collared moose locations from the winters of 1995–2001. You are to test the hypothesis that mature bull moose spend most of their time in large riparian *Salix alexensis* (willow) stands whenever there is deep snow. The winters of 1995, 1997, and 2000 were deep snow winters while the winters of 1996,1998,1999, and 2001 were light snow winters. The moose biologist wants the following GIS analyses:

1) For mature bull moose, compute the mean distance to riparian *Salix alexensis* (riparian is defined as within 100 meters of a river) for light versus deep snow years.
2) Compute percentage of mature bull moose locations that were within 100 meters of riparian *Salix alexensis* for light versus deep snow years.
3) For the *Salix alexensis* polygons that had mature bull moose locations in them, compute the mean *Salix alexensis* polygon area in hectares for light versus deep snow years.

Fill in the following flowchart with the appropriate vector GIS tools to solve your problem.

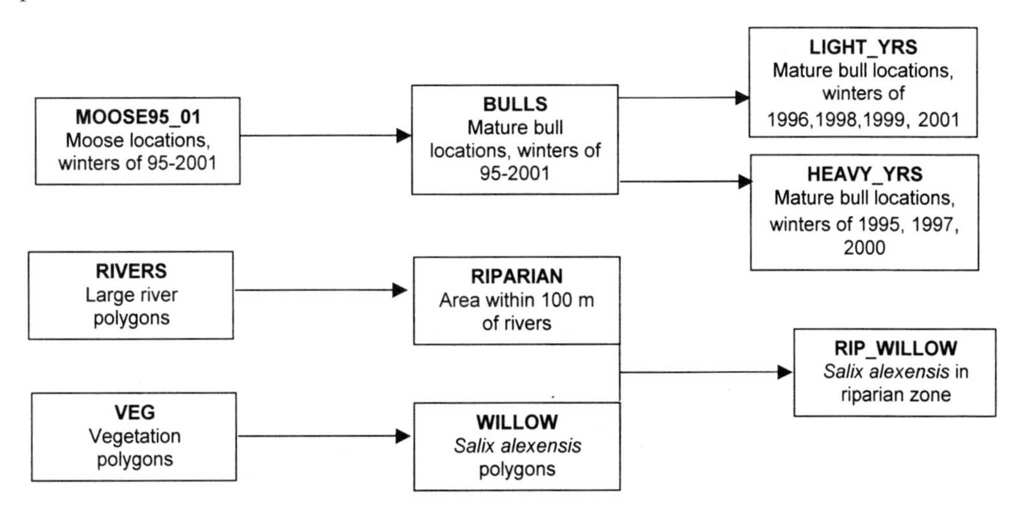

For mature bull moose, compute the mean distance to riparian *Salix alexensis* (riparian is defined as within 100 meters of a river) for light versus deep snow years.

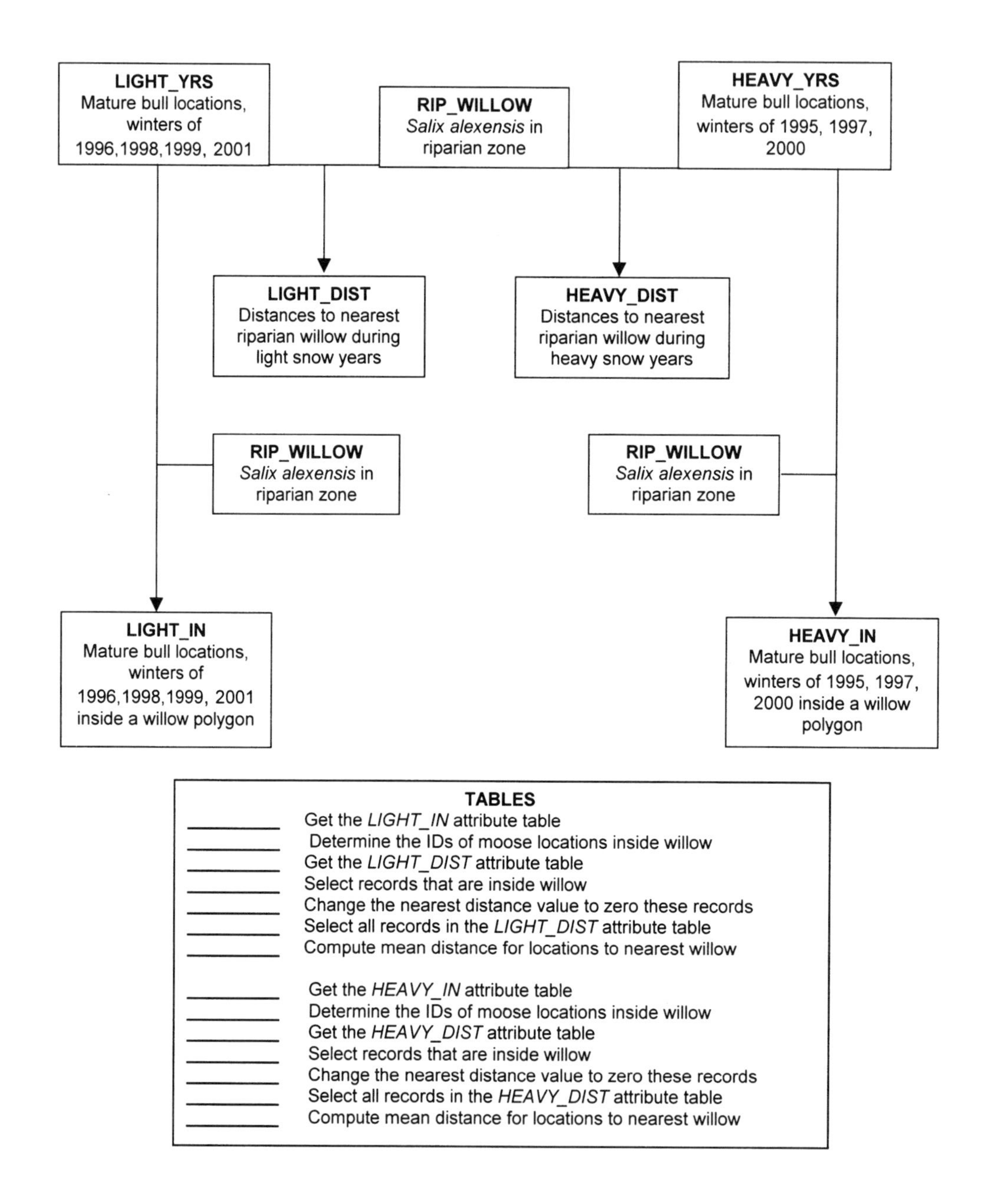
LIGHT_YRS
Mature bull locations, winters of 1996,1998,1999, 2001
RIP_WILLOW
Salix alexensis in riparian zone
HEAVY_YRS
Mature bull locations, winters of 1995, 1997, 2000
LIGHT_DIST
Distances to nearest riparian willow during light snow years
HEAVY_DIST
Distances to nearest riparian willow during heavy snow years
RIP_WILLOW
Salix alexensis in riparian zone
RIP_WILLOW
Salix alexensis in riparian zone
LIGHT_IN
Mature bull locations, winters of 1996,1998,1999, 2001 inside a willow polygon
HEAVY_IN
Mature bull locations, winters of 1995, 1997, 2000 inside a willow polygon
TABLES
Get the LIGHT_IN attribute table
Determine the IDs of moose locations inside willow
Get the LIGHT_DIST attribute table
Select records that are inside willow
Change the nearest distance value to zero these records
Select all records in the LIGHT_DIST attribute table
Compute mean distance for locations to nearest willow
Get the HEAVY_IN attribute table
Determine the IDs of moose locations inside willow
Get the HEAVY_DIST attribute table
Select records that are inside willow
Change the nearest distance value to zero these records
Select all records in the HEAVY_DIST attribute table
Compute mean distance for locations to nearest willow

Compute percentage of time that mature bull moose locations were within 100 meters of riparian *Salix alexensis* for light versus deep snow years.

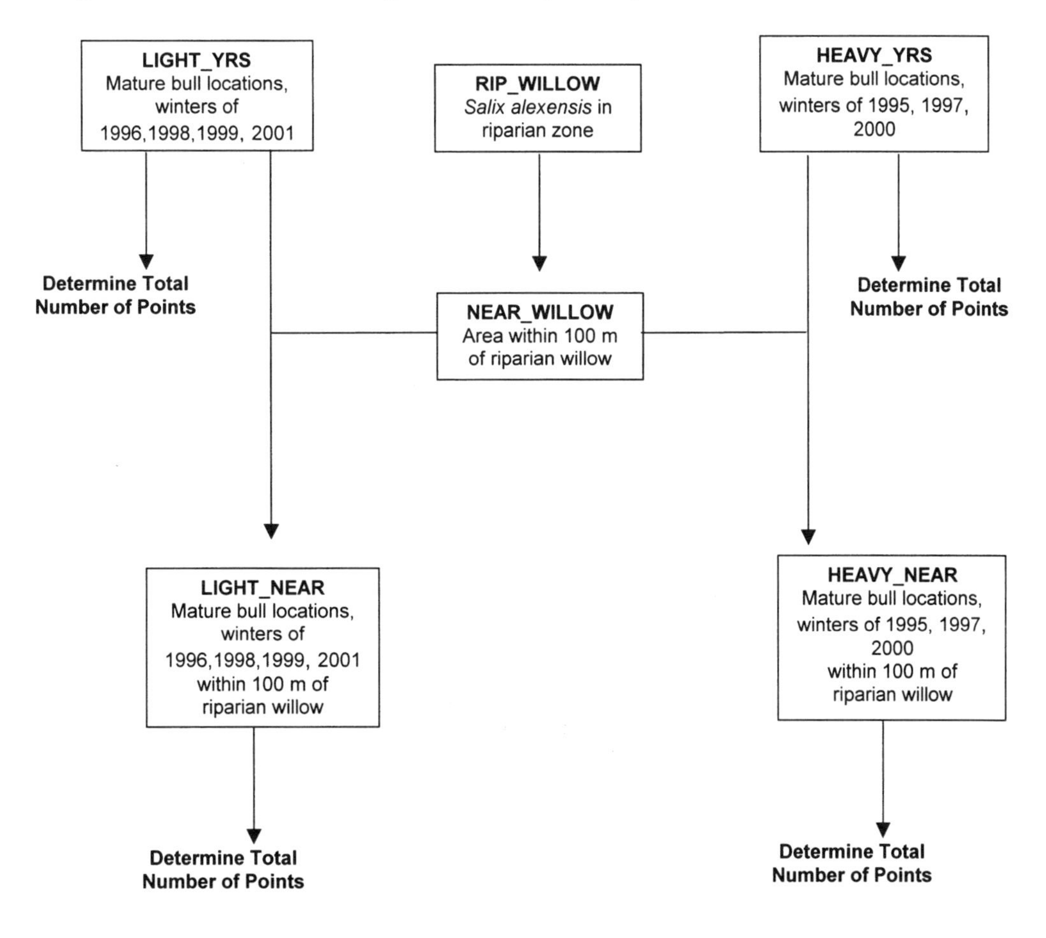

For the *Salix alexensis* polygons that had mature bull moose locations in them, compute the mean *Salix alexensis* polygon area in hectares for light versus deep snow years.

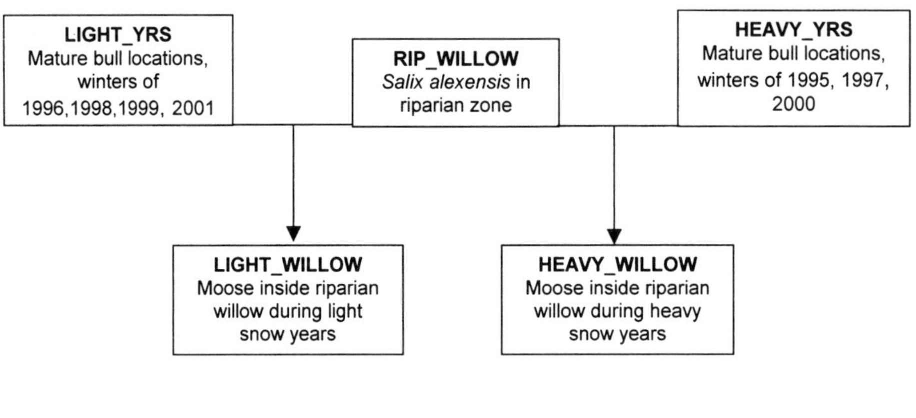

TABLES

________ Add a column called Ha to the *LIGHT_WILLOW* table
________ Get the *LIGHT_WILLOW* attribute table
________ Compute the hectares for the willow each point is in
________ Compute mean hectares for willow polygons that the points are in

________ Add a column called Ha to the *HEAVY_WILLOW* table
________ Get the *HEAVY_WILLOW* attribute table
________ Compute the hectares for the willow each point is in
________ Compute mean hectares for willow polygons that the points are in

FIRE HYDRANT INSPECTION APPLICATION

You have a table of county-owned fire hydrants and their mileage recorded by driving along all county roads. The table contains information such as date of last inspection, age of hydrant, maximum flow rate, and so on. You also have a county-wide roads line theme and a parcels polygon theme. You also have a table of parcel owners; the table contains parcel-IDs, owner names, addresses, and phone numbers. All hydrants in the county are inspected on a five-year basis. Your job is to create a text file of names and addresses of parcel owners with property within 100 meters of a fire hydrant that is scheduled to be inspected in 2001.

Fill in the following flowchart with the appropriate vector GIS tools to solve your problem.

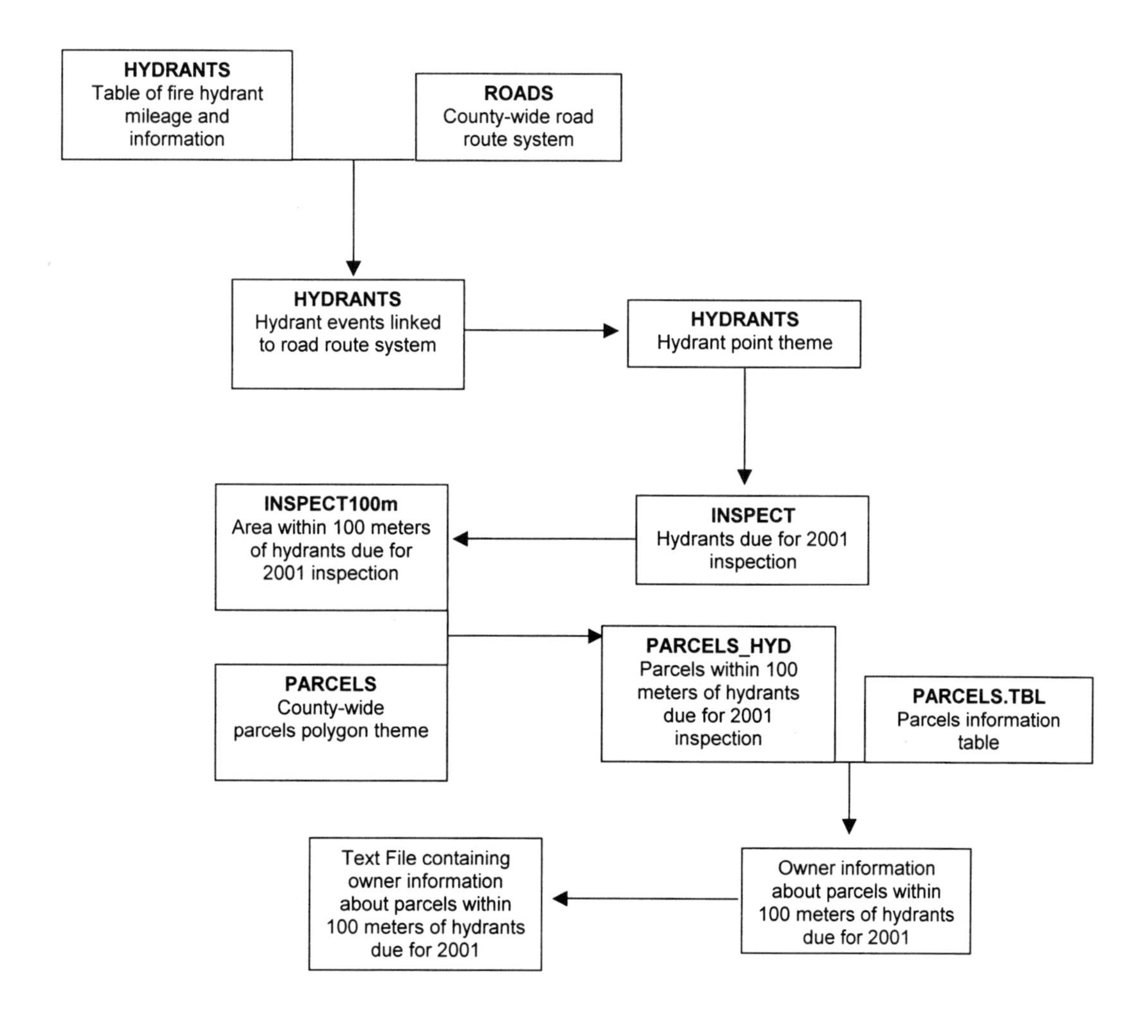

FOREST FIRE APPLICATION

You have a line theme of flight lines over the center of lakes, a polygon theme of lakes, and a point theme of the centers of five currently burning wildfires. In order for tankers to reload water, they need a lake that is at least 3 km long. Develop a map showing all lakes available for tanker reloading and which fire they are the closest to.

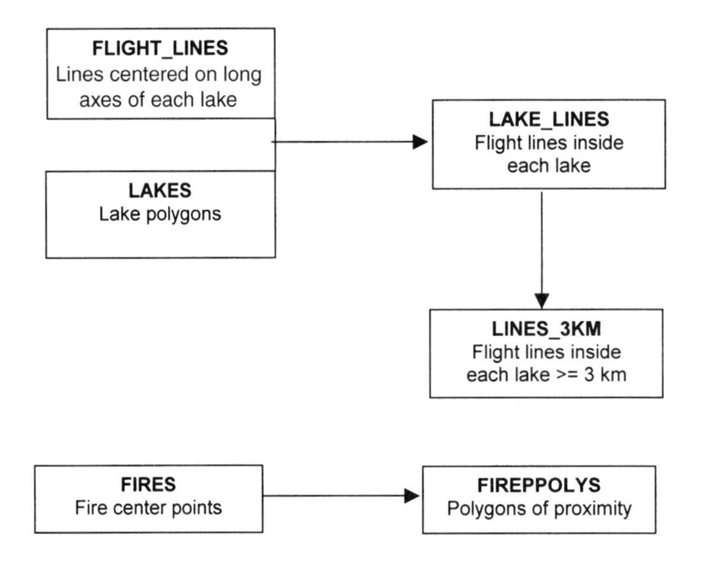

EMERGENCY PHONE APPLICATION

You are hired as a GIS guru for the facilities management department of a large university. You have a polygon theme of buildings, a line theme of sidewalks, and a point theme of outside emergency phones. Your job is to show the area of the campus that is most lacking in emergency phone service. Fill in the following flowchart with the appropriate GIS tools to solve your problem.

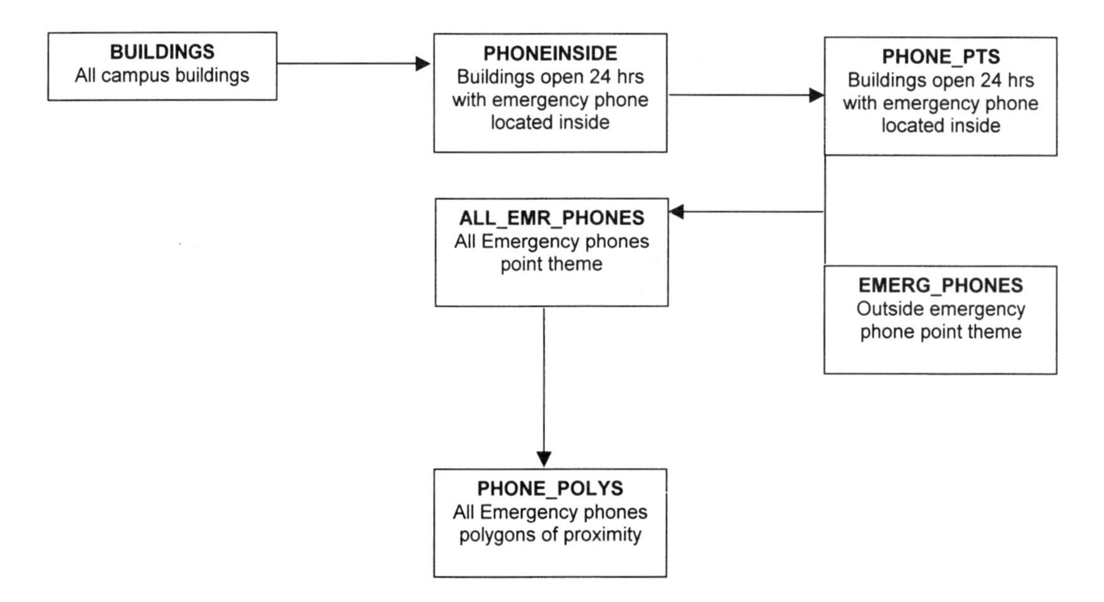

Chapter 11

Grid Exercises

INTRODUCTION

In this chapter, you can test your understanding of GIS tools for application of problem-solving in the Grid GIS world. The following table summarizes the Grid GIS tools that were covered in Chapter 8.

Grid Arithmetic Operators:	
+	Addition.
-	Subtraction.
*	Multiplication.
/	Division.
Selection Tools:	
TEST	Returns 1 / 0 based on logical query.
SELECT	Returns original / NODATA based on logical query.
CON	Returns user-specified values based on logical query.
Grouping Tools:	
RECLASS	Change cell values based on remap table assignments.
REGIONGROUP	Groups cells of the same value that touch each other.
Distance Tools:	
EXPAND	Expands user-specified cell values (analogous to BUFFER).
EUCDISTANCE	Computes the distance to closest non-NODATA cell for each cell.
Optimal Path Tools:	
COSTDISTANCE	Estimates the minimum accumulative cost to resource cells.
COSTPATH	Estimates the least cost path from source to destination cells.
Topographic Tools:	
SLOPE	Estimates rate of maximum slope change for each cell.
ASPECT	Estimates the slope direction for each cell.

REAL ESTATE APPLICATION

You have grids of elevation, roads, and ownership. You want to find land owned by the Fairbanks Northstar Borough that is above 500 feet elevation (out of the ice fog), on a easterly, southerly, or westerly slope, and within 100 meters of a road. Your elevation values is in meters.

Fill in the following flowchart with the appropriate Grid tools:

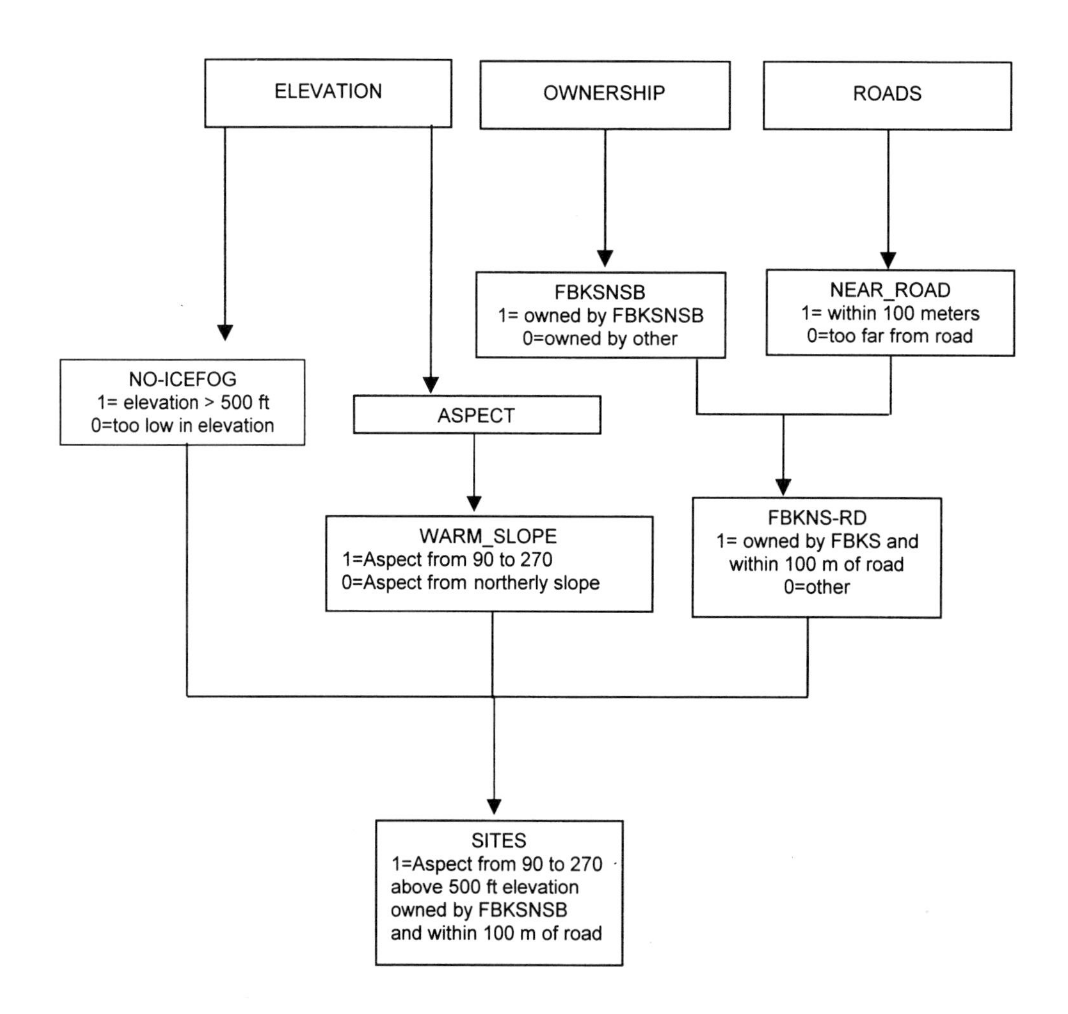

WATERSHED APPLICATION

You have a grid of vegetation types, and a grid of watersheds. You want a table showing the area of each watershed and the area of black spruce in each watershed. Each grid cell is 25 by 25 meters.

Fill in the following flowchart with the appropriate grid operations to solve your GIS problem:

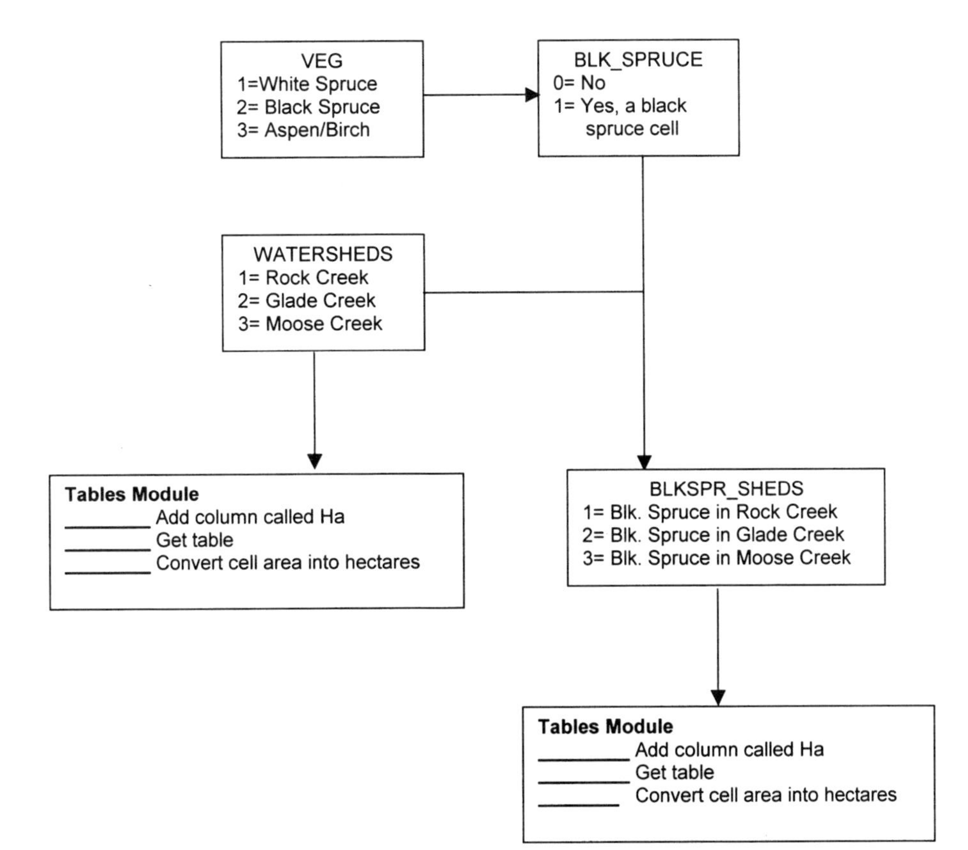

WATERFOWL HABITAT ANALYSIS

You have grids of lakes. One grid is from early summer and the other grid is from late summer. You want to create a grid of the best lakes for waterfowl habitat . . . lakes that had at least 1 hectare of exposed mudflats in late summer due to drawdown of the lake water level. The cell size of each grid is 10 meters. Fill in the following flowchart to solve your GIS problem.

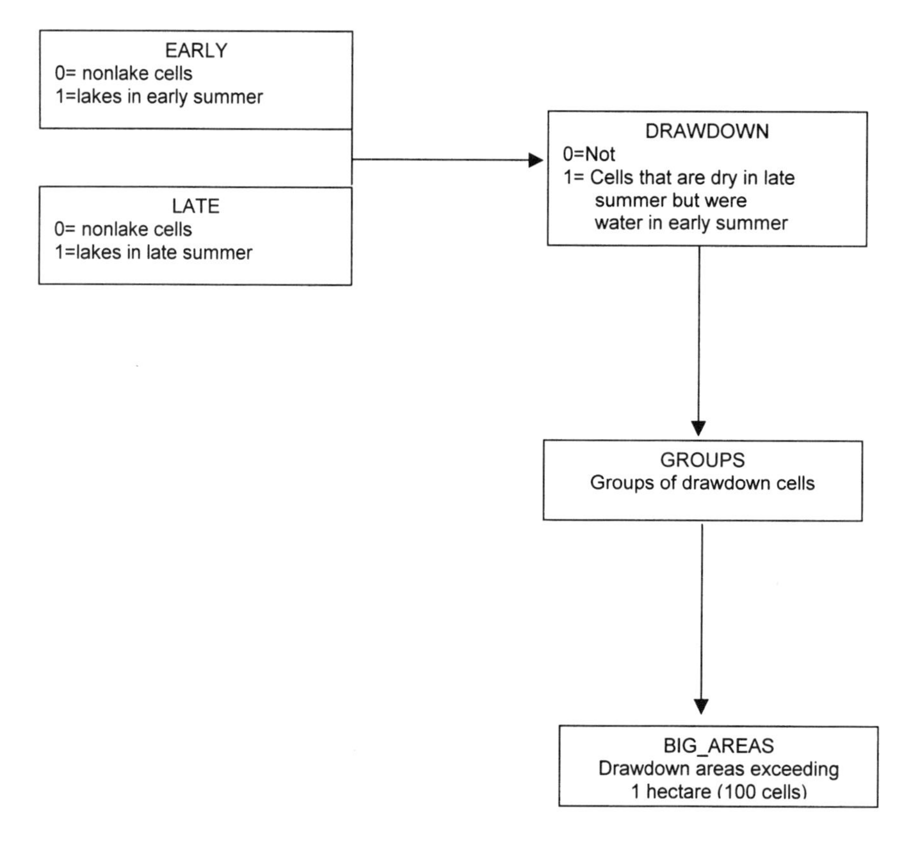

GROUNDWATER APPLICATION

You have a grid of well locations, gas station locations, soils, and parcel ownership that contains the address of the owner of each parcel.

Outline what tools you would use to develop a text file of all well owners that have wells on sandy textured soils within a kilometer of a gas station.

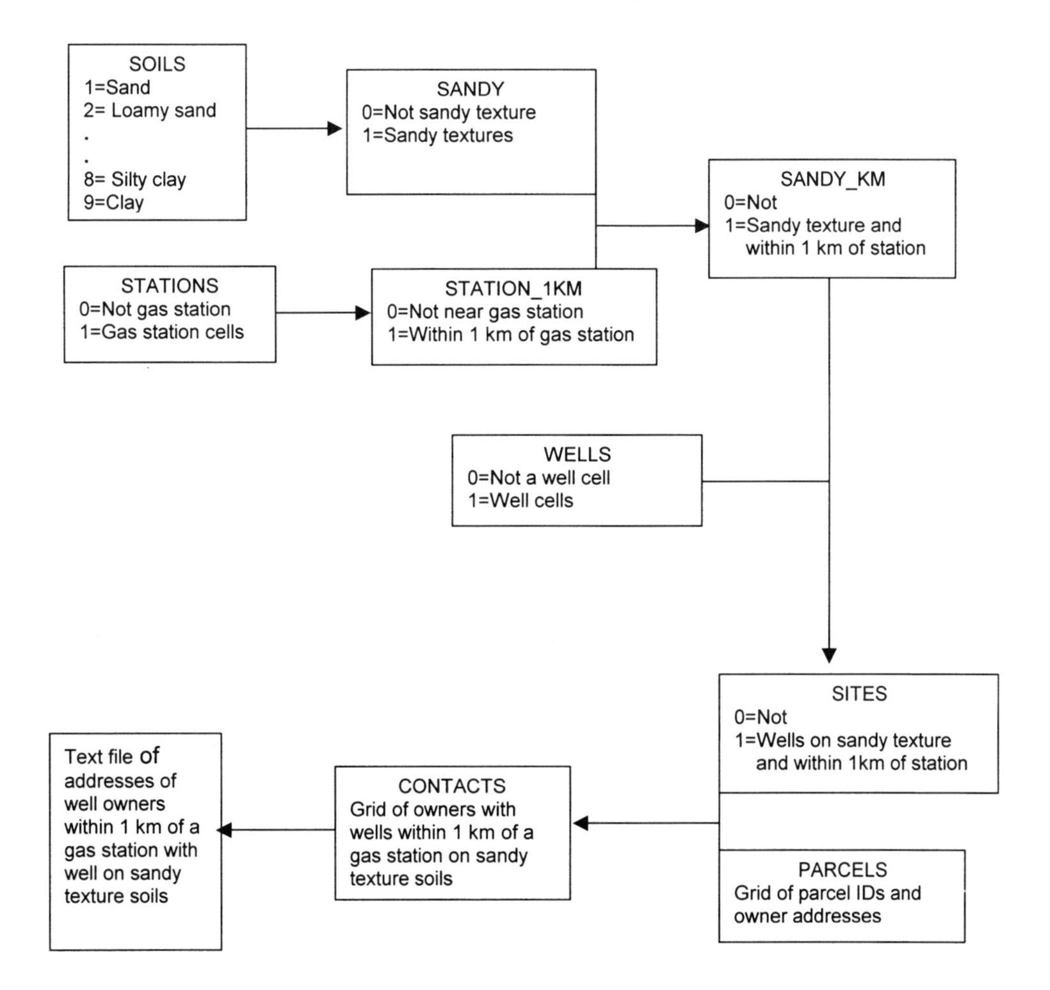

MOOSE HABITAT ANALYSIS

You have grids of moose locations, random locations and vegetation types. You want to estimate the mean distance of moose and random locations to the nearest willow cell. All grid cells are 10 meters in size.

Fill in the following flowchart with the appropriate grid operations to solve your GIS problem:

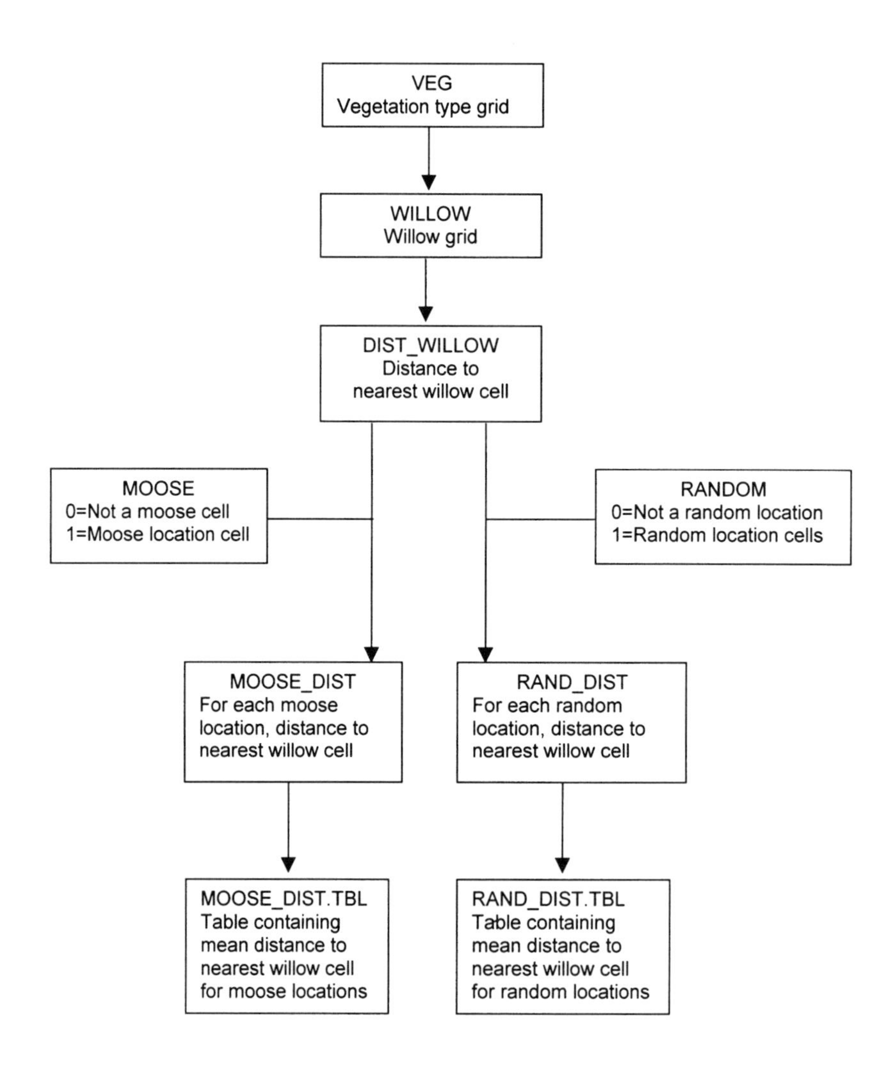

Chapter 12

Saving Time In GIS Analysis

INTRODUCTION

In this chapter are tips on how to improve efficiency in using GIS. Although the examples are from using Arc/Info, many of these ideas can be applied to any GIS.

WORK WHILE YOU PLAY

Some GIS processes such as image rectification, spatial join, and buffering operations take considerable processing time. You can run these as batch jobs to be excuted while you sleep or play. In the unix environment, it is a three-step process:

1. Edit a file telling the batch processor the type of unix shell script being used. Set your ARC environment variables by sourcing your .cshrc and then list your arc commands. For example:

   Edit rectify.batch

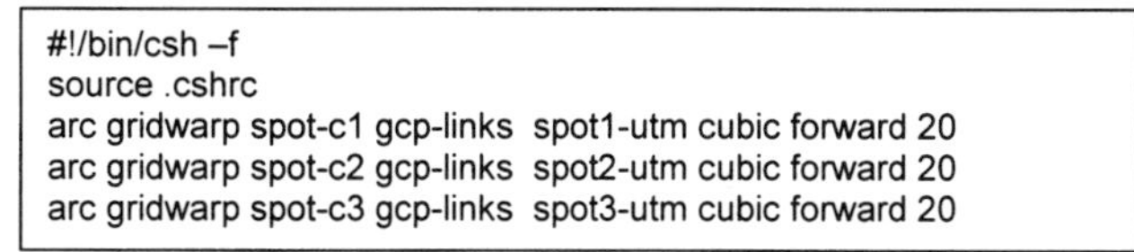

```
#!/bin/csh –f
source .cshrc
arc gridwarp spot-c1 gcp-links  spot1-utm cubic forward 20
arc gridwarp spot-c2 gcp-links  spot2-utm cubic forward 20
arc gridwarp spot-c3 gcp-links  spot3-utm cubic forward 20
```

2. Make your batch file executable using the unix chmod command. For example:

   chmod u+x rectify.batch

3. Use the unix at command to submit your batch job to start at 10 p.m.

   at -f rectify.batch 10:00pm Feb 10

If you work in the windows NT or windows 2000 environment, there is a similar **AT** command available for batch processing.

AVOID TELEPHONE TAG

We live in a digital world. E-mail instead of phoning probems to GIS customer support. This allows you to:

1) Avoid telephone tag.
2) Avoid time-zone differences.
3) Attach sample error messages and data.
4) Store a file detailing your question and customer support's solution.
5) Document poor customer support, if you feel justified in complaining.

PAINLESS DOCUMENTATION

It is easy to talk about good documentation and how important it is. But it is human nature to put off documentation until the current crisis is over—and in many cases GIS work can be crisis management where clear, thorough documentation may be put on the back burner for a long, long time.

Documentation should never be on the back burner because it is painless! The trick is to document as you work, through the use of documentation files. A documentation file is a edited file where all GIS commands are entered into and then copied to the GIS command window as in the following example:

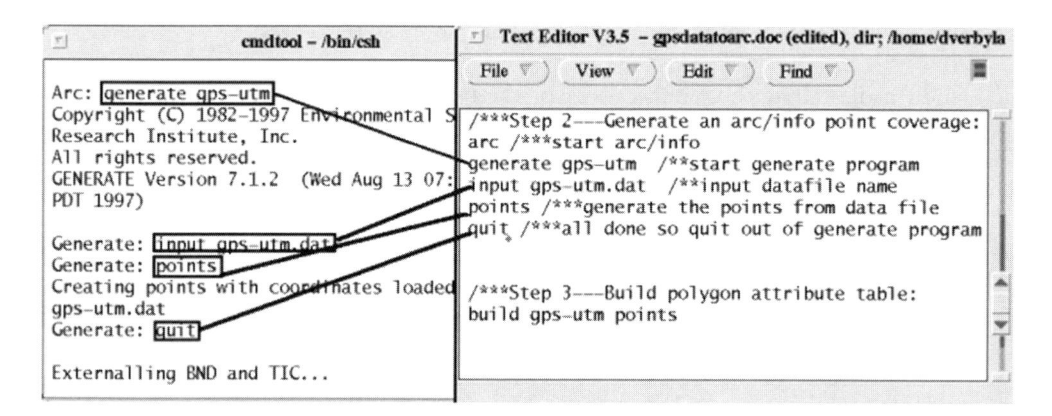

Documentation files are a good practice for the following reasons:

1) You can work on your commands while the GIS is currently executing.

Why wait for a GIS to execute a command? In the above example, arc may take several minutes to generate the point coverage. During that time you could enter into your documentation file comments and the next commands (build gps-utm points).

2) Trivial facts that may become important later are captured.

For example, what workspace was the coverage in? Was the datum NAD27 or NAD83? What was fuzzy tolerance? etc.

```
&show &workspace; describe roads-utm

/home/dverbyla/nrm338
          Description of SINGLE precision coverage roads-utm

                      FEATURE CLASSES
                      Number of   Attribute      Spatial
Feature Class         Subclass    Features   data (bytes)  Index?  Topology?
-------------    --------   ---------  ------------  -------  ---------
ARCS                              78          30
NODES                             82

                      SECONDARY FEATURES
Tics                          4
Arc Segments                      1017

                        TOLERANCES
Fuzzy  =             1.608 V        Dangle =              0.000 N

                      COVERAGE BOUNDARY
Xmin =             428499.844          Xmax =             442600.594
Ymin =             7171804.500         Ymax =             7181133.000

                          STATUS
The coverage has not been edited since the last BUILD or CLEAN.

                    COORDINATE SYSTEM DESCRIPTION
Projection   UTM   Zone 6   Datum NAD27   Units METERS   Spheroid CLARKE1866
```

3) Repeating operations is easy.

Repetitive commands can easily be copied and modified as a block of commands to be pasted to the GIS command window. For example, just before you take a coffee break, you could copy and paste the following to be excuted while you are gone . . .

```
/***generate 95 well point coverages
generate gpspts95
input gpsutm95.txt
points
quit
build gpspts95 point

/***generate 96 well point coverages
generate gpspts95
input gpsutm96.txt
points
quit
build gpspts96 point

/***generate 97 well point coverages
generate gpspts97
input gpsutm97.txt
points
quit
build gpspts97 point
```

4) How did you do that last year?

I don't remember exactly, but I can easily and efficiently E-mail you the documentation file . . .

ASSUME YOUR GIS LIES

You can save yourself time by being skeptical about the results from GIS operations. Sometimes a lack of understanding can be the source of trouble. Fortunately, GISs allow you to visually check out the results of most analysis operations. Here are a few examples where a lack of understanding is the source of trouble, but visual checks saved the day:

1) Moose habitat analysis. A biologist has a point theme of moose locations and a polygon theme of a series of wildfire burns of various ages. He uses the **NEAR** tool to estimate the distance to the nearest wildfire polygon for each moose observation. He then computes the mean distance to each age class burn and gets the following table:

Burn Age Class	Mean Distance (m)	Area of Burn (Ha)
<10 years	58	32
10-20 years	419	889
>20 years	33	43

It seems strange that the mean distance for the burn 10-20 years old is so much larger than the other burns. The biologist visually checks out moose locations relative to that burn . . . and selects the moose location with the greatest distance from the 10-20 year old burn.

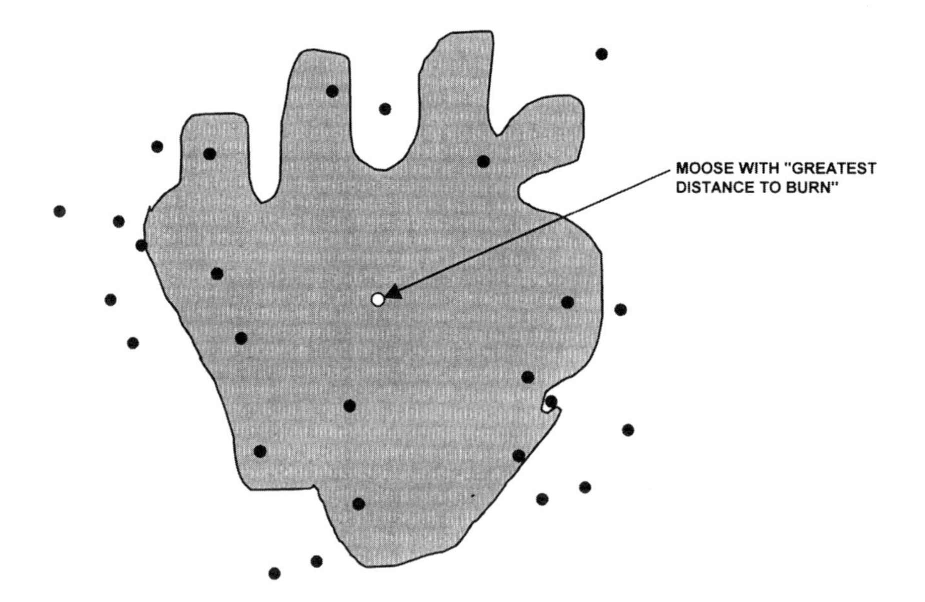

By visually checking out the results of the analysis, the biologist discovers that for moose inside the burn, the distance values were all greater than zero and not zero as expected. The problem was easily solved once the biologist recognized that there was a flaw in his interpretation of the analysis.

2) Groundwater well analysis. A series of shallow wells have been established to monitor possible contamination from a chemical spill. All wells are spaced at least 10 km away from each other. For this application, the wells to be used for a special chemical analysis have to be at least 100 meters away from any oak stand to avoid tannins associated with oak leaves. The analyst buffers the wells by 100 meters and intersects the buffered theme with a theme of nonoak polygons. The analyst then reselects polygons that represent wells that are at least 100 meters away from any oak forest (polygons with an area equal to pi $*100^2$). However, no intersected buffers meet this criteria.

   By visually checking out the results of the analysis, the analyst finds that there are indeed wells that meet this criteria.

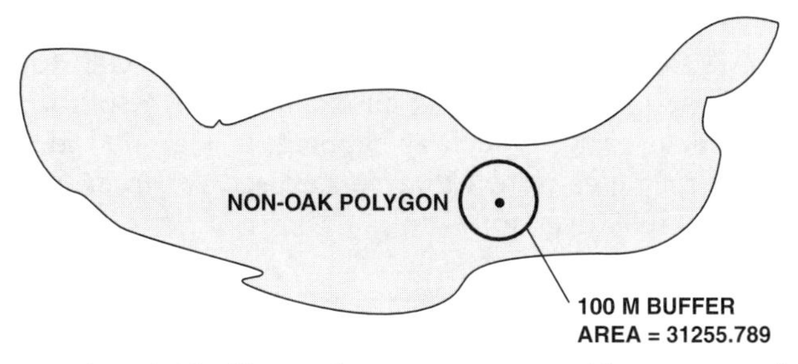

   The problem was that the buffer circle is approximated by a series of arcs and thus the area is less than pi $*100^2$ (31415.9265). So the analyst recognizes the problem and easily solves it by selecting buffered wells with areas greater than or equal to 31255.

3) Vegetation Index Analysis. A common vegetation index used in remote sensing is the Normalized Difference Vegetation Index (NDVI) which is:

   (NIR Reflectance – Red Reflectance) / (NIR Reflectance + Red Reflectance)

   where the index ranges from –1 to +1 with values less than 0.10 representing sparsely vegetated or unvegetated grid cells.

   A remote sensing analyst wants to select all vegetated pixels prior to a classification by selecting those pixels with an NDVI greater than 0.10. Using two grids, he does the analysis as follows:

```
Grid: NDVI = (NIR_Grid  –  Red_Grid) / (NIR_Grid  +  Red_Grid)
Grid: Veg_pixels = Select( NDVI, Value gt 10)
```

   Where the NIR_Grid, Red_Grid are integer grids of percent reflectance, ranging from 0 to 100.

   Instead of boldly going straight to the Veg_pixels = Select(NDVI, Value gt 10) the analyst should first check to make sure the NDVI values seem reasonable. He could do this by sampling some pixels and looking at the input Red_Grid, NIR_Grid, and output NDVI values . . .

Red_Grid	NIR_Grid	NDVI
5	25	0

Red_Grid	NIR_Grid	NDVI
2	10	0

Red_Grid	NIR_Grid	NDVI
10	5	0

All the grid cells have an NDVI value of zero! The problem is that the two input grids are integer and therefore the output grid is integer by default. Once you recognize that this is the problem, the solution is easy to figure out. You would have to use the **FLOAT** function to specify that you want a floating point calculation.

Grid: NDVI = FLOAT(NIR_Grid – Red_Grid) / FLOAT(NIR_Grid + Red_Grid)

As a check we could sample some pixels to make sure the calculation was correct:

Red_Grid	NIR_Grid	NDVI		Red_Grid	NIR_Grid	NDVI		Red_Grid	NIR_Grid	NDVI
5	24	0.655		11	19	0.267		15	25	.250

So far we have seen examples where the GIS really was not lying . . . poor understanding of the GIS tools was the problem. Sometimes, the GIS really does lie . . . especially with the release of new tools. As an example, the GeoProcessing Wizard in ArcView 3.2 allows for operations such as clipping. It is a good idea to always test new tools to make sure they give results that you expect. As an example, we generate two polygon themes with 50% overlap:

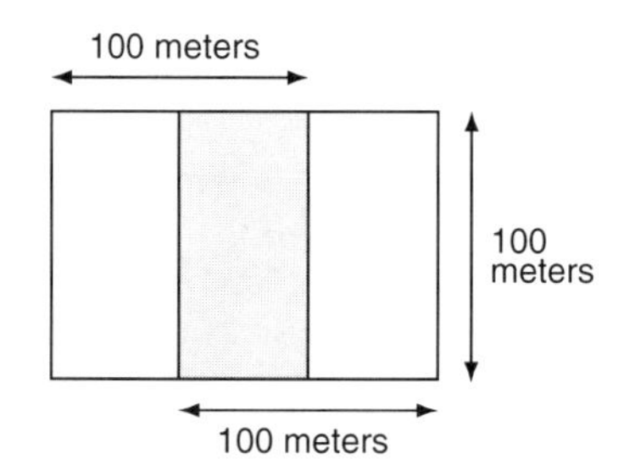

The correct area of the clip of these 2 test polygons should be 100 X 50 = 5000 m^2 and the correct perimeter should be 100+100+50+50 = 300 m. We test-drive the Geoprocessing Wizard to see if it gives us the correct answer..

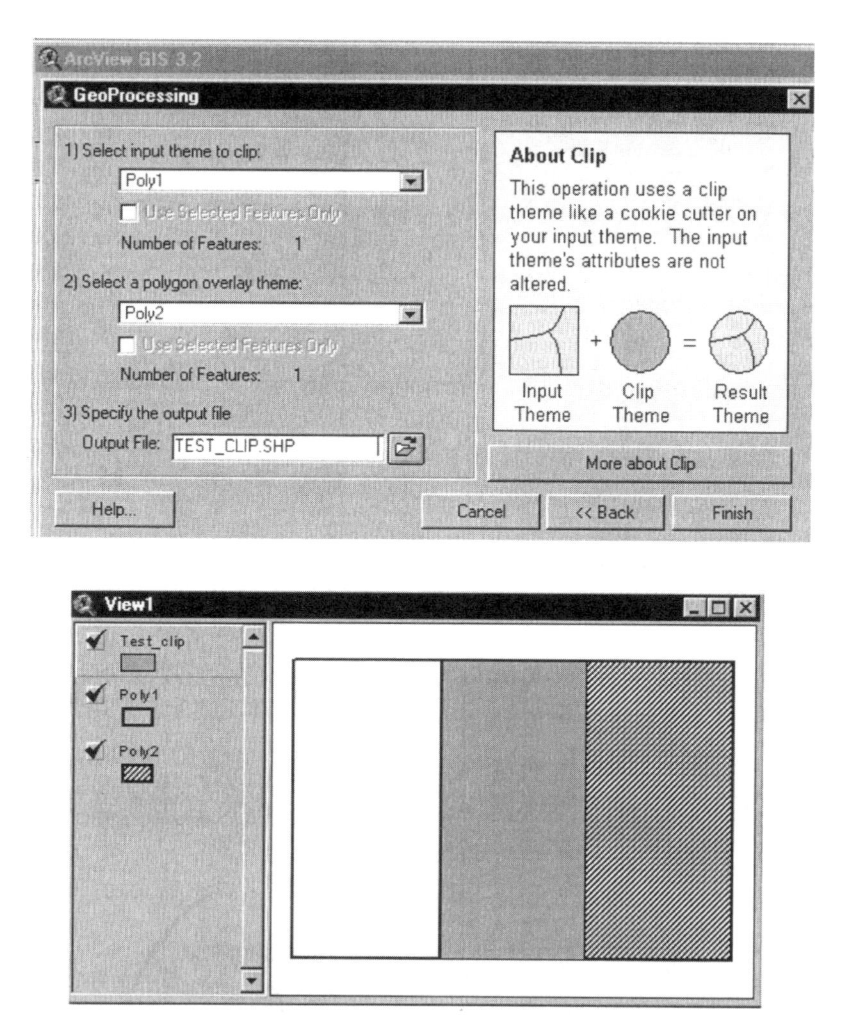

The resulting Test_clip theme looks reasonable . . . but did the area and perimeter get correctly computed?

Attributes of Test_clip .shp

Shape	Area	Perimeter	Poly1#	Poly1-id
Polygon	10000.000	400.000	0	0

The GIS gives the wrong Area and Perimeter! Instead of calculating the correct area after a clipping operation, it copies the original input polygon area and perimeter!

Once you recognize the problem, you can solve it by using Avenue **[Shape].ReturnArea** and **[Shape].ReturnLength** requests with the table calculator as follows:

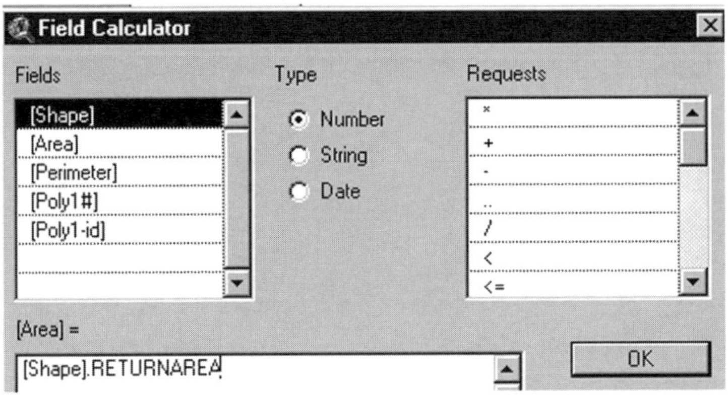

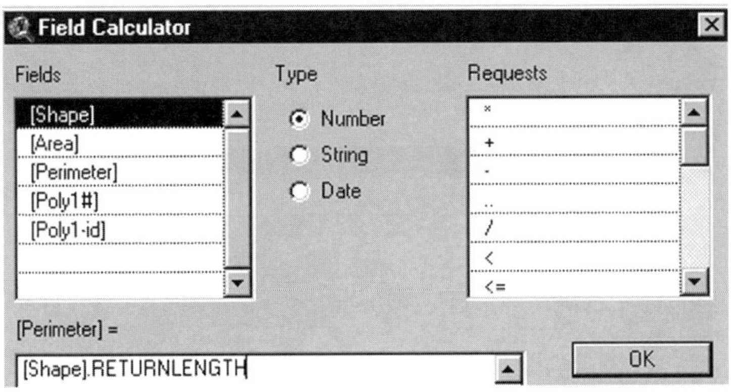

Field Calculator

Fields: [Shape], [Area], [Perimeter], [Poly1#], [Poly1-id]
Type: Number, String, Date
Requests: *, +, -, .., /, <, <=

[Area] =
[Shape].RETURNAREA

Field Calculator

Fields: [Shape], [Area], [Perimeter], [Poly1#], [Poly1-id]
Type: Number, String, Date
Requests: *, +, -, .., /, <, <=

[Perimeter] =
[Shape].RETURNLENGTH

And the correct Area and Perimeter are calculated for the output polygons.

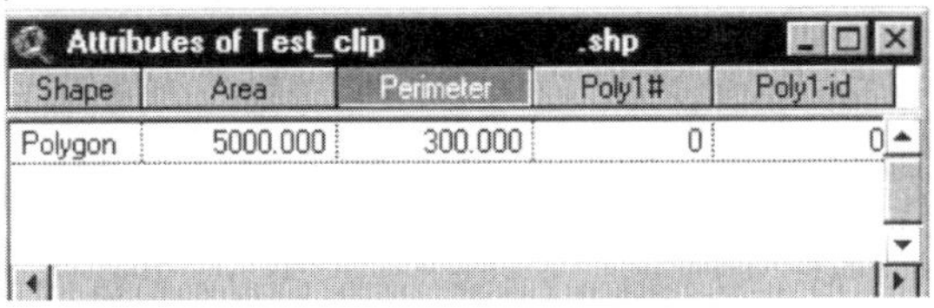

Attributes of Test_clip.shp

Shape	Area	Perimeter	Poly1#	Poly1-id
Polygon	5000.000	300.000	0	0

LESS CAN BE BETTER

GIS work is fun because it is challenging—you have to figure things out! Often frustration can be minimized and efficiency maximized if you work with a reduced spatial data set.

For example, a Landsat Thematic Mapper scene contains over 40 million grid cells. You could wait for 40 million grid cells to be processed before you find out you made an error. It would be more efficient to subset a small area from the scene to debug all your image processing methods with. In ARC/INFO Grid, you can do this with the SETWINDOW command. Quickly figure out how to correctly do your analysis on a 10 row by 10 column grid. After you are successful, you can then apply the same analysis operations to your huge grid.

Another example is using digital orthos for on-screen digitizing of wildfire burn scars. If you find your digital ortho display too slow (and you can live with 2 -meter pixels), you could reduce a 200 mb 1-meter cell size image to a 50 mb 2-meter cell size image by using the Grid RESAMPLE function.

In the vector environment, instead of wasting time struggling with tens of thousands of points, lines or polygons, first use the arc RESELECT command to create a small subset to work with. Once you have successfully figured out the correct analysis operations on a few features, then apply the same operations to your theme that contains tens of thousands of features.

GIS TABLES ISSUES

Consider keeping some of your attributes external to the GIS. You can always relate these attributes to your feature attribute table. Why would you want to do this? First, have you ever deleted the wrong coverage by mistake? If you had all your attributes stored in the feature attribute table, you lost them along with the geography. When you use external tables that can be linked, your attribute data are preserved when the coverage is accidentally deleted. You may have to redigitize the coverage features, but your attribute table is already constructed.

Secondly, related tables can make quality assurance a much easier process. Consider a coverage with 40,000 polygons, each of which is attributed with one of 22 soil types. You could have the actual name of the soil type entered for each polygon. Or you could have integer soil codes for each polygon and then use a related Look-up table where the actual names for the soil types exist. If you discover that some polygons are mislabeled, it is relatively simple to find and correct them, because the integer codes are short and errors will tend to stand out when the values are listed. If you go the other route, and choose to store the soil names in the polygon attribute table, *'hystic pergellic cryoquept'* may be entered 27,000 times, *'hystic pergellic cryoqupt'* may be entered 110 times, *'hystc pergellic cryoquept'* 2 times, *'hystic pergelic cryoquept'* 10 times, and *'hystic pergellic c8ryoquept'* 4 times. When you query for polygons where soil_type = *'hystic pergellic cryoquept'* you will miss 126 polygons that should have been selected. When you do recognize that there are errors in the database, correcting them will be time-consuming because the errors are scattered among 40,000 records and the exact spelling errors are unknown.

CONSIDER ABANDONING YOUR GIS

GIS is great software for spatial analysis. It can be pretty lame however for tabular analysis, graphics, and statistical analysis beyond very simple applications. Therefore, you should always keep your options open for using other software when you need the tools. Consider exporting your GIS table into a real spreadsheet program if you need sophisticated spreadsheet operations. Export your data into a statistical package if you need powerful statistical analysis. And if you require graphics beyond basic charts and graphs, think about exporting your GIS data into a graphics program. Use the best suite of tools available to you, and your work will be more fun and efficient.

DO NOT REPEAT YOURSELF

GIS work should not be a mundane process of repeating commands; there is usually a way to minimize repetition.

1) Startup files. Use a startup file to automatically execute common tasks. For example, the arc processor executes a startup file associated with each module when you start that module. For example, the *.arc* startup file is executed when you start Arc, while the *.arcedit* startup file is executed when you start Arcedit. By creating a .startup file, you can avoid repeating yourself every time you start a module. Here is an example .arcedit startup file for the Arcedit module. Note that the user does not have to specify any of this information as these commands are automatically executed from the .arcedit file.

```
/***.arcedit startup file to avoid specify session settings:

/**hardware specifications:
&terminal 9999 /***Xwindows
&display 9999 4 /***fullscreen X-windows canvas
digitizer 9100 /dev/ttyb /**calcomp 9100 digitizer to serial port b

/***set snapping tolerances :
&echo &on /***echo back the settings
nodesnap closest 50 /***snap to closest node within 50 meters
arcsnap on 50 /***snap to an arc within 50 meters
intersectarcs add /***automatically put a node when arcs intersect
weedtolerance 2 /***vertex minimum spacing is 2 meters

/***set drawing rules:
mapextent %.nrm338%/studyarea
drawenvironment arcs nodes tics ids links /***drawing rules
nodecolor node red /***color nodes red
&echo &off /***turn the echo off

/****user selects coverage to edit:
&sv cover = [getcover * -all 'Select a coverage to edit' ]
editcoverage %cover%

&return /***done with .arcedit startup so return to arcedit
```

2) Use AML Repeat and Expansion Character

In ARC/INFO, the AML processor interprets the (! as the beginning of a repetitive process. For example,

```
projectcopy cover /net/shemp/export/data/veg-utm newveg-utm
projectcopy cover /net/shemp/export/data/veg-utm newsoils-utm
projectcopy cover /net/shemp/export/data/veg-utm newburns-utm
```

Can be replaced by the following:

```
projectcopy cover /net/shemp/export/data/veg-utm (! newveg-utm newsoils-utm newburns-utm !)
```

3) Overwriting Temporary Grids

In Grid, by default you cannot overwrite an existing grid. For example:

```
Grid: temp = select (ndvi, 'Value gt 100')
GRD ERROR —Unable to create the output grid: temp
Output exists

Grid: kill temp
Killed TEMP with the ARC option
Grid: temp = select (ndvi, 'Value gt 100')
Running...
```

You can tell Grid that you want to overwrite existing grids by using the VERIFY OFF command. That way you do not have to constantly delete old grids before you write to them:

```
Grid: verify off
Grid: temp = select (ndvi, 'Value gt 100')
Running...
```

INDEXING FEATURES AND ATTRIBUTES

Imagine that you just picked a gallon of blueberries and you pull your favorite cookbook off the shelf. To find a recipe for blueberry pie, you could start at the first page of the cookbook and sequentially search page by page through the book until you find what you are looking for. A more efficient approach would be to look in the book index under "Blueberry." In an analogous matter, it is more efficient to search GIS features and attributes if they are indexed.

Attributes can be indexed in ARC/INFO by using the **INDEXITEM** command and in ArcView by first making the table field active, and then selecting Field → Create Index. By indexing attributes, you can speed up operations that select or relate features by attribute values.

Spatial features can be indexed in ARC/INFO by using the **INDEX** command and in ArcView by first making the Shape field of the attribute table active, and then selecting Field → Create Index. By indexing spatial features, you can speed up operations that retrieve features by spatial location such as drawing and graphical selection operations.

Appendix

GIS DATA MODEL EXERCISE SOLUTIONS

1) You have the following line theme:

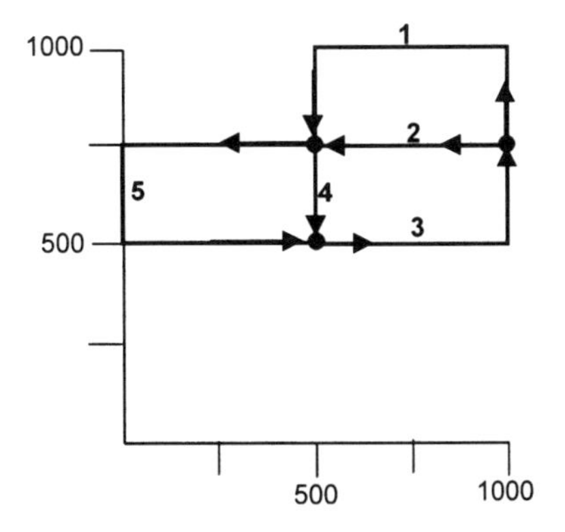

Fill in the following information from using the DESCRIBE operation.

There are three nodes. An arc starts at a from-node and ends at a to-node. Therefore there are 5 arcs . . . the arrows in the figure are arbitrary directions the arcs were originally digitized as. Arc#5 has two vertices at 0,500 and 0,750 while Arc#1 has a vertex at 1000,1000. Therefore the XMIN is 0, YMIN is 500, XMAX is 1000, and YMAX is 1000.

Number of Arcs	Number of Nodes	XMIN	YMIN	XMAX	YMAX
5	3	0	500	1000	1000

2) You have the following points. Point 100 is a well that is 200 feet deep, on parcel 356, while point 101 is a well that is 300 feet deep on parcel 387. Fill in the correct values for the point attribute table associated with this theme.

By definition, a point has no area or perimeter so the attribute table contains zeros for those values. The Well-ID is the user-specified identification tag while the Well# is generated by the GIS as a unique ID for each point.

+100

+101

Well.PAT

Area	Perimeter	Well#	Well-ID	Depth	Parcel
0	0	1	100	200	356
0	0	2	101	300	387

3) You have the following lines. Arc#1 has a length of 100 meters.

Fill in the correct values for the arc attribute table associated with this theme.

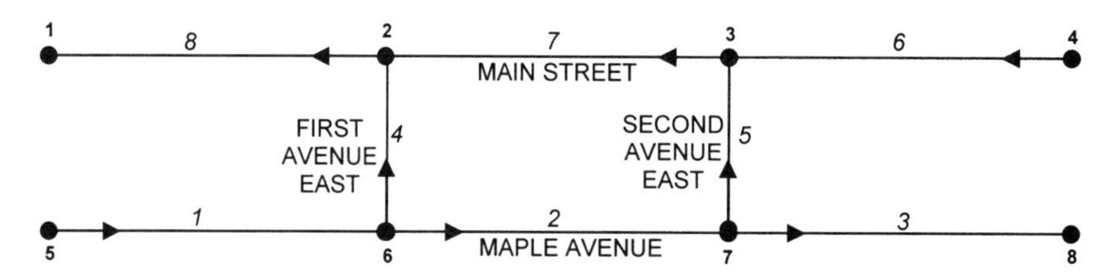

Fnode#	Tnode#	Length	Street#	Street-ID	Name
5	6	100	1	1	MAPLE AVENUE
6	7	100	2	2	MAPLE AVENUE
7	8	100	3	3	MAPLE AVENUE
6	2	50	4	4	FIRST AVENUE EAST
7	3	50	5	5	SECOND AVENUE EAST
4	3	100	6	6	MAIN STREET
3	2	100	7	7	MAIN STREET
2	1	100	8	8	MAIN STREET

4) You have the following polygons. These parcels have the following tax information:

	Parcel 203	Parcel 208	Parcel 209
Property Tax	$1,500	$3,000	$2,000
Owner	Mr. John Smith PO Box 100 Fairbanks, AK	Ms. Jane Doe PO Box 75 Deerville, CA	Mr. Roger Rabbit 235 Rabbit Hole Dr Carrotville, AK

Fill in the correct values for the polygon attribute table associated with this theme.

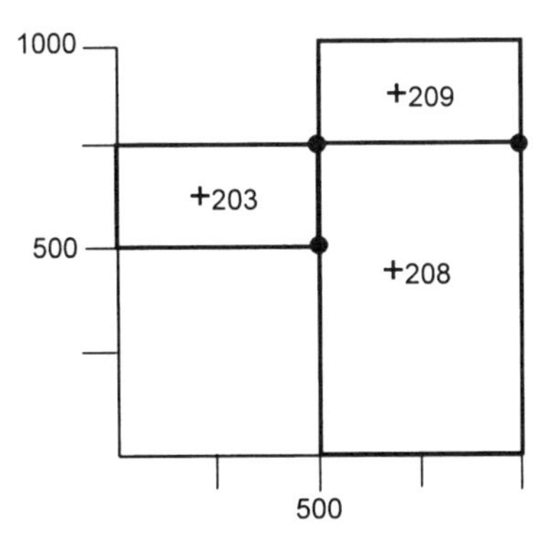

The GIS assigns a unique ID to each polygon, including the universe polygon which is assigned an area as negative of the summed area. The area of parcel 203 is 500 X 250 = 125000 and the perimeter is 500+250+500+250 = 1500. The area of parcel 208 is 500 X 750 = 375000 and the perimeter is 500 + 750 + 500 + 750 = 2500. The area and perimeter of parcel 209 is the same as parcel 203. The universe polygon has an area of the sum of the three parcel polygons (125000 + 375000 + 125000) and is assigned a negative value to indicate that it is an artificial polygon.

Area	Perimeter	Parcel#	Parcel-ID	Tax	Name	Street	City	State
-625000	5500	1	0	0				
125000	1500	2	203	1500.0	Mr. John Smith	PO Box 100	Fairbanks	AK
375000	2500	3	208	3000.0	Ms. Jane Doe	PO Box 75	Deerville	CA
125000	1500	4	209	2000.0	Mr. Roger Rabbit	235 Rabbit Hole Dr	Carrotville	AK

5) You have the following grid of slope classes. Build a Value Attribute Table for your grid.

For each unique value, count the total number of grid cells and fill in the Value Attribute Table. There are 11 cells with 0, 41 cells with 1, and 48 cells with 2 as values. As a check we have 10 rows and 10 columns and all cells contain valid values, therefore the total count should be 100.

Slope_Class Grid:

0	0	1	1	2	2	2	2	2	1
0	0	1	1	2	2	2	2	2	1
0	0	1	1	2	2	2	2	2	1
1	1	1	1	2	2	2	2	2	1
1	1	1	1	2	2	2	2	2	1
1	1	1	1	2	2	2	2	2	1
1	1	1	1	2	2	2	2	2	1
1	1	1	1	2	2	2	2	2	0
1	1	1	1	2	2	2	2	0	0
1	1	1	1	2	2	2	2	0	0

Slope_Class Value Attribute Table:

Value	Count
0	11
1	41
2	48

6) You have the following grid of soil pH. Build a Statistics Table for your grid.

The descriptive statistics of minimum, maximum, mean, and standard deviation are computed using all grid values:

Soil_PH

6.6	6.3	6.2	5.9
6.5	6.7	6.2	6.1
6.5	6.6	6.4	6.3
6.6	6.5	6.5	6.4

Soil_PH Statistics Table

Min	Max	Mean	STDV
5.9	6.7	6.394	0.208

TABULAR ANALYSIS EXERCISE SOLUTIONS

TABULAR ANALYSIS EXERCISES

1) You have an attribute table about soil polygons. You run the **statistics** program to create a new table summarizing the area of soil polygons by texture class. The output table is as follows:

Texture	Frequency	Sum-Area
1	21389	3371.357086
2	40987	6671.368010
3	*****	27204.001052
4	81298	20271.315022
5	92381	25244.364040

Why is the **Frequency** for texture class 3 not displayed in the table? **The value for the attribute is larger than the output width allocated for that column (output width of 5).**

How can you solve the problem so that **Frequency** for texture class 3 is displayed in the table? **Use ALTER to make the output width of frequency larger than 5 . . . try an output width of 10 to accommodate values up to 9,999,999,999.**

2) You have the following arc attribute table:

Stream#	Length	Ownership	Trout_count
1	3371.357086	1	156
2	6671.368010	2	354
3	27204.001052	1	45
4	20271.315022	1	98
5	25244.364040	3	322

Which records would be selected in the following expression:

Tables: **SELECT** stream.aat

Tables: **RESELECT** Ownership in {1,3} AND Trout_count GT 200

AND means the conditions of ownership of 1 or 3, and trout_count >200 both have to be true. Only the 5th record satistifies those criteria:

Stream#	Length	Ownership	Trout_count
1	3371.357086	1	156
2	6671.368010	2	354
3	27204.001052	1	45
4	20271.315022	1	98
5	25244.364040	3	322

Which records would be selected in the following expression:

Tables: **SELECT** stream.aat

Tables: **RESELECT** Ownership in {1,3} OR Trout_count GT 200

OR means that either one side or the other side or both sides of the expression need to be true. Therefore all 5 records are selected since they either have an ownership of 1 or 3, or they have a trout count > 200.

Stream#	Length	Ownership	Trout_count
1	3371.357086	1	156
2	6671.368010	2	354
3	27204.001052	1	45
4	20271.315022	1	98
5	25244.364040	3	322

3) You have a street arc attribute table containing two attributes: **Speed_Limit** which is the maximum allowable speed in miles per hour and **Length** which is the length of each arc in meters. You add another attribute column called **Time** . How would you use calculate the time in minutes it would take to travel across each arc at the maximum speed limit? There are 5280 feet in a mile and 3.281 feet in a meter.
You can convert Speed_Limit in miles/hour to feet/hour by multiplying by 5280 feet/ 1 mile. You can convert feet/hour to feet/minute by multiplying by 1 hour/60 minutes. And you can convert Length in meters to feet by multiplying by 3.281 feet/ 1 meter.

CALCULATE FT_PER_MIN = Speed_Limit * 5280 * (1.0 / 60.0)

CALCULATE Length_FT = Length * 3.281

CALCULATE TIME = Length_FT / FT_PER_MIN

As a check we could create a test table with one record having a speed limit of 25 miles per hour and the arc length of 1 mile. The second record could have a speed limit of 60 miles per hour and an arc length of 10 miles. We do the calculations and then list the table:

Speed_Limit	Length	Length_FT	FT_PER_MIN	TIME
25.0	1609.2655	5280.00	2200.0	2.400
60.0	16092.6543	52800.00	5280.0	10.000

Our test table checks out . . . it takes 10 minutes to travel a 10 mile (52800 ft) arc at a speed of 60 minutes per hour. And it takes 2.4 minutes at 25 miles per hour to cross a mile-long arc (5280 feet). (Check 25 miles/hr * 2.4 minutes * 1 hour/60 minutes = 1 mile).

4) Correct the following logical expression:

RESELECT SPECIES CN 'KING SALMON' OR CN 'SOCKEYE SALMON'
The problem with the above expression is that the GIS does not know which attribute to search to see if a record contains 'SOCKEYE SALMON'
The correct expression is:
RESELECT SPECIES CN 'KING SALMON' OR SPECIES CN 'SOCKEYE SALMON'

5) Correct the following calculation:

CALCULATE ACRES = HECTARES X 2.471
The problem is that the operator for multiply is * and not X
The correct expression is:
CALCULATE ACRES = HECTARES * 2.471

6) Correct the following calculation:

CALCULATE ACRES = AREA / 43,560
The problem is that the character, is unknown to the expression translator
The correct expression is:
CALCULATE ACRES = AREA / 43560

7) Correct the following logical expressions:

The problem is that the all records need to be selected before each record selection.
Otherwise, you would get a message like "0 Records selected"
RESELECT VEGCODE = 1
CALCULATE SHADECOLOR = 27
RESELECT VEGCODE = 2
0 Records selected
CALCULATE SHADECOLOR = 35
RESELECT VEGCODE = 3
0 Records selected
CALCULATE SHADECOLOR = 67

The correct expression is:
ASELECT
RESELECT VEGCODE = 1
CALCULATE SHADECOLOR = 27
ASELECT
RESELECT VEGCODE = 2
CALCULATE SHADECOLOR = 35
ASELECT
RESELECT VEGCODE = 3
CALCULATE SHADECOLOR = 67

8) You have selected a **FOREST.PAT** polygon attribute table. You create a new attribute called **SITE_CLASS** based on an existing **SITE_INDEX** attribute. **SITE_CLASS** of 1 would be any polygon with **SITE_INDEX** less than 50, **SITE_CLASS** of 2 would be any polygon with **SITE_INDEX** between 50 and 75, and **SITE_CLASS** of 3 would be any polygon with a **SITE_INDEX** greater than 75.

Fill in the appropriate TABLES commands to do the following:
ADDITEM /*Add a new attribute column called Site_class**

SELECT FOREST.PAT /**Select the forest polygon attribute table**

RESELECT SITE_INDEX LT 50 /*Select all records with site_index less than 50**

CALCULATE SITE_CLASS = 1 /*Fill in the Site_class attribute with a value of 1**

ASELECT /*Select all records in the table**

RESELECT SITE_INDEX GE 50 AND SITE_INDEX LE 75 /*Select site_index 50- 75**

```
CALCULATE SITE CLASS = 2  /***Fill in the Site_class attribute with a value of 2
ASELECT                   /***Select all records in the table
RESELECT SITE INDEX GT 75 /***Select all records with site_index greater than 75
CALCULATE SITE CLASS = 3  /***Fill in the Site_class attribute with a value of 3
ASELECT                   /***Select all records in the table
RESELECT AREA LT 0        /***Select the universe polygon record
CALCULATE SITE CLASS = 0  /***Fill in the Site_class attribute with a value of 0
```

9) You have a point attribute table of waterfowl nests containing the following items:

UNIT	The management unit the nest is in
X-COORD	The GIS X-coordinate of each nest location
Y-COORD	The GIS Y-coordinate of each nest location
NEST-ID	The Identification Number of each nest
SPECIES	The species code (1=mallard, 2=pintail, 3=widgeon,4=green wing teal)
AGECLASS	The age class of the nesting duck (1=first year, 2=older than first year)
CLUTCH_SIZE	The number of eggs in each nest

You want to produce a table with the following information:

Unit	Species	Age Class	Total Number of Eggs	Total Number of Nests
1	1	1	121	12
1	1	2	345	42
1	2	1	32	7
1	2	2	213	19
1	3	1	267	22
2	1	1	465	54
2	1	2	132	12
2	3	1	197	15

What would you use for frequency and summary items to generate this information?

Frequency item(s): These are the categories that the GIS summarizes by: UNIT, SPECIES, AGECLASS

Summary item(s): These are the summed or totaled attributes: CLUTCH_SIZE

This would give you the total number of eggs for each category. Since each record represents a nest, FREQUENCY is always an output attribute from the program and will represent the total number of nests in each category.

10) There is a proposal to purchase some land for an experimental forest research site. Your job is to produce a table listing the hectares and percent of area for each vegetation class in this area. You have a vegetation polygon attribute table and another table of vegetation names as follows:

Vegetation Polygon Attribute Table

Area	Veg#	Veg-ID	Size-class	Type
-929,7919.191	1	0	0	0
3447.094	2	101	P	1
7017.024	3	102	S	7
and so on...	and so on...	and so on...	and so on...	and so on...

type_names.tbl

-1	'Cutover'
0	'Universe polygon'
1	'Black Spruce'
2	'White Spruce
3	'Aspen'
4	'Birch'
5	'Open Water'
6	'Willow'
7	'Alder'
8	'Dwarf Birch'
9	'Sedge Meadow'
10	'Calamagrostis Grass'

Fill in the appropriate tools in the following flowchart that would produce a sorted table of total hectares and percent of area for each vegetation class.

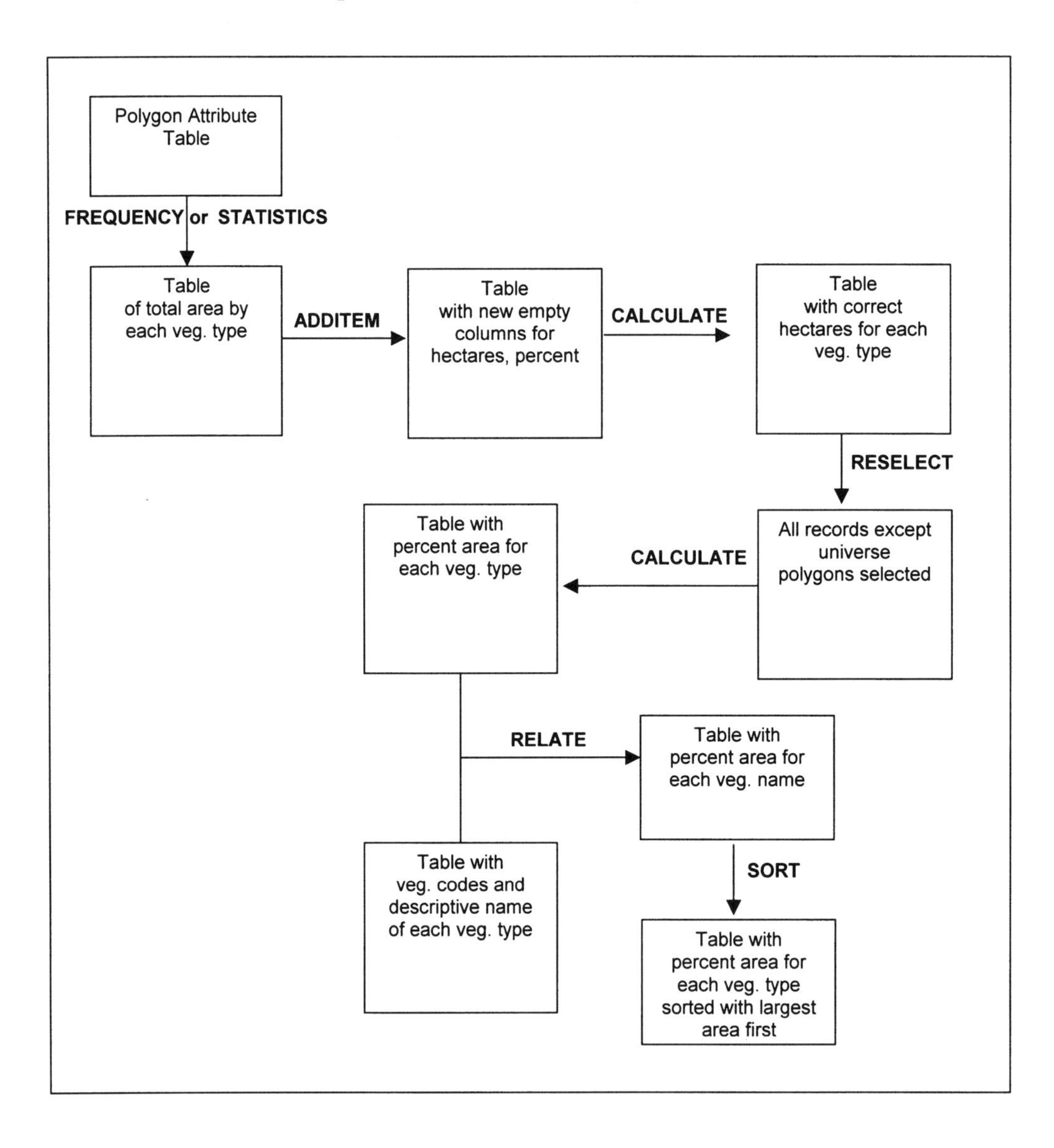

POINT ANALYSIS EXERCISE SOLUTIONS

1) You have a point coverage of manhole covers and a line coverage of streets. You want to determine the street name for each manhole cover.

Fill in the following flowchart with the appropriate GIS tools to solve your problem:

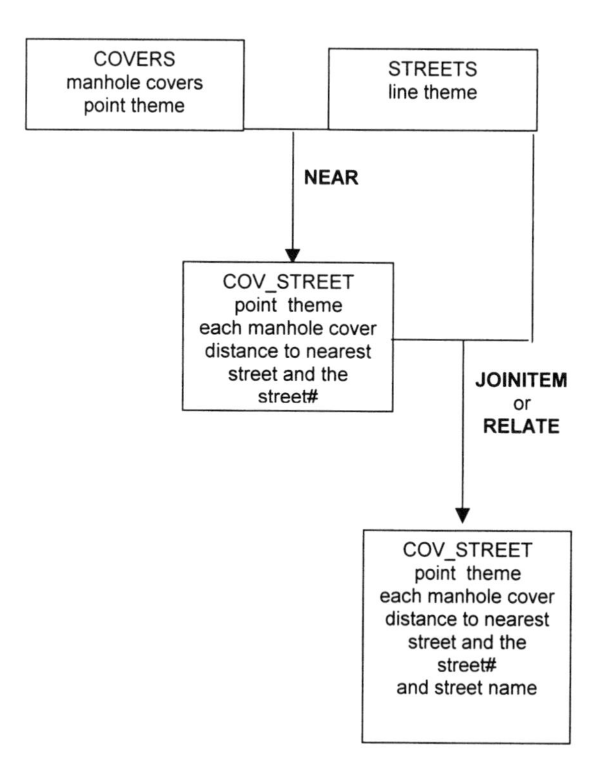

2) You have a point theme of wells to sample groundwater. You want to calculate the mean distance for each well to all other wells. Fill in the following flowchart to solve your problem:

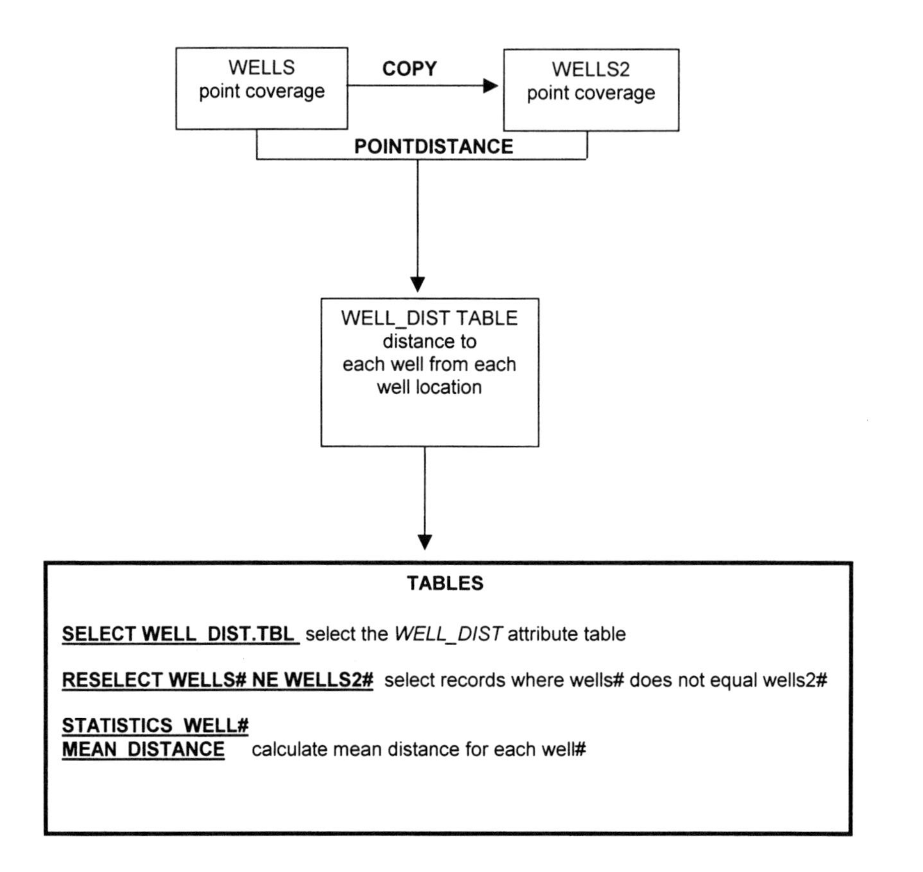

3) You have a point theme of lightning strike locations and a polygon theme of elevation classes. Your goal is to produce a table like the following:

Elevation Zone	Total Area (ha)	Total Lightning Strikes	Lightning Strikes per million ha
1			
2			
3			
4			
5			

Fill in the following flowchart with the appropriate GIS tools to solve your problem:

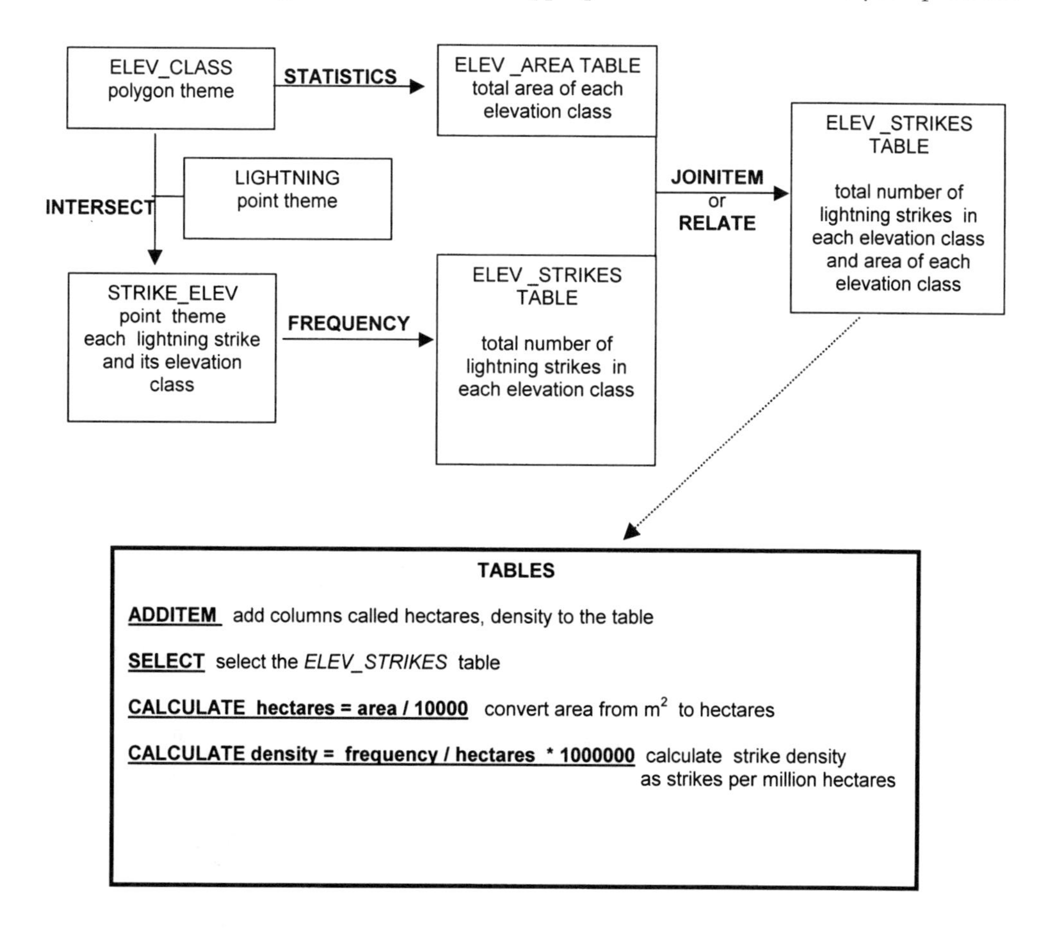

4) You have 5 GPS survey monuments set-up in the Tanana Flats area. Draw 5 Thiessen polygons associated with these GPS monuments:

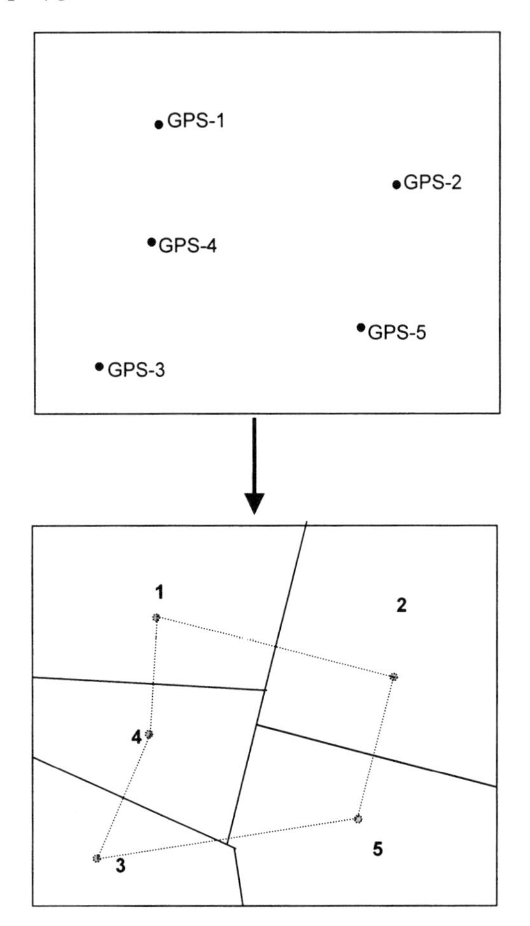

5) You have a point coverage of bufflehead nest box locations along the Chena river. Outline the GIS analysis tools that you would use to select all nest box points that are at least 1 km away from any other nest box.

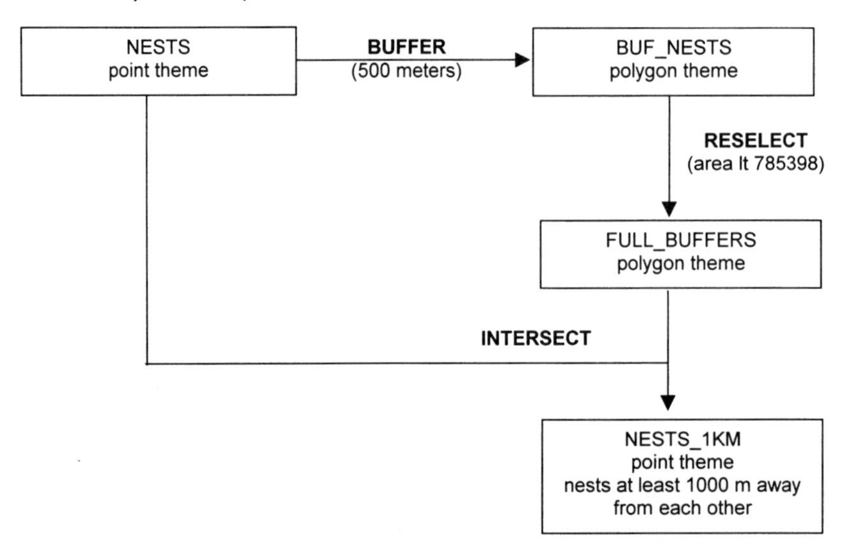

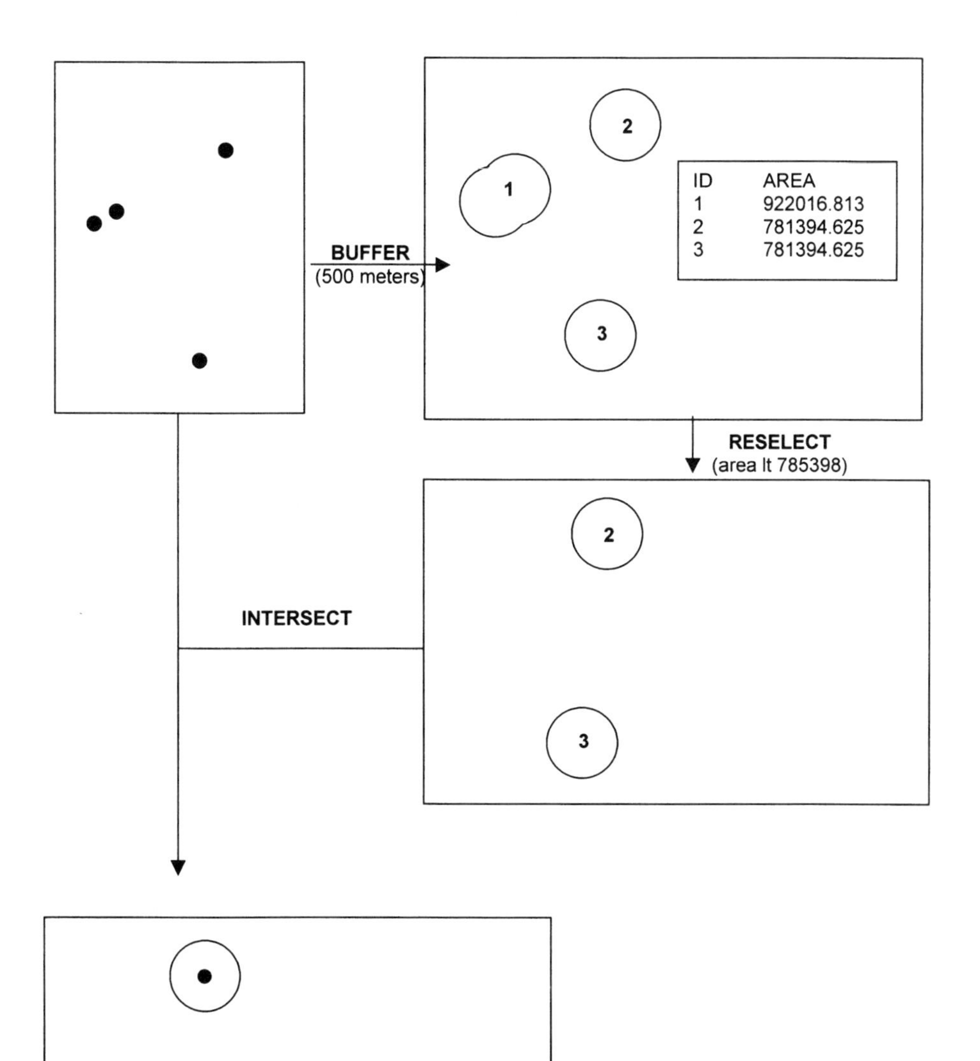

2
1
ID AREA
1 922016.813
2 781394.625
3 781394.625
BUFFER
(500 meters)
3
RESELECT
(area lt 785398)
2
INTERSECT
3

6) You have a point coverage in meters of snowshoe hare locations. You want to know how many of these locations are within 50 meters of a willow polygon . . . .

list hares.pat

HARES#	HARES-ID
1	1
2	2
3	3
4	4
5	5
6	6
7	7
8	8
9	9
10	10

buffer hares buf_50m # # 50 # point
list buf_50m.pat

AREA	PERIMETER	BUF_50M#	BUF_50M-ID	INSIDE
-78123.359	3137.729	1	0	1
7812.336	313.773	2	1	100
7812.336	313.773	3	2	100
7812.336	313.773	4	3	100
7812.336	313.773	5	4	100
7812.336	313.773	6	5	100
7812.336	313.773	7	6	100
7812.336	313.773	8	7	100
7812.336	313.773	9	8	100
7812.336	313.773	10	9	100
7812.336	313.773	11	10	100

From the buffer table above, we can infer that all points had full buffers since the areas are all the same and slightly less than pi*50^2. Notice also that the attribute BUF_50M-ID contains the original hare ID numbers.

list willow.pat

AREA	PERIMETER	WILLOW#	WILLOW-ID
-680190.375	7027.910	1	0
151522.984	1548.660	2	102
392388.719	2466.836	3	103
47269.211	1157.825	4	101
53964.820	990.133	5	104
35044.672	864.457	6	105

intersect buf_50m willow hares_50m
list hares_50m.pat

Area	Perimeter	HARES _50M#	HARES _50M-ID	BUF _50M#	BUF _50M-ID	INSIDE	WILLOW #	WILLOW -ID
-31906.	1735.	1	0	1	0	1	1	0
7812.	313.	2	1	3	2	100	2	102
7812.	313.	3	2	4	3	100	3	103
1293.	175.	4	3	5	4	100	4	101
560.	138.	5	4	5	4	100	4	101
7812.	313.	6	5	6	5	100	3	103
2196.	212.	7	6	7	6	100	5	104
4417.	266.	8	7	11	10	100	6	105

How many of the ten hare locations are within 50 meters of a willow polygon?

The attribute BUF_50M-ID contains the original hare IDs. Therefore the following hares are within 50 meters of a willow polygon: Hare#2, Hare#3, Hare#4, Hare#5, and Hare#6 and Hare#10 . . . so a total of 6 Hares.

Notice that Hare#4 is within 50 meters of willow 101 twice . . .

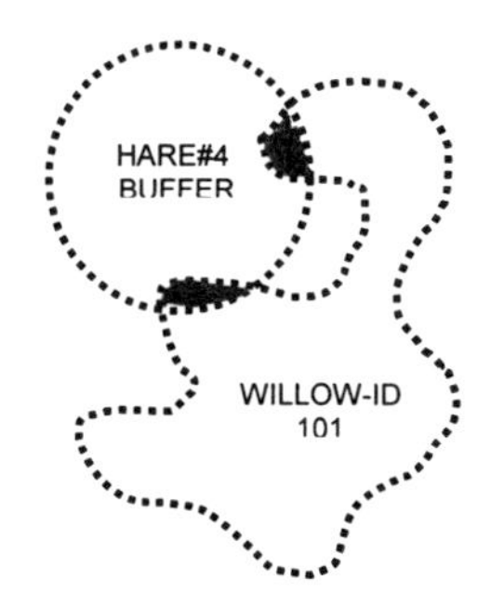

7) You have a point coverage of radio-collared moose. The point attributes include Moose-ID, sex, and age. You also have a line coverage of rivers. You want to produce a table as follows:

Moose-ID	Average Distance to River

Fill in the following flowchart to produce the above table:

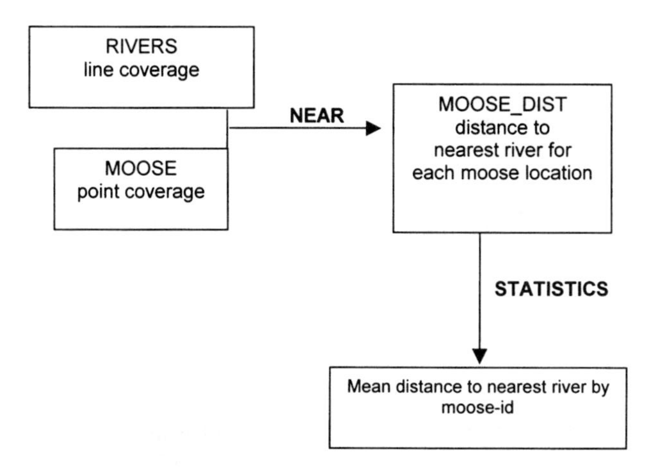

8) You have a point coverage of Polar Bear locations stored in longitude/latitude. You want to create a new theme of bear locations north of 70 degrees and between –170 and –100 degrees of latitude. Fill in the following flowchart to produce this new theme.

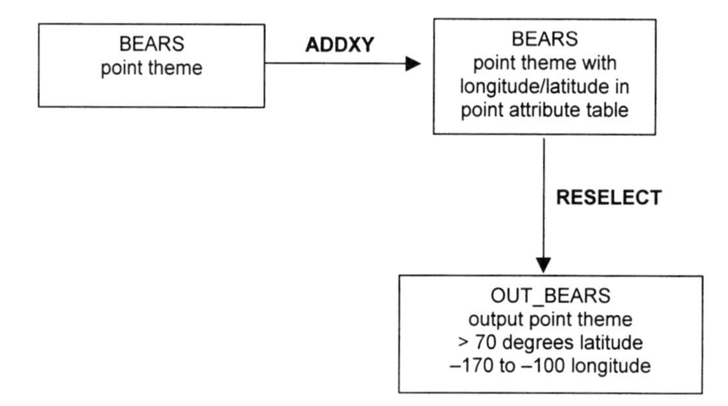

9) You have a theme of endangered plant species locations and a theme of randomly located points. You want to know what percent of plant points are within clay loam soil polygons compared to the percent of random points in clay loam polygons. Fill the following flowchart with the appropriate GIS tools to solve this problem

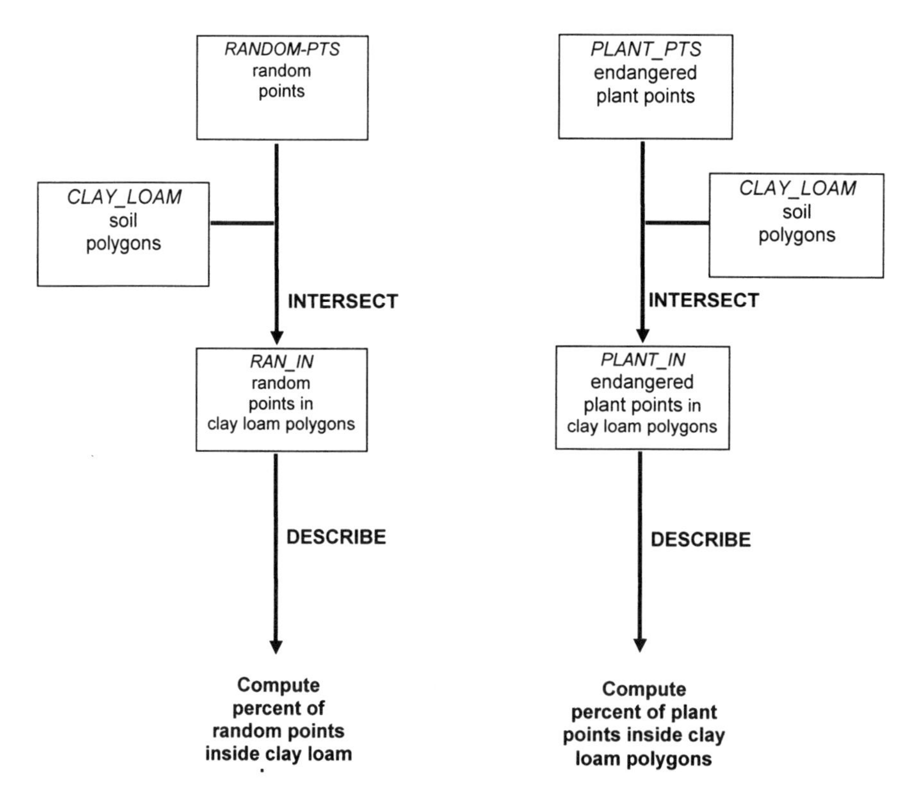

In the final step, you would use DESCRIBE to determine the total number of original points and the total number of points in clay loam polygons. Then you could calculate the percentage of points inside the clay loam polygons using a calculator.

10) You have a theme of endangered plant species locations. You want the mean, minimum, and maximum distance of the plant locations to the nearest clay_loam polygon. Fill the following flowchart with the appropriate GIS tools to solve this problem.

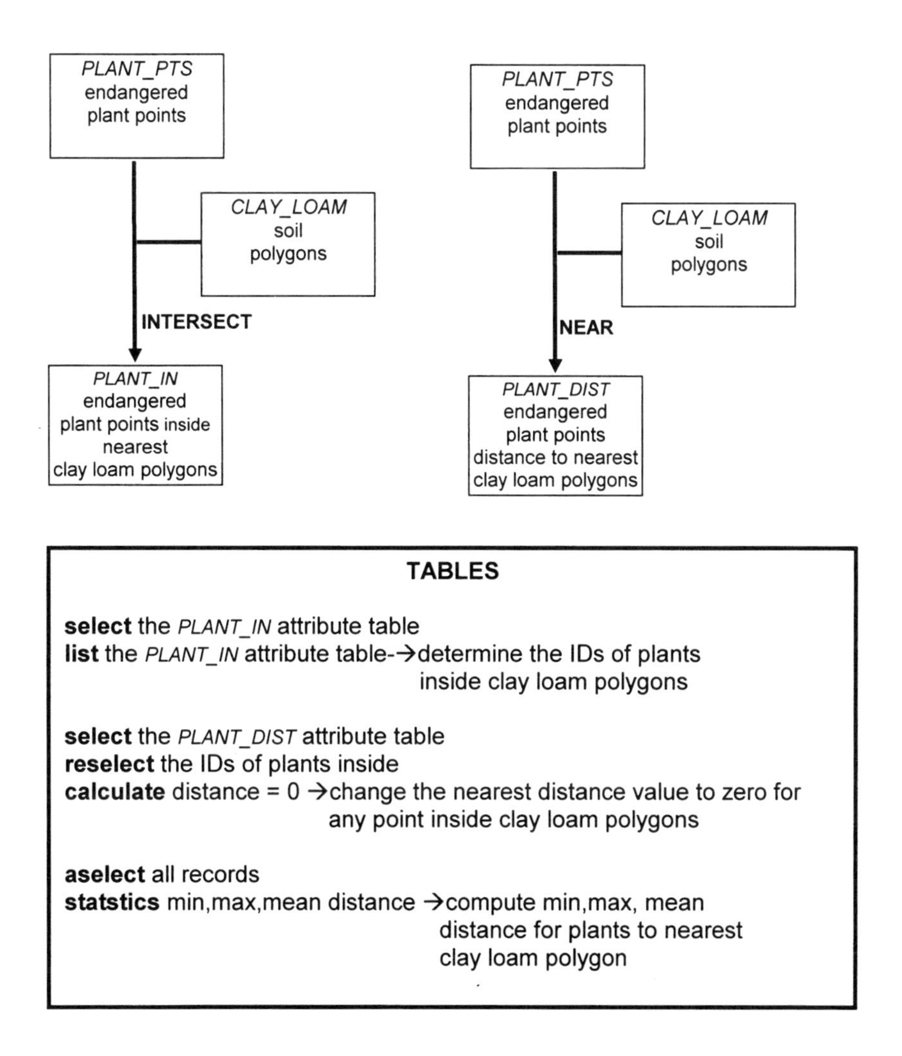

LINE ANALYSIS EXERCISE SOLUTIONS

1) You have the following line coverage of pipes. You run **DISSOLVE** using *Pipe_Class* as the dissolve item. What will your output theme look like?

Pipe Arc Attribute Table

Pipe#	Length	Pipe_Class	Diameter	Flow
1	1000	1	36	1500
2	300	1	36	1400
3	700	1	36	1300
4	1000	1	36	1200
5	300	1	36	1100
6	600	1	36	0
7	800	3	12	800
8	1000	2	12	700
9	800	3	10	600
10	800	3	10	600
11	800	3	12	800
12	800	3	12	800
13	1000	2	18	700
14	800	4	4	600

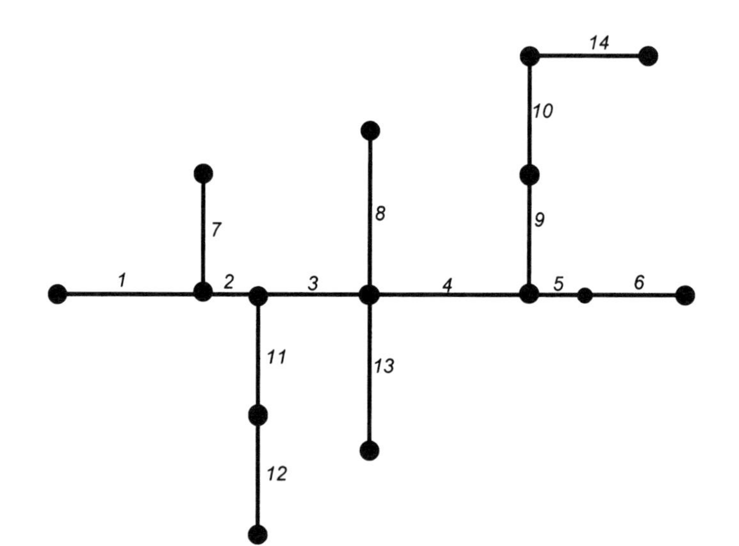

The only arcs that you can dissolve that contain *Pipe_Class* of 1 are arcs 5 and 6. They get dissolved to arc# 5 which is 300 + 600 = 900 in length. The arcs that contain *Pipe_Class* of 2 are arcs 8 and 13. They intersect the main pipe arcs 3 and 4 and therefore cannot be dissolved. The arcs that contain *Pipe_Class* of 3 are arcs 7, and 9-12. Of these, arcs 9 and 10 can be dissolved together and arcs 11 and 12 can be dissolved together. And there is only one arc with *Pipe_Class* of 4, so it cannot be dissolved with any other arcs.

Dissolved_Pipes#	Length	Pipe_Class
1	1000	1
2	300	1
3	700	1
4	*1000*	*1*
5	**900**	**1**
7	800	3
8	1000	2
9	***1600***	***3***
11	***1600***	***3***
13	1000	2
14	800	4

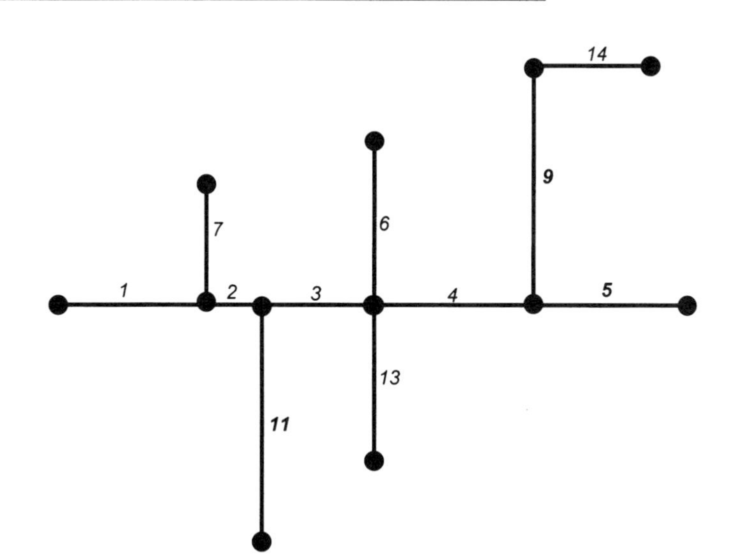

2) You have a line coverage of roads and a point coverage of cabin locations. You want to find all cabins that are within 1 mile of a road. Your GIS coordinate system is in meters.

Fill in the following flowchart to solve the problem:

Your GIS coordinate system is in meters. There are 1609.344 meters per mile, therefore your buffer distance should be 1609.344

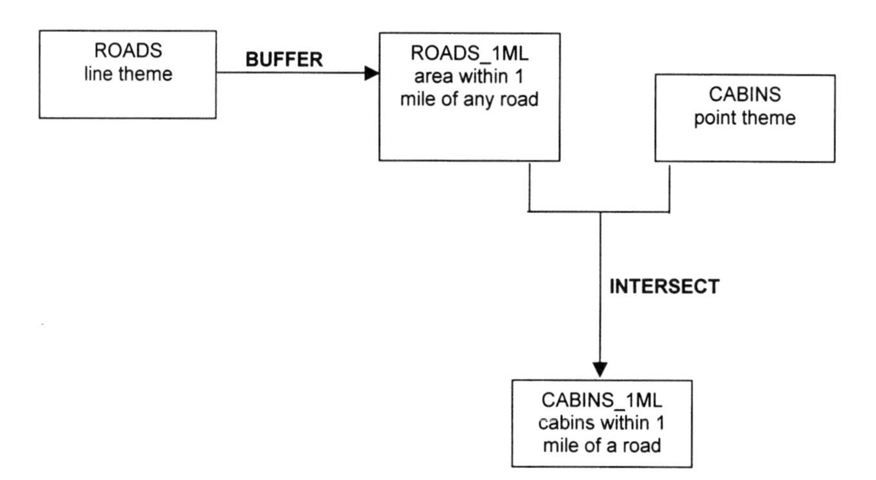

3) You have a line coverage of streams with an arc attribute called *King_count* representing the count of king salmon observed along each arc. For each stream, determine the total density of king salmon per mile. Your GIS coordinate system is in meters.

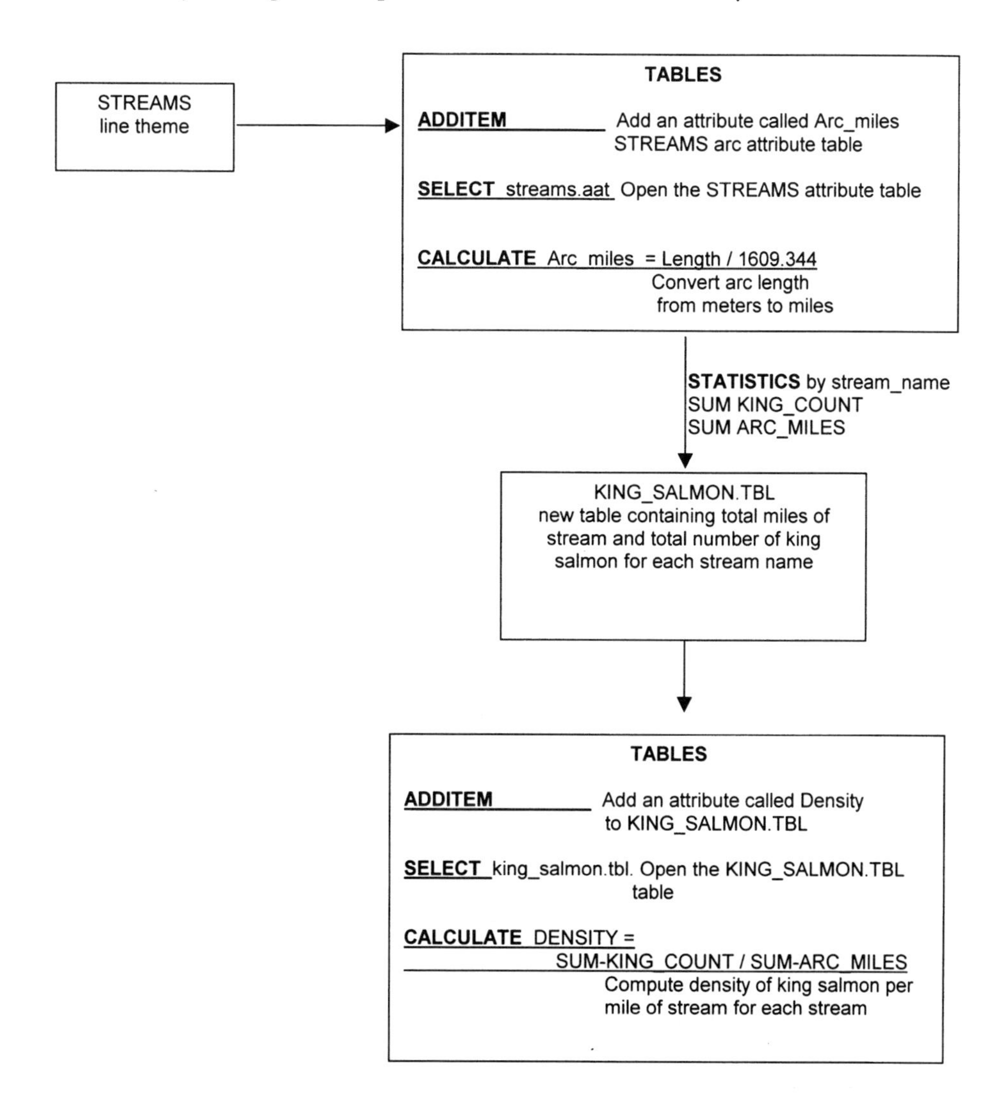

4) You have a line theme where each line represents the border between two vegetation types as follows:

Vegetation Arc Attribute Table						
Fnode#	Tnode#	Lpoly#	Rpoly#	Length	Arc#	Arc-ID
		1	2	7409.314	1	2
		3	2	4061.212	2	2
		1	3	231.273	3	3
		1	4	8.255	4	1
		3	1	236.307	5	4
		3	4	4057.586	6	1
		1	4	7073.343	7	1

Vegetation Polygon Attribute Table					
Area	Perim.	Vegetation#	Vegetation-ID	Veg_Class	Veg_Name
		1	0	0	
		2	1	3	Pin Oak
		3	2	2	Sweet Gum
		4	3	1	Red Maple

From searching the polygon attribute table, you know there is one stand of Sweet Gum and one stand of Pin Oak in the theme. What is the length of the border between the Sweet Gum and Pin Oak stand? Fill in the appropriate **TABLES** tools to solve the problem:

From the polygon attribute table, you can see that the Pin Oak polygon internal ID is Vegetation#2 and the Sweet Gum polygon internal ID is Vegetation#3. Therefore, find any arcs where the left polygon is Pin Oak (Lpoly#2) and the right polygon is Sweet Gum (Rpoly#3) or the right polygon is Pin Oak (Rpoly#2) and the left polygon is Sweet Gum (Lpoly#3)

TABLES

SELECT VEGETATION.AAT Get the arc attribute table

RESELECT (LPOLY# EQ 2 AND RPOLY# EQ 3) OR
(LPOLY# EQ 3 AND RPOLY# EQ 2) Find the arc(s) that make up the border

STATISTICS
SUM LENGTH Determine the total length of the border

5) You have a line theme of roads and a point theme of auto accidents. Each road arc has a highway code attribute. You want to store all the accidents that occurred on the Parks Highway (highway code= 2) as a new point theme.

Fill in the following flowchart to solve your GIS problem:

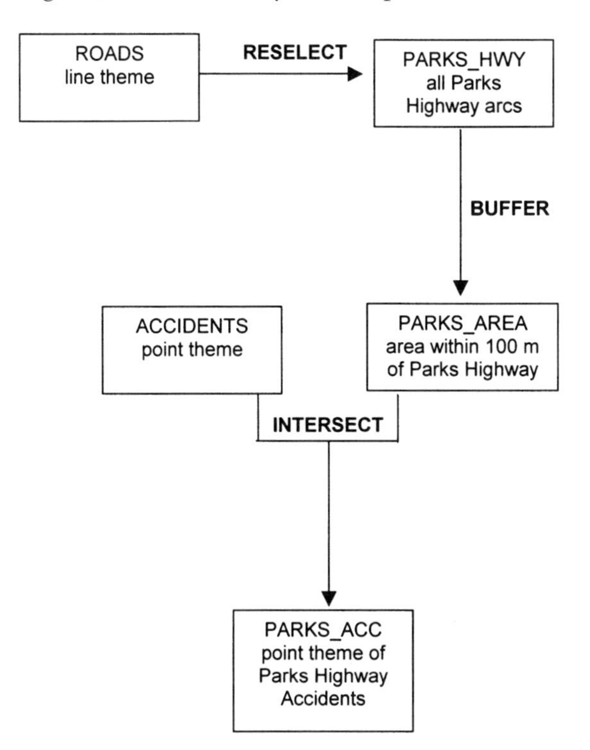

6) You have a polygon theme of parcels and a line coverage representing a utility right of way. Each parcel polygon has the owner's phone number and address stored in the polygon attribute table. You want to generate a text file of owner phone numbers and addresses for all parcels within the right of way.

Fill in the following flowchart to solve your GIS problem:

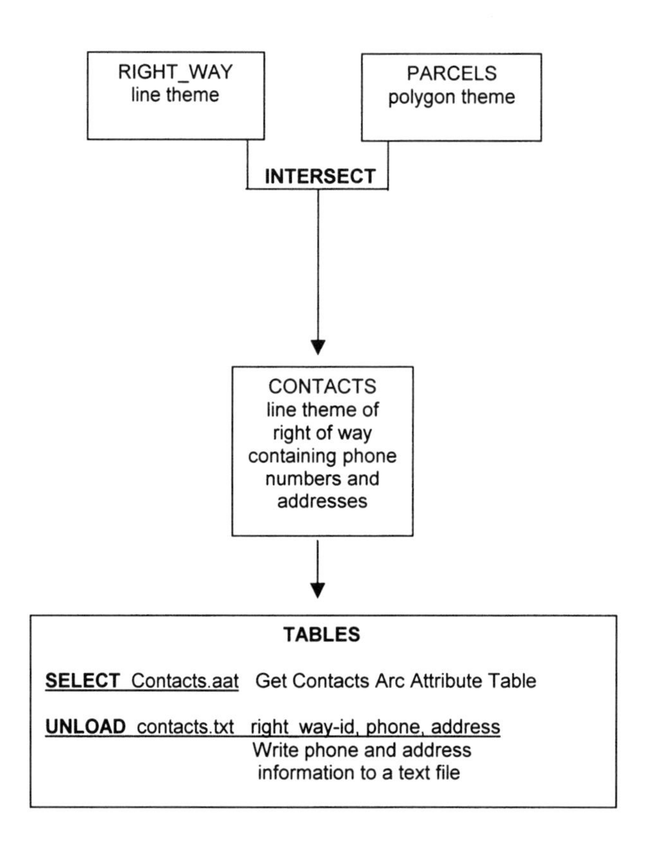

7) You have a line theme of gas pipelines and a polygon theme of ownership. You want to produce a table showing for each ownership, the total length of pipeline by pipe diameter class and by pipe age class.

Fill in the following flowchart to solve your GIS problem:

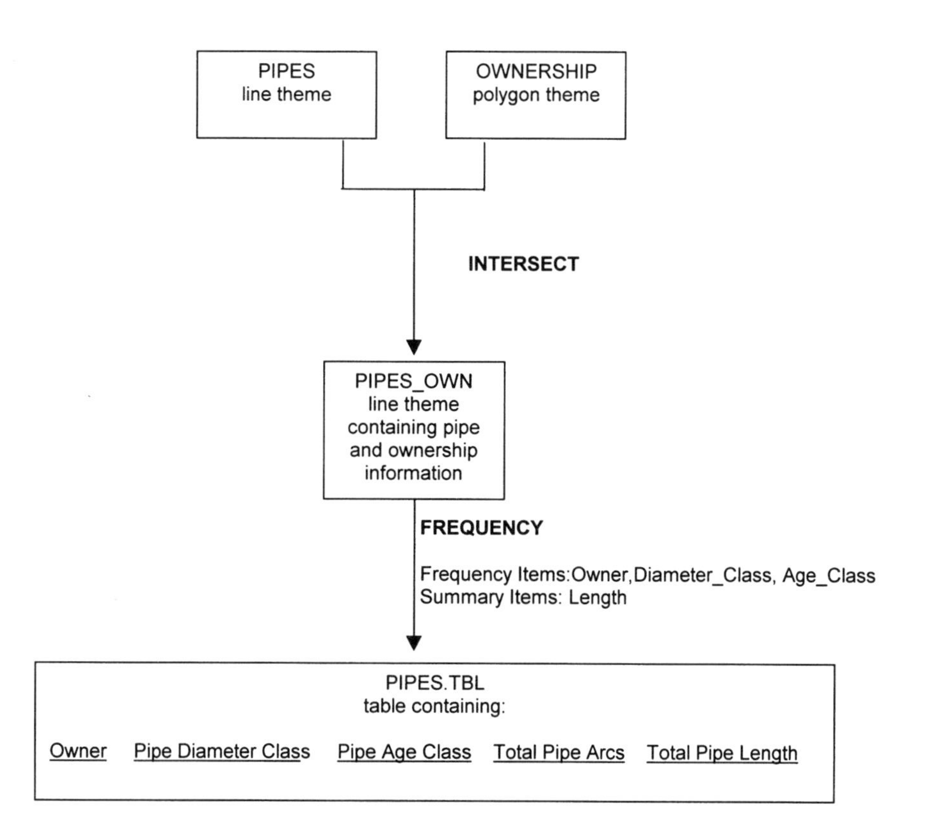

8) You have a line theme of contour lines with an attribute of elevation. Elevations are recorded at contour intervals of 10 meters, ranging from 10 to 980 meters. You want to assign a dashed line to minor contour elevations that are tens of meters (for example: 10,20,30,40, etc.) and a solid line to major contour elevations that are hundreds of meters (for example 100,200,300).

Fill in the following flowchart to solve your GIS problem:

TABLES

ADDITEM contours.aat contour_class 2 4 B
Add a column called *Contour_Class* to the arc attribute table

SELECT contours.aat Get the arc attribute table

RESELECT elev in {100,200,300,400,500,600,700,800,900}
Get the major elevation contour records

CALCULATE Contour_Class = 2 Assign a *Contour_Class* value of 2

NSELECT ______________ Get the minor elevation contour records

CALCULATE Contour_Class = 1 Assign a *Cuntour_Class* value of 1

9) You have a line coverage of streets and a point coverage of fire hydrants. You have a line coverage of streets and a point coverage of fire hydrant locations. You want to find all hydrants where the distance between fire hydrants is greater than 1 km and the hydrants are within 10 meters of a street.

Fill in the following flowchart to solve the problem:

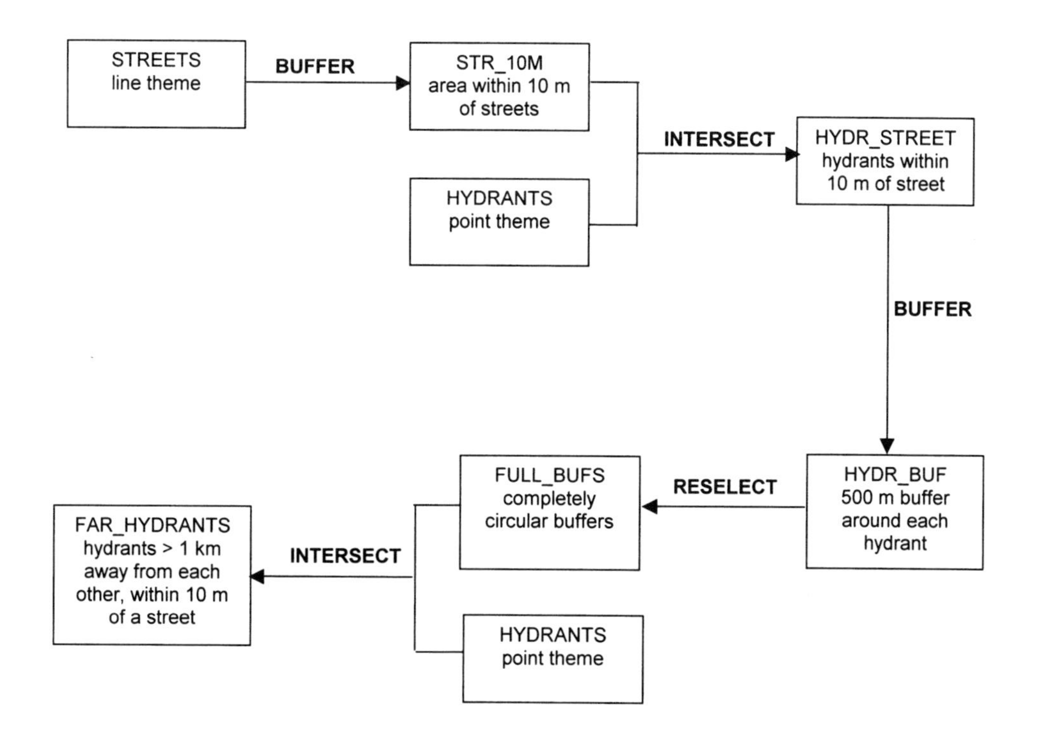

10) A grizzly bear biologist wants you to use a line theme of streams and a point theme of bears to produce the following three tables. The GIS coordinate system is in meters. Your job is to figure out the total area within 1 km of each stream and the total number of bears within these 1 km areas

Stream Name	**Number of Bears** (within 1 km of stream)	**Area (ha)** within 1 km of stream
Clear Creek		

Stream Name	**Number of Bears** (within 1 km of stream)	**Area (ha)** within 1 km of stream
Moose River		

Stream Name	**Number of Bears** (within 1 km of stream)	**Area (ha)** within 1 km of stream
Rapid River		

Fill in the appropriate tools to solve the problem on the next page:

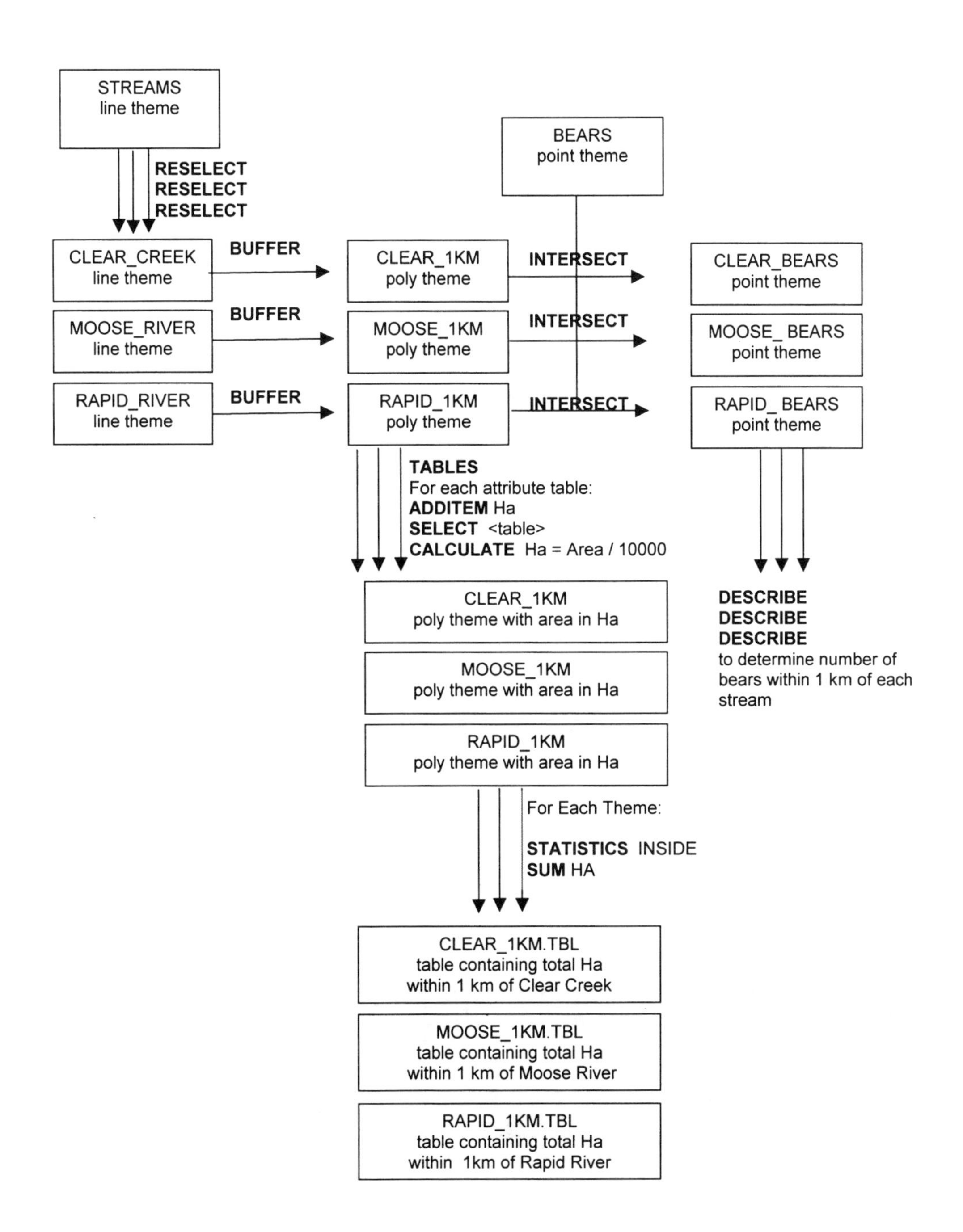
STREAMS
line theme
RESELECT
RESELECT
RESELECT
BEARS
point theme
CLEAR_CREEK
line theme
BUFFER
CLEAR_1KM
poly theme
INTERSECT
CLEAR_BEARS
point theme
MOOSE_RIVER
line theme
BUFFER
MOOSE_1KM
poly theme
INTERSECT
MOOSE_ BEARS
point theme
RAPID_RIVER
line theme
BUFFER
RAPID_1KM
poly theme
INTERSECT
RAPID_ BEARS
point theme
TABLES
For each attribute table:
ADDITEM Ha
SELECT <table>
CALCULATE Ha = Area / 10000
CLEAR_1KM
poly theme with area in Ha
MOOSE_1KM
poly theme with area in Ha
RAPID_1KM
poly theme with area in Ha
DESCRIBE
DESCRIBE
DESCRIBE
to determine number of bears within 1 km of each stream
For Each Theme:
STATISTICS INSIDE
SUM HA
CLEAR_1KM.TBL
table containing total Ha within 1 km of Clear Creek
MOOSE_1KM.TBL
table containing total Ha within 1 km of Moose River
RAPID_1KM.TBL
table containing total Ha within 1km of Rapid River

NETWORK ANALYSIS EXERCISE SOLUTIONS

1) You have a report of a fire at 1721 First Avenue West. Draw the location of that address on the following streets theme. **The first step is to search the arc attribute table and match the address components to the street arc attributes Name=First, Type=Avenue, Suffix= West . . .**

Arc#	Length	F_node	T_node	Left_from	Left_to	Right_from	Right_to	Name	Type	Suffix
1	400	1	2	200	300	201	299	Main	Street	
2	400	2	3	302	400	301	399	Main	Street	
3	400	3	4	402	500	401	499	Main	Street	
4	400	5	1	100	2	99	1	Third	Avenue	
5	400	6	2	100	2	99	1	Second	Avenue	West
6	400	8	3	100	2	99	1	First	Avenue	West
7	400	11	4	100	2	99	1	Center	Street	
8	400	6	5	1814	1998	1813	1997	First	Avenue	West
9	200	7	6	1744	1812	1743	1811	First	Avenue	West
10	200	8	7	1692	1742	1691	1741	First	Avenue	West
11	100	9	8	1612	1690	1611	1689	First	Avenue	West
12	200	10	11	1552	1610	1551	1609	First	Avenue	West
13	100	11	10	1500	1550	1549	1551	First	Avenue	West
14	400	12	5	200	102	199	101	Third	Avenue	West
15	400	13	6	200	102	199	101	Second	Avenue	West

Then find the record that has the correct address range.

Arc#	Length	F_node	T_node	Left_from	Left_to	Right_from	Right_to	Name	Type	Suffix
6	400	8	3	100	2	99	1	First	Avenue	West
8	400	6	5	1814	1998	1813	1997	First	Avenue	West
9	200	7	6	1744	1812	1743	1811	First	Avenue	West
10	200	8	7	1692	1742	1691	1741	First	Avenue	West
11	100	9	8	1612	1690	1611	1689	First	Avenue	West
12	200	10	11	1552	1610	1551	1609	First	Avenue	West
13	100	11	10	1500	1550	1549	1551	First	Avenue	West

Then find Arc#10 and determine which side is the right side of the arc:

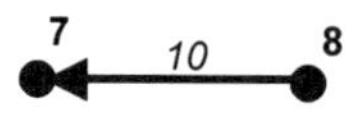

And then interpolate between the arc addresses to estimate the location of 1721.

1741 – 1691 = address range of 50 over a distance of 200.

1721 – 1691 / 50 = 60 percent of the arc or .60 * 200 = distance of 120 from starting node.

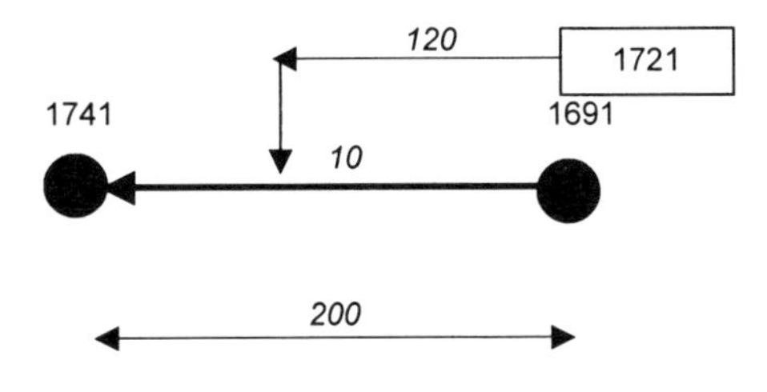

2) Given the following network theme, determine the path that is the quickest to get from point A to point B. First, we label the time it takes to travel across each arc from the information in the arc attribute table

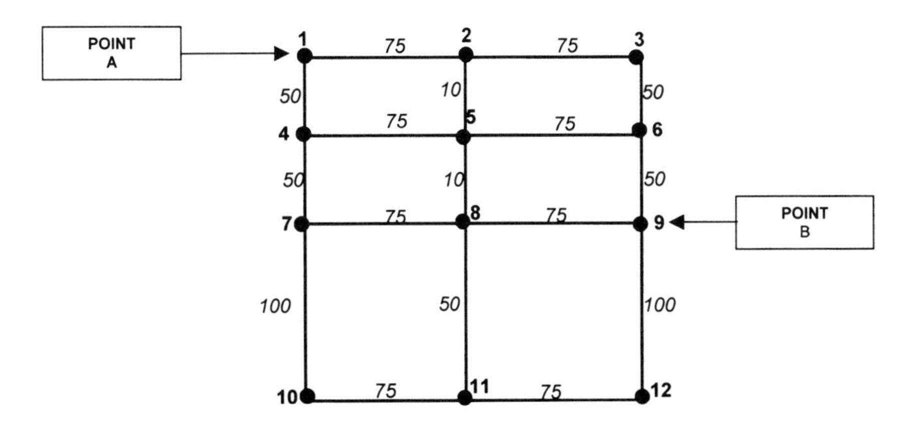

We build two tables, one of nodes that have already been processed, and one of adjacent nodes to process. We start at Point A, node#1:

Processed Nodes

Node	Cumulative Cost	Previous Node
1	0	none

Adjacent Nodes

Node	Cumulative Cost	Previous Node
2	75	1
4	50*	1

We then pick the adjacent node with the least cumulative cost (node#4) and add that to the processed nodes list. Then add the nodes adjacent to the last processed node to the adjacent nodes list. . . nodes 5 and 7.

Processed Nodes

Node	Cumulative Cost	Previous Node
1	0	none
4	50	1

Adjacent Nodes

Node	Cumulative Cost	Previous Node
2	75*	1
5	125	4
7	100	4

We then pick the adjacent node with the least cumulative cost (node#2) and add that to the processed nodes list. Then add the nodes adjacent to the last processed node to the adjacent nodes list.

Processed Nodes

Node	Cumulative Cost	Previous Node
1	0	none
4	50	1
2	75	1

Adjacent Nodes

Node	Cumulative Cost	Previous Node
5	125	4
7	100	4
3	150	2
5	85*	2

We then pick the adjacent node with the least cumulative cost (node#5) and add that to the processed nodes list. Then add the nodes adjacent to the last processed node to the adjacent nodes list.

Processed Nodes

Node	Cumulative Cost	Previous Node
1	0	none
4	50	1
2	75	1
5	85	2

Adjacent Nodes

Node	Cumulative Cost	Previous Node
7	100	4
3	150	2
6	160	5
8	95*	5

We then pick the adjacent node with the least cumulative cost (node#8) and add that to the processed nodes list. Then add the nodes adjacent to the last processed node to the adjacent nodes list.

Processed Nodes

Node	Cumulative Cost	Previous Node
1	0	none
4	50	1
2	75	1
5	85	2
8	95	5

Adjacent Nodes

Node	Cumulative Cost	Previous Node
7	100*	4
3	150	2
6	160	5
7	170	8
9	170	8
11	145	8

We then pick the adjacent node with the least cumulative cost (node#7) and add that to the processed nodes list. Then add the nodes adjacent to the last processed node to the adjacent nodes list.

Processed Nodes

Node	Cumulative Cost	Previous Node
1	0	none
4	50	1
2	75	1
5	85	2
8	95	5
7	100	4

Adjacent Nodes

Node	Cumulative Cost	Previous Node
3	150*	2
6	160	5
7	170	8
9	170	8
11	145	8
10	200	7

We then pick the adjacent node with the least cumulative cost (node#3) and add that to the processed nodes list. Then add the nodes adjacent to the last processed node to the adjacent nodes list.

Processed Nodes

Node	Cumulative Cost	Previous Node
1	0	none
4	50	1
2	75	1
5	85	2
8	95	5
7	100	4
3	150	2

Adjacent Nodes

Node	Cumulative Cost	Previous Node
6	160*	5
9	170	8
11	145	8
10	200	7
6	200	3

We then pick the adjacent node with the least cumulative cost (node#6) and add that to the processed nodes list. Then add the nodes adjacent to the last processed node to the adjacent nodes list.

Processed Nodes

Node	Cumulative Cost	Previous Node
1	0	none
4	50	1
2	75	1
5	85	2
8	95	5
7	100	4
3	150	2
6	160	5

Adjacent Nodes

Node	Cumulative Cost	Previous Node
9	170	8
11	145*	8
10	200	7
9	210	6

We then pick the adjacent node with the least cumulative cost (node#11) and add that to the processed nodes list. Then add the nodes adjacent to the last processed node to the adjacent nodes list.

Processed Nodes

Node	Cumulative Cost	Previous Node
1	0	none
4	50	1
2	75	1
5	85	2
8	95	5
7	100	4
3	150	2
6	160	5
11	145	8

Adjacent Nodes

Node	Cumulative Cost	Previous Node
9	170*	8
10	200	7
9	210	6
10	270	11
12	220	11

We then pick the adjacent node with the least cumulative cost (node#9) and add that to the processed nodes list. Then add the nodes adjacent to the last processed node to the adjacent nodes list.

Processed Nodes

Node	Cumulative Cost	Previous Node
1	0	none
4	50	1
2	75	1
5	85	2
8	95	5
7	100	4
3	150	2
6	160	5
11	145	8
9	170	8

Adjacent Nodes

Node	Cumulative Cost	Previous Node
10	200*	7
10	270	11
12	220	11
12	270	9

We then pick the adjacent node with the least cumulative cost (node#10) and add that to the processed nodes list. Then add the nodes adjacent to the last processed node to the adjacent nodes list. All the nodes adjacent to node 10 (nodes 7,11) have already been processed. So we add node 12 as the final entry to the processed nodes table.

Processed Nodes

Node	Cumulative Cost	Previous Node
1	0	none
4	50	1
2	75	1
5	85	2
8	95	5
7	100	4
3	150	2
6	160	5
11	145	8
9	170	8
10	200	7
12	220*	11

Adjacent Nodes

Node	Cumulative Cost	Previous Node
12	220*	11
12	270	9
		10

The quickest path from node 1 to node 9 is 9←8←5←2←1 and the optimal route would take 170 seconds.

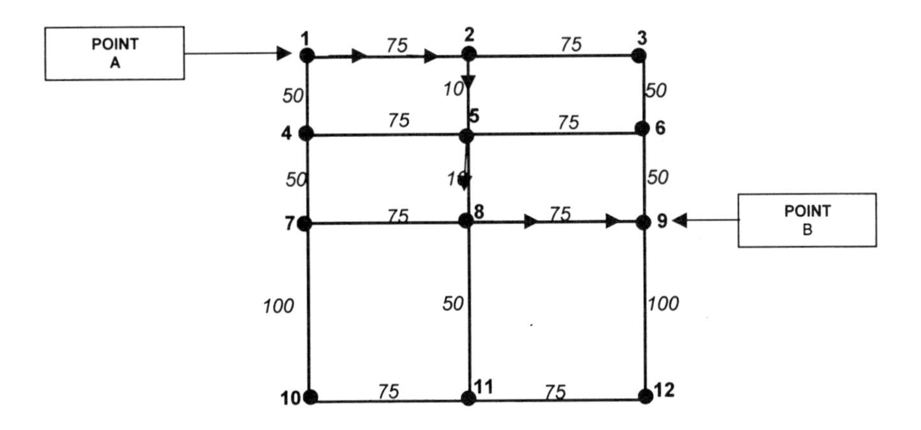

3) You are managing wilderness cabins on a network of backcountry ski trails. The average skiing time per 100 meters of trail for various slopes is as follows:

Slope Class	-Down Slope Time (seconds per 100 m)	+Up Slope Time (seconds per 100 m)
Level	40	40
0-5 percent	30	60
6-15 percent	20	120
>15 percent	15	200

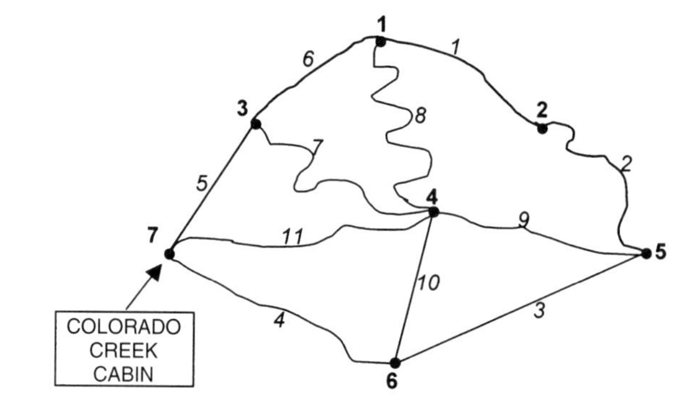

Delineate service network that represents all trails that are within a half hour from the Colorado Creek Cabin.

The first step is to compute the time in seconds it would take to travel across each arc. For example, arc#1 would take : 40 seconds/100m * 1100 m = 440 seconds. Arc#3 would take 20 seconds/100 m * 1200 m = 240 seconds to travel from node#6 to node#5 and 120 seconds/100 m * 1200 m = 1440 seconds to travel up slope from node#5 to node#6. Draw these times on the network figure.

Arc#	Length_meters	From_node	To_node	Slope	To_secs	Back_secs
1	1100	1	2	0	440	440
2	1000	2	5	0	400	400
3	1200	6	5	-15	240	1440
4	1800	7	6	+15	2160	360
5	1100	7	3	0	440	440
6	1000	1	3	0	400	400
7	1400	3	4	-18	210	2800
8	2200	1	4	-20	330	4400
9	1300	4	5	-10	260	1560
10	1200	4	6	-25	180	2400
11	2000	7	4	-12	400	2400

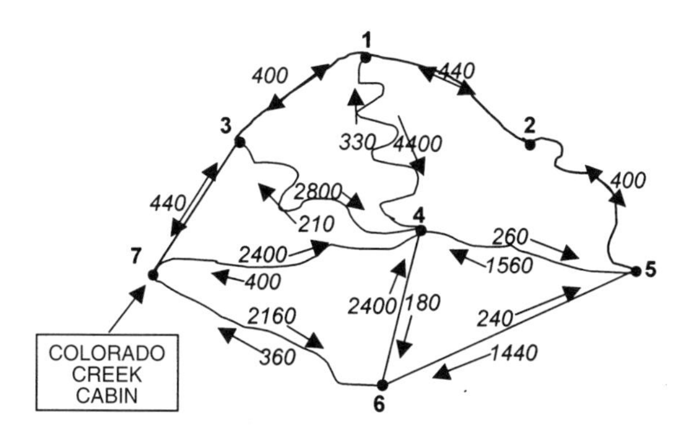

The time limit is 30 minutes * 60 seconds/minute = 1800 seconds. Starting at node#7, allocate network arcs until you reach that limit.

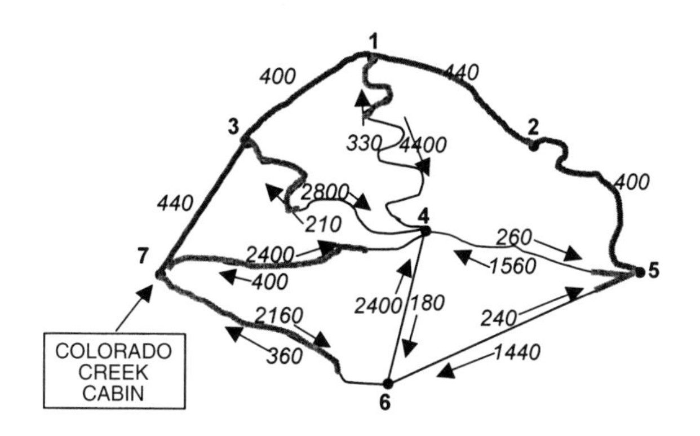

DYNAMIC SEGMENTATION EXERCISE SOLUTIONS

1) You have a route system of irrigation canals as follows. Circle the first section of route#1 and the second section of route#3. **The first section of route#1 is from 2/3s of the arc to the end of arc#1 (66.66 percent to 100 percent). The second section of route#3 is from 0 to 50 percent of arc#9.**

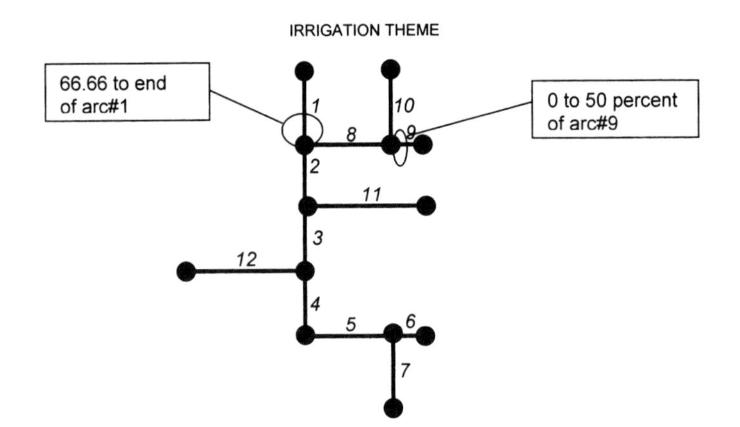

IRRIGATION SECTION TABLE:

Arc#	Routelink#	F-Meas	T-Meas	F-Pos	T-Pos
1	1	10	15	66.66	100
2	1	15	25	0	100
3	1	25	35	0	100
4	1	35	45	0	100
5	1	45	60	0	100
6	1	60	65	0	100
7	2	0	15	0	100
8	3	0	15	0	100
9	3	15	17.5	0	50
10	4	0	15	0	100
11	5	0	20	0	100
12	6	0	20	0	100

2) You have the following point theme of auto accidents and network of streets.

Accidents Point Attribute Table

Accident#	DATE	Num_Vehicles
1	01/01/2000	1
2	01/01/2000	3
3	01/02/2000	2
4	01/04/2000	2
5	01/07/2000	1
6	01/07/2000	1
7	01/08/2000	2

Streets Section Table:

Arc#	Routelink#	F-Meas	T-Meas	F-Pos	T-Pos
1	1	0	15	0	100
2	1	15	25	0	100
3	1	25	35	0	100
4	1	35	45	0	100
5	1	45	60	0	100
6	1	60	65	0	100
7	2	0	15	0	100
8	3	0	15	0	100
9	3	15	20	0	100
10	4	0	15	0	100
11	5	0	20	0	100
12	6	0	20	0	100

STREETS THEME

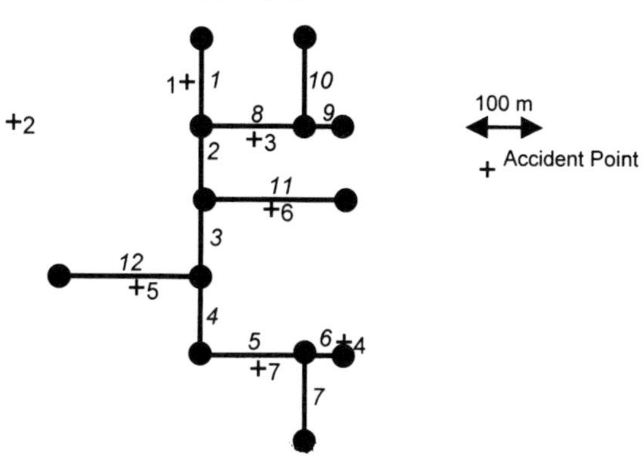

You enter the **ADDROUTEMEASURE** command and specify *Accidents* as the name of your point theme, *Streets* as the name of your route theme and route system, and *acc_events.tbl* as the name of your output event table. You specify a search radius of 100 meters. Fill in your output event table:

Accident number 2 is not within 100 m of any street so it is not included in the output event table. Accident#1 is closest to halfway down arc#1 → route 1, measure 7.5, Accident#7 is halfway down arc#5 → route1, measure 52.5. Accident#4 is at the end of arc#6 → route 1, measure 65. Accident#3 is halfway down arc#8 → route 3, measure 7.5. Accident#6 is halfway down arc#11 → route 5, measure 10. And Accident #5 is halfway down arc#12 → route 6, measure 10.

acc_events.tbl

Accident#	Route#	Measure	Date	Num_Vehicles
1	1	7.5	01/01/2000	1
7	1	52.5	01/08/2000	2
4	1	65	01/04/2000	2
3	3	7.5	01/02/2000	2
6	5	10	01/07/2000	1
5	6	10	01/07/2000	1

3) You have a route system of streams as follows.

Streams arc attribute table

Streams#	Length	Name
1	195	Moose Creek
2	295	Poplar River
3	210	Poplar River
4	65	Poplar River
5	310	Sheep Creek

Streams route system tables:

Route-ID
1
2
3
4

Arc#	Routelink#	F-Meas	T-Meas	F-Pos	T-Pos	Section#
1	1	0	195	0	100	1
3	2	0	210	0	100	2
4	2	210	275	0	100	3
2	3	0	295	0	100	4
5	4	0	310	0	100	5

Salmon Event Table

Route#	From	To	Sockeye_Count	King_Count
2	100	120	30	0
2	140	160	40	10
2	180	190	20	5
2	190	195	18	6
2	200	210	0	5

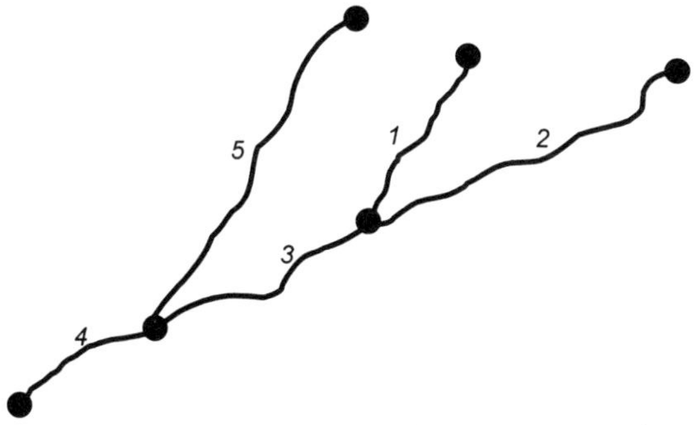

Circle the stream section that has the highest density of Sockeye Salmon.

Circle the stream section that has the highest density of King Salmon.

The first step is to add 2 columns to your event table for density of Sockeye Salmon and density of King Salmon. Density is calculated as count / length where length is the length of each section (To – From):

Salmon Event Table

Route#	From	To	To-From	Sockeye _Count	Density _Sockeye	King _Count	Density _King
2	100	120	20	30	1.5	0	0
2	140	160	20	40	2.0	10	0.5
2	180	190	10	20	2.0	5	0.5
2	190	195	5	18	**3.6**	6	**1.2**
2	200	210	10	0	0	5	0.5

The highest density for both Sockeye and King Salmon is the section from 190 to 195 on route#2. So we circle that section.

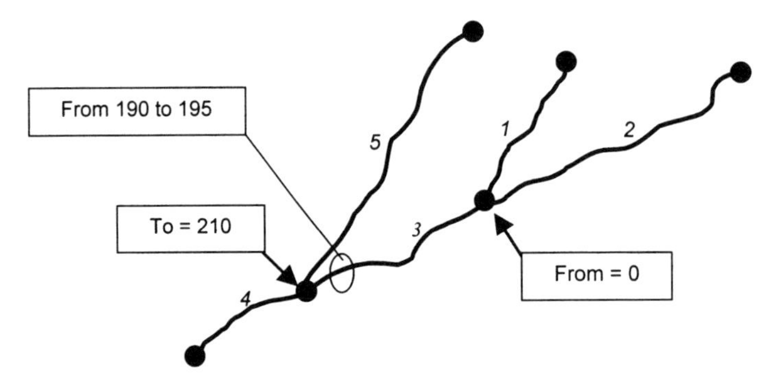

4) You have a route system of streams as follows. What would be the output if you run **RESELECT** on your streams with the following expression:

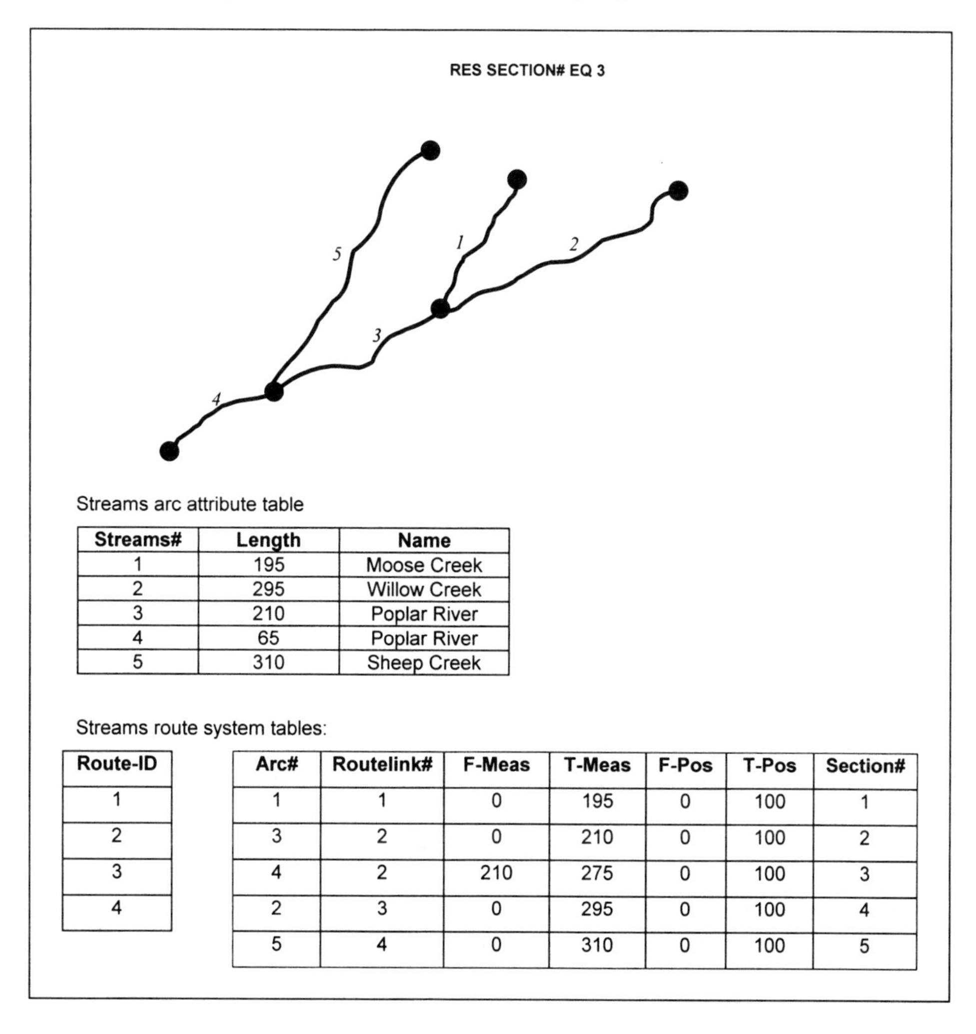

Streams arc attribute table

Streams#	Length	Name
1	195	Moose Creek
2	295	Willow Creek
3	210	Poplar River
4	65	Poplar River
5	310	Sheep Creek

Streams route system tables:

Route-ID
1
2
3
4

Arc#	Routelink#	F-Meas	T-Meas	F-Pos	T-Pos	Section#
1	1	0	195	0	100	1
3	2	0	210	0	100	2
4	2	210	275	0	100	3
2	3	0	295	0	100	4
5	4	0	310	0	100	5

The output would be section# = 3 and would look like:

Arc#	Routelink#	F-Meas	T-Meas	F-Pos	T-Pos	Section#
4	2	210	275	0	100	3

5) You have the following route system. Draw what the output theme would look like if you applied the **ROUTEARC** command to this route system.

ARC #5 ARC #6

Route Attribute Table

Route#	Route-ID
1	19

Section Attribute Table

Arc#	Routelink#	F-Meas	T-Meas	F-Pos	T-Pos	Section#	Surface
5	1	0	1.5	0	33..33	1	Dirt
5	1	1.5	3.5	33.33	77.77	2	Gravel
5	1	3.5	4.5	77.77	100	3	Paved
6	1	4.5	10.0	0	100	4	Paved

INPUT THEME

ROUTEARC

6) You have the following route system. Draw what the output theme would look like if you applied the **SECTIONARC** command to this route system.

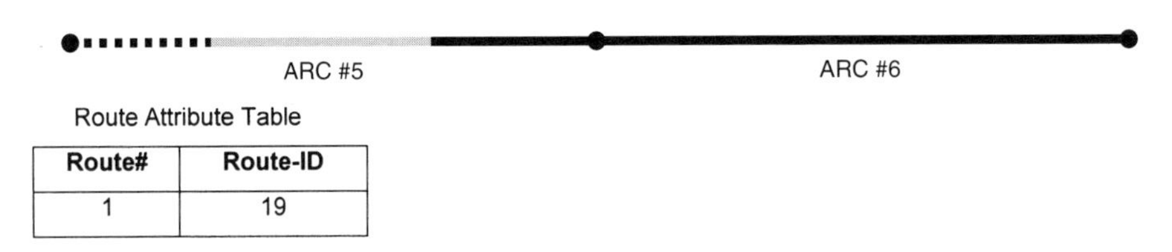

Route Attribute Table

Route#	Route-ID
1	19

Section Attribute Table

Arc#	Routelink#	F-Meas	T-Meas	F-Pos	T-Pos	Section#	Surface
5	1	0	1.5	0	33..33	1	Dirt
5	1	1.5	3.5	33.33	77.77	2	Gravel
5	1	3.5	4.5	77.77	100	3	Paved
6	1	4.5	10.0	0	100	4	Paved

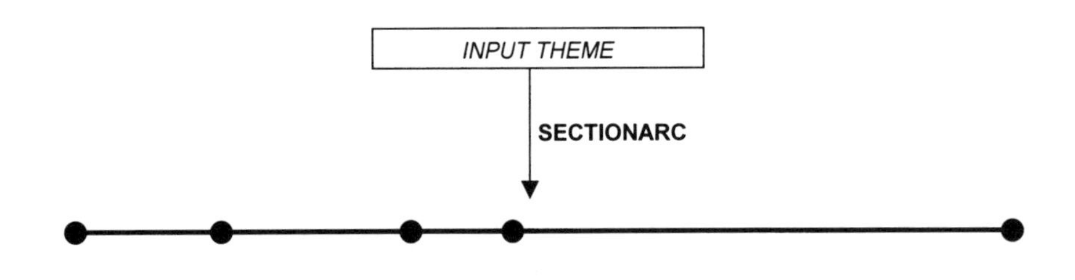

7) You have the following bus route system:

Arc Attribute Table

Arc#	Length
1	500
2	500
3	1000
4	500
5	1000
6	500
7	1000
8	500
9	1000
10	500
11	1000

Bus route system tables:

Route-ID
1
2
3

Arc#	Routelink#	F-Meas	T-Meas	F-Pos	T-Pos	Section#
1	1	0	50	0	100	1
2	1	50	100	0	100	2
3	1	100	200	0	100	3
1	2	0	50	0	100	4
2	2	50	100	0	100	5
4	2	100	150	0	100	6
5	2	150	250	0	100	7
6	2	250	300	0	100	8
7	2	300	350	0	50	9
1	3	0	50	0	100	10
2	3	50	100	0	100	11
8	3	100	150	0	100	12
9	3	150	250	0	100	13
10	3	250	300	0	100	14
11	3	300	325	0	25	15

What would be the results if you run **ROUTESTATS** on your route system with the following options?

RouteStats: ***ARCLENGTH***
RouteStats: ***MEASURELENGTH***

ARCLENGTH will return the total length of each route in GIS coordinate units. For example, route 3 is composed of arc# 1,2,8,9,10 and 25 percent of arc#11. Therefore the ARCLENGTH returned for route 3 is: 500 + 500 + 500 + 1000 + 500 + 0.25(1000)= 3,250.
MEASURELENGTH will return the total length of each route in measure units. For example, route 1 has a total of 200 measure units.

Table Output from ROUTESTATS:

Route#	Arclength	Measurelength
1	2000	200
2	3500	350
3	3250	325

8) You have the following point and linear events along a street route system. Draw the output themes that would result from using **EVENTPOINT** and **EVENTARC.**

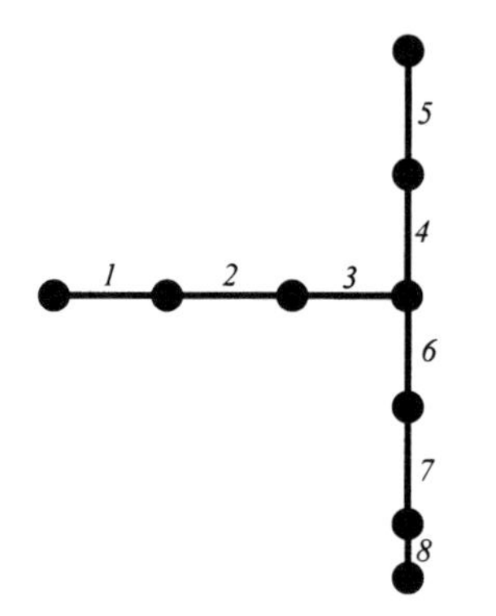

Route-ID
1
2
3

Arc#	Routelink#	F-Meas	T-Meas	F-Pos	T-Pos	Section#
1	1	0	100	0	100	1
2	1	100	200	0	100	2
3	1	200	300	0	100	3
4	2	0	100	0	100	4
5	2	100	200	0	100	5
6	3	0	100	0	100	6
7	3	100	200	0	100	7
8	3	200	250	0	100	8

Poles Event Table

Route-ID	Pole#	Location
1	110	50
1	111	150
2	245	50
2	246	150
3	247	75
3	248	200

Speed Limit Event Table

Route-ID	Speed_Limit	End
1	45	300
2	35	200
3	35	100
3	25	125
3	35	250

EVENTPOINT would take the pole events and determine their correct location in the GIS coordinate system based on the event location attribute. The output point theme would look like the following:

Point#	Route-ID	Pole#	Location
1	1	110	50
2	1	111	150
3	2	245	50
4	2	246	150
5	3	247	75
6	3	248	200

+245

+246

110 111
+ +

+247

+248

EVENTARC would take the speed limit events and determine their correct location in the GIS coordinate system based on the event location attribute. The output line theme would look like the following:

Arc Attribute Table:

Arc#	Length	Route-ID	Speed_Limit	End
1	300	1	45	300
2	200	2	35	200
3	100	3	35	100
4	25	3	25	125
5	125	3	35	250

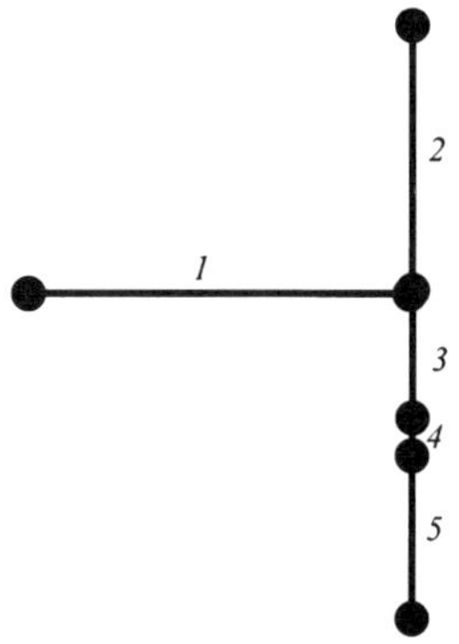

1
2
3
4
5

9) You have two event tables of King Salmon counts along a stream route. Fill in the output table from executing **OVERLAYEVENTS** with these two event tables.

1999 King Salmon Counts

Route#	From	To	Count_1999
1	60	65	3
1	110	130	18
1	165	175	14

2000 King Salmon Counts

Route#	From	To	2000_Count
1	60	75	1
1	110	130	3
1	165	175	2

Stream Section Table:

Arc#	Routelink#	F-Meas	T-Meas	F-Pos	T-Pos
1	1	0	200	0	200

With the *UNION* option, the first step is to combine the *From*- and *To*- segments of the section: 60 to 65, 65 to 75, 110 to 130, 165 to 175. Then tally up the counts for these stream stretches . . .

OVERLAYEVENTS OUTPUT TABLE (*UNION OPTION*)

Route#	From	To	Count_1999	Count_2000
1	60	65	**3**	**1**
1	65	75	**0**	**1**
1	110	130	**18**	**3**
1	130	165	**0**	**0**
1	165	175	**14**	**2**
1	175	175	**0**	**0**

With the *INTERSECT* option, the first step is to find the stretches that completely overlap in both event tables: 60 to 65, 110 to 130, and 165 to 175. Then tally up the counts for these stream stretches . . .

OVERLAYEVENTS OUTPUT TABLE (*INTERSECT OPTION*)

Route#	From	To	Count_1999	Count_2000
1	60	65	**3**	**1**
1	110	130	**18**	**3**
1	165	175	**14**	**2**

Notice that the output may be misleading due to some stretches not perfectly overlapping. For example, for 60 to 65, the table contains 1 for Count_2000. This may be true, but the original stretch was 60 to 75 that year so the King may have been counted anywhere in that stretch. To avoid this type of possible error, it would be best to have the *From-* and *To-* measures consistent from year to year.

10) You have the following route system and irrigation event table. Fill in the output event table resulting from using **DISSOLVEEVENTS** with ***Pipe_Type*** as the dissolve item.

Input Route System

Route#	Arclink#	F-Meas	T-Meas	F-Pos	T-Pos
1	1	0	100	0	100
1	2	100	200	0	100
1	3	200	250	0	100
1	4	250	300	0	100
2	1	0	100	0	100
2	2	100	200	0	100
2	5	200	250	0	100
2	6	250	300	0	100
3	1	0	100	0	100
3	2	100	200	0	100
3	7	200	250	0	100
3	8	250	300	0	100

Input Irrigation Pipe Event Table

Route#	From	To	Pipe_Type	Date_Installed
1	0	65	1	07/29/2000
1	65	100	1	07/30/2000
1	100	175	1	08/01/200
1	175	200	1	08/02/2000
1	200	300	2	08/03/2000
2	0	65	1	07/29/2000
2	65	100	1	07/30/2000
2	100	175	1	08/01/200
2	175	200	1	08/02/2000
2	200	300	2	08/03/2000
3	0	65	1	07/29/2000
3	65	100	1	07/30/2000
3	100	175	1	08/01/200
3	175	200	1	08/02/2000
3	200	250	2	08/03/2000
3	250	300	2	08/04/2000

Output Table from DISSOLVEEVENTS

Route#	From	To	Pipe_Type
1	0	200	1
1	200	300	2
2	0	200	1
2	200	300	2
3	0	200	1
3	200	300	2

POLYGON ANALYSIS EXERCISE SOLUTIONS

1) Polygons are assumed to be pure in their content and discrete and absolute in their boundary locations. In other words, they are assumed to be homogeneous with sharp, distinct boundaries. This is rarely the case. Rank the following polygon themes in terms of their likely purity and boundary sharpness.

Theme	Purity Rank	Boundary Sharpness Rank
Wetland Polygons	**Medium (some other smaller wetland types can be inside a wetland polygon)**	**Medium (changes with water level)**
Tax Parcels	**Highest (Each polygon has properties that do not naturally change)**	**Highest (Surveyed with good accuracy)**
Soil Polygons	**Lowest (Undersurface ..inclusions are unknown)**	**Lowest (Undersurface ..boundary interpreted by geomorphology)**

2) You have polygon themes of a wildfire burn from 1963 and a burn from the year 2000. Fill in the following flowchart to determine the total number of hectares from the 1963 burn that burned in 2000.

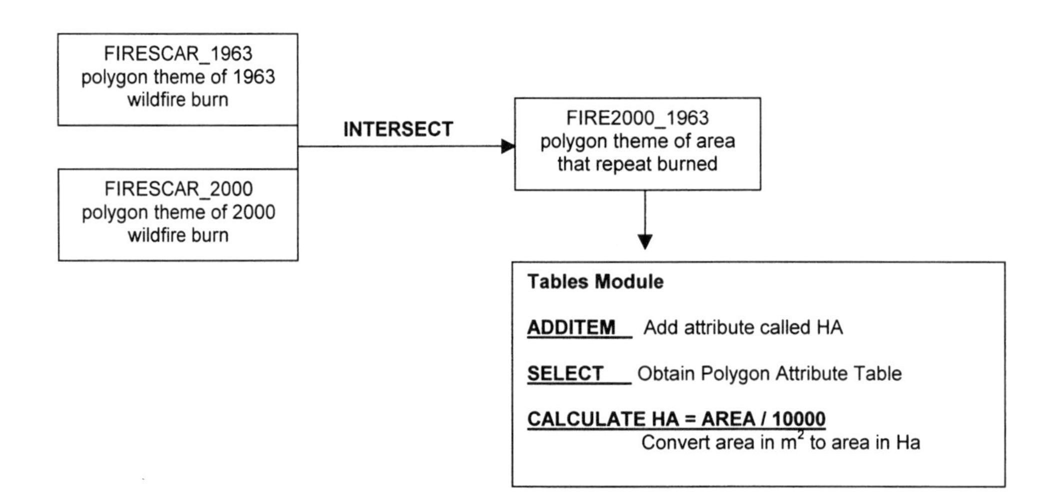

3) You have a line theme of rivers and streams and a polygon theme of three watershed areas (1= Rock Creek Watershed, 2=Clear Creek Watershed, 3=Willow Creek Watershed). For each watershed, you want to know the density of rivers and streams in meters per hectare of the watershed. Your GIS coordinate system is in meters.

Fill in the following flowchart to solve your GIS problem.

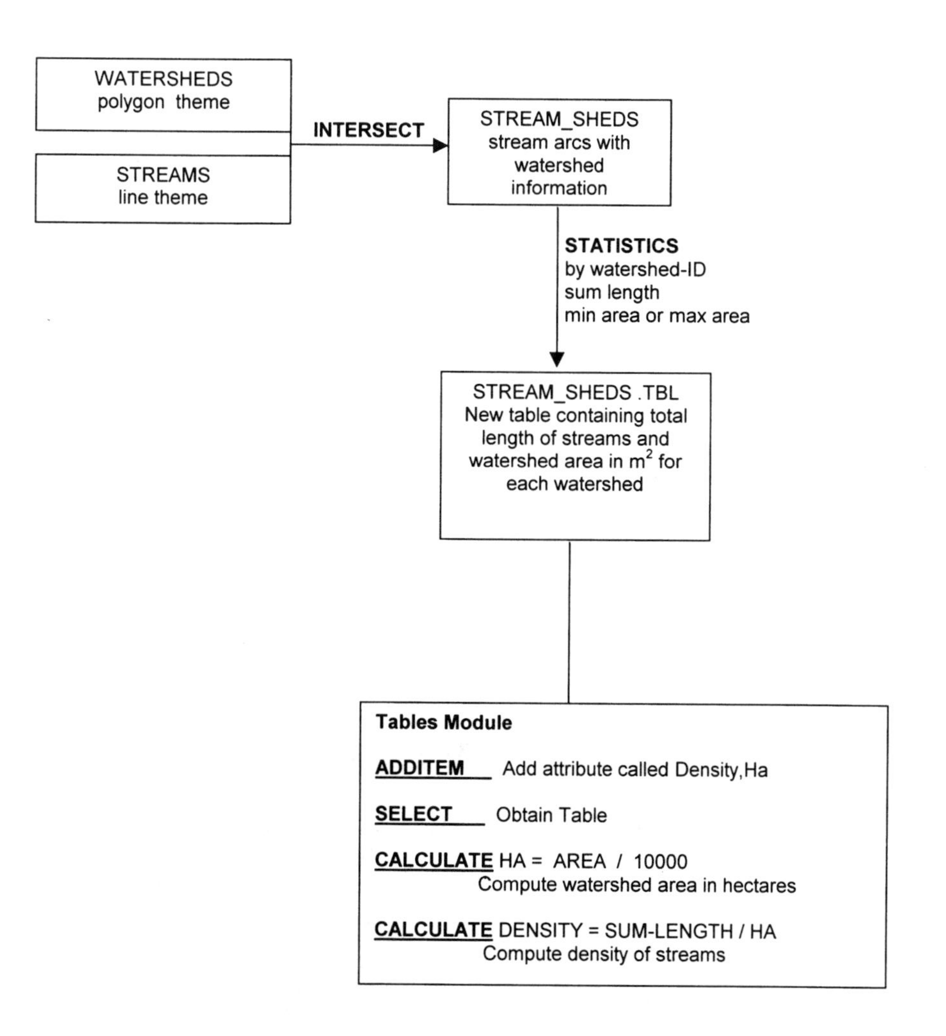

4) You have a point theme of mallard nest locations with three age classes of nesting mallards. You also have a polygon theme of land cover types (1= tussock 2=low shrub 3=high shrub 4=barren). Your GIS coordinate system is in meters.

You want to estimate the density of nests in each land cover type by mallard age class as follows:

Tussock Type (hectares)	Low Shrub Type (hectares)	High Shrub Type (hectares)	Barren Type (hectares)

Mallard Age Class	Tussock Type (total nests)	Low Shrub Type (total nests)	High Shrub Type (total nests)	Barren Type (total nests)
1				
2				
3				

Fill in the following flowchart to solve your GIS problem.

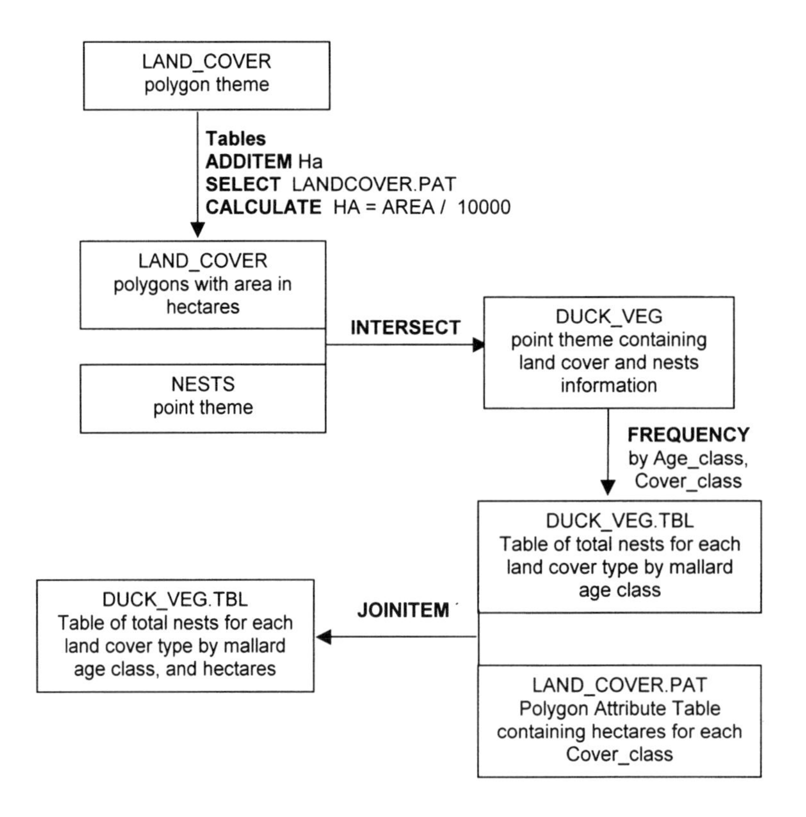

5) You have a polygon theme of vegetation types *(vegpolys)* and a polygon theme of watershed areas *(watersheds)*. You want a table showing the area of each watershed and the area of black spruce in each watershed.

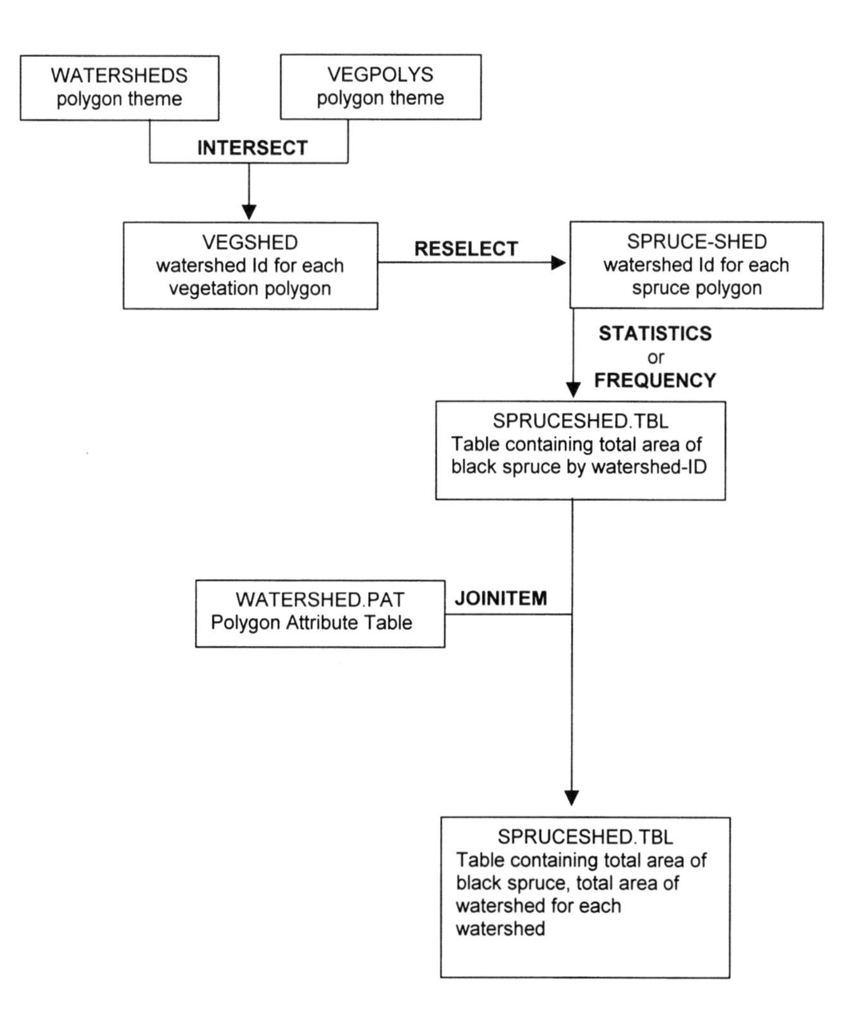

6) You have three polygon themes: SOIL, VEG, and TINPOLYS. SOIL contains soil information, VEG contains vegetation information, and TINPOLYS are triangles where the slope and aspect of each triangle has been estimated.

You are checking for possible errors in the vegetation theme. You want to find if there are any aspen polygons that occur over poorly drained soils or on northerly facing slopes that are over 10 percent in slope gradient. These are most likely errors in vegetation mapping, because aspen typically grows on relatively dry, warm sites.

Fill in the following flowchart to solve your GIS problem.

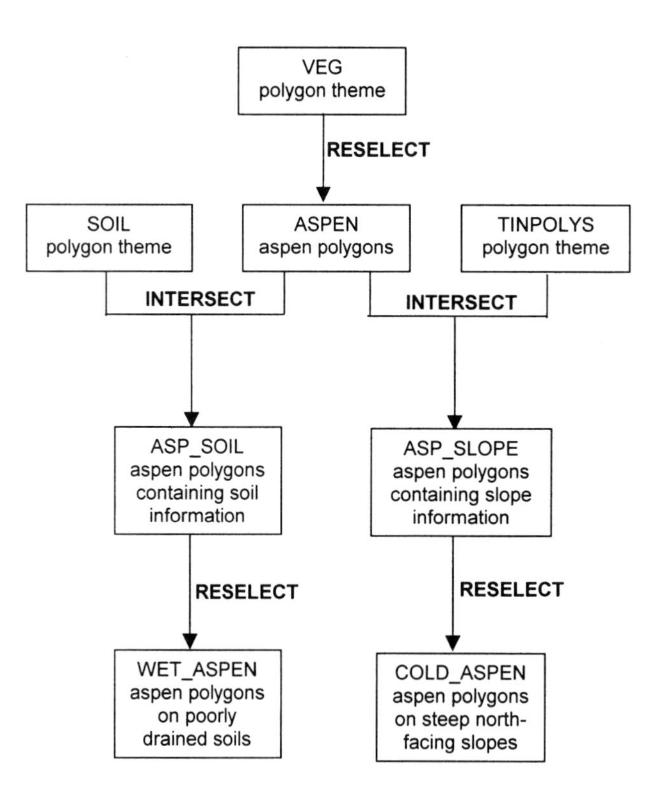

7) You have three candidate routes for a new power line. The cost of building the line varies by amount of land owned privately, by soil type to build on, and by amount of wetlands to cross, and also total distance. You want to know this type of information for the three routes. Fill in the following flowchart with the most appropriate GIS tools:

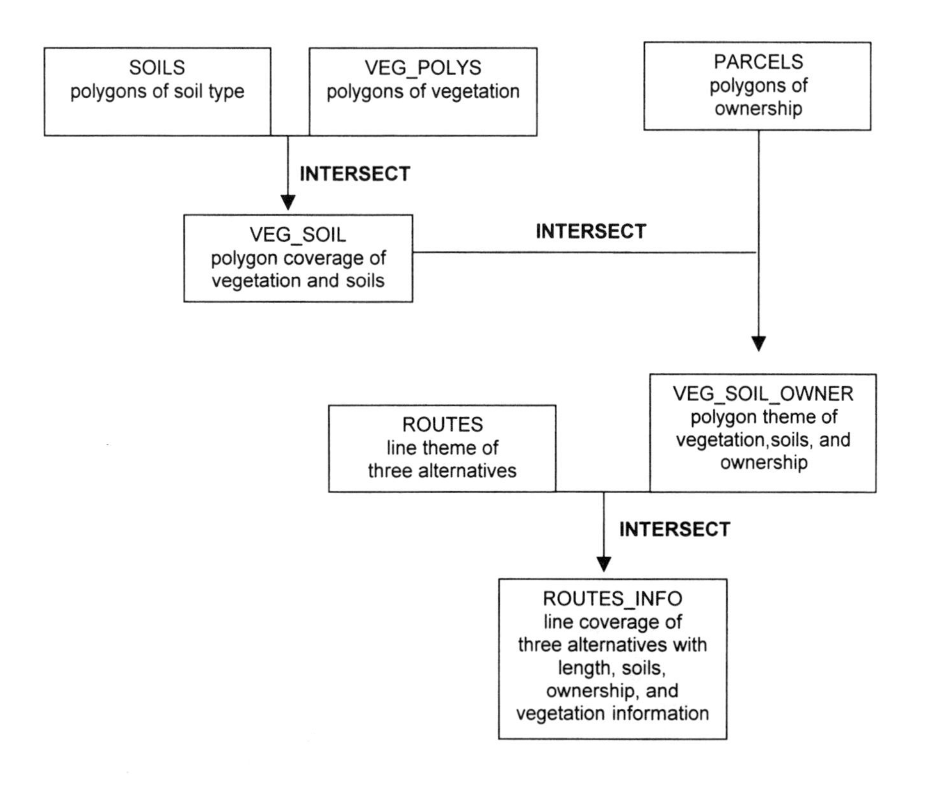

8) You have a pretimber harvest polygon theme of alder clumps and a posttimber harvest point theme of planted spruce seedlings. You want to find the mean height growth of the spruce seedlings within 10 meters of preharvest alder, versus seedlings that were at least 10 meters away from the preharvest alder clumps. Fill in the following flowchart with the most appropriate GIS tools:

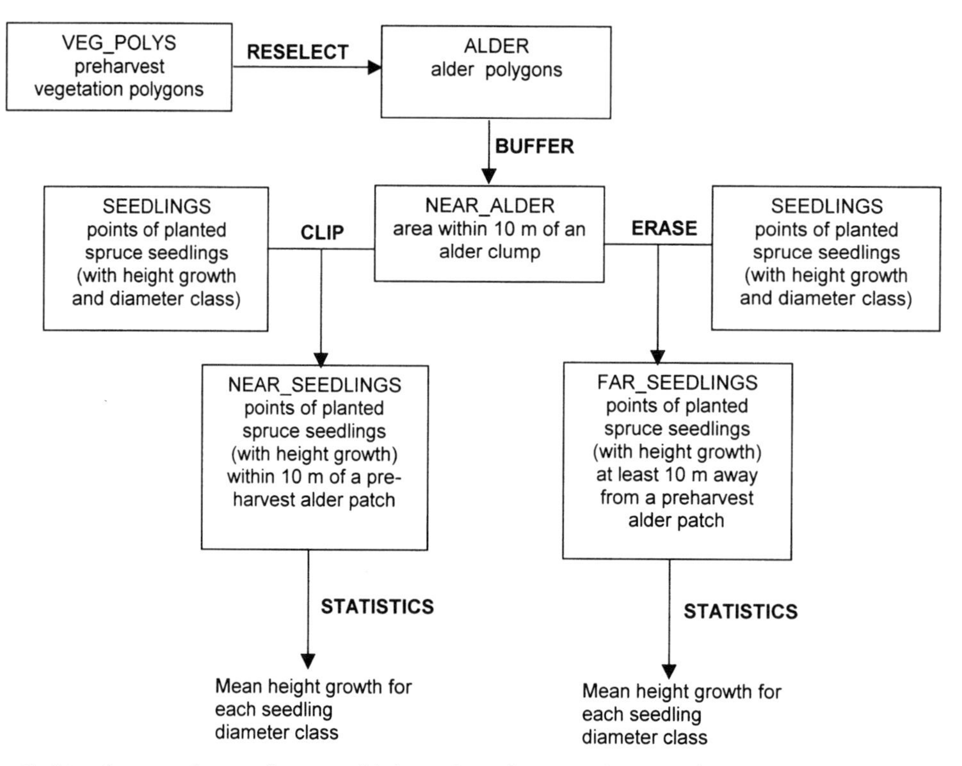

9) You have polygon themes of lakes. One theme is from early summer and the other theme is from late summer. You want to create a new theme of the best lakes for waterfowl habitat . . . lakes that had at least 1 hectare of exposed mudflats in late summer due to drawdown of the lake water level. Fill in the following flowchart to solve your GIS problem.

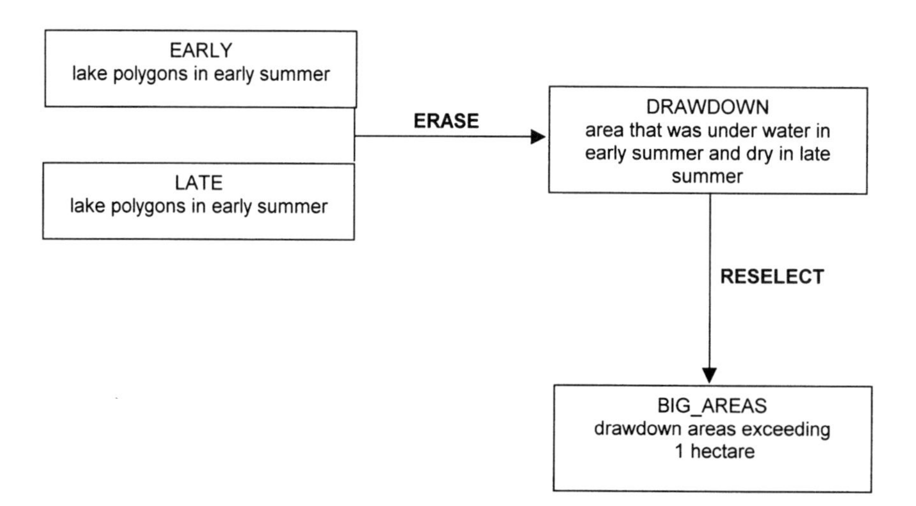

10) You have a point theme of well locations, a point theme of gas station locations, a soils polygon theme, and a parcels polygon theme containing the address of the owner of each parcel.

Outline what tools you would use to develop a text file of all well owners that have wells on sand, loamy sand, or sandy loam soils within a kilometer of a gas station.

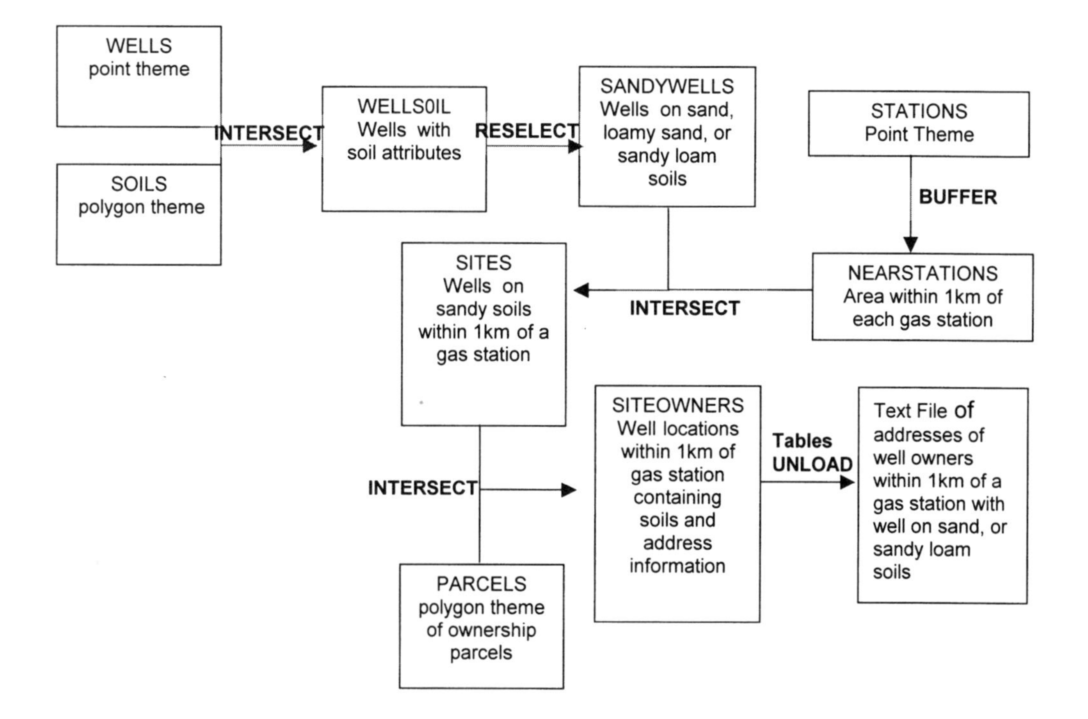

GRID ANALYSIS EXERCISE SOUTIONS

1) You run **REGIONGROUP** on the following grid. Fill in the output grid:

Start with the upper left cell, and assign all cells that have a value of zero and are connecting as group 1, proceed to the next cell on the row and assign all cells that have a value of 90 and are connecting as group 2, and so on. Once you have completed your group assignments, you can build your output value attribute table.

Input Grid

0	90	90	270	270	90
0	90	90	270	270	90
90	90	90	270	270	90
90	90	90	270	270	90
180	180	180	180	180	180
180	180	180	180	180	180

Output Grid

1	2	2	3	3	4
1	2	2	3	3	4
2	2	2	3	3	4
2	2	2	3	3	4
5	5	5	5	5	5
5	5	5	5	5	5

Output Grid Value Attribute Table

Value	Count
1	2
2	10
3	8
4	4
5	12

The output grid has 36 cells, so as a check, add up your count values to make sure they sum to 36.

2) What would the output grid contain if you use **TEST** with the following logical expression: **Value in {1,3,8,9}**

Test returns a 1 if the cell meets the logical expression criteria, 0 if it does not. Therefore the cells containing 1,3,8, or 9 receive 1s and the other cells receive 0s.

Input Grid

0	1	2	2	3	1
0	1	2	2	2	8
1	2	3	8	2	1
2	2	3	7	7	3
4	4	3	6	8	9
4	5	5	6	8	0

Output Grid

0	1	0	0	1	1
0	1	0	0	0	1
1	0	1	1	0	1
0	0	1	0	0	1
0	0	1	0	1	1
0	0	0	0	1	0

Output Grid Value Attribute Table

Value	Count
0	21
1	15

As a check, the total count should sum to 36 . . . and it does.

3) What would the output grid contain if you use **SELECT** with the following logical expression: **Value in {1,3,8,9}**

SELECT returns the cell value if the cell meets the logical expression criteria; NODATA if it does not. Therefore the cells not containing 1,3,8, or 9 receive NODATA.

Input Grid

0	1	2	2	3	1
0	1	2	2	2	8
1	2	3	8	2	1
2	2	3	7	7	3
4	4	3	6	8	9
4	5	5	6	8	0

Output Grid

	1			3	1
	1				8
1		3	8		1
		3			3
		3		8	9
				8	

Output Grid Value Attribute Table

Value	Count
1	5
3	5
8	4
9	1

As a check, the total of number of valid (not NODATA) cells is 15. Therefore the sum of count should be 15 and it is (5 + 5 + 4 + 1 = 15).

4) You have a grid of willow cells. You want to buffer all willow areas by 1 cell. Fill the output grid with the appropriate values after the **EXPAND** function is executed.

BUFWILLOW = EXPAND (WILLOW, 1, LIST, 4, 5, 6,)

WILLOW Grid

0					5				
0	4					5			
	4						5		
							5		
			6	6					
			6	6					
					6				
								19	
							19	19	

BUFWILLOW Grid

4	4	4		5	5	5	5		
4	4	4		5	5	5	5	5	
4	4	4			5	5	5	5	
4	4	4	6	6	6	5	5	5	
		6	6	6	6	5	5	5	
		6	6	6	6	6			
		6	6	6	6	6			
				6	6	6			
								19	
							19	19	

5) You have the following grids output from running the **COSTDISTANCE** function.

Accumulative Cost Grid

0	200	350	650	1183	1045
150	283	483	633	833	1033
350	574	566	866	1216	1045
750	845	**786**	612	716	750
1300	1257	745	362	450	300
1200	950	550	250	150	0

Backlink Grid

0	5	5	5	3	4
7	6	6	6	5	5
7	6	6	5	3	6
7	6	2	3	4	3
7	2	2	2	3	3
1	1	1	1	1	0

First draw the direction arrows associated with the codes in the Backlink Grid.

Then use the Backlink Grid and Accumulative Cost Grid to determine the optimal path starting at cell (3,4). The optimal path has a cost of 786

Fill in the output grid resulting from estimating an optimal path using the COST-PATH function:

From Cell Grid

		1			

Optimal Path Grid

	3				
		3			
			3	3	1

6) You have a grid of forest types and a grid of three watersheds. You want to produce the following information.

	Number of Grid Cells in Watershed by Vegetation Type				
Watershed	Black/White Spruce	Aspen-Birch	Riparian Shrub	Other	Total
1					
2					
3					

Fill in the following flowchart with the appropriate grid tools to solve your problem.

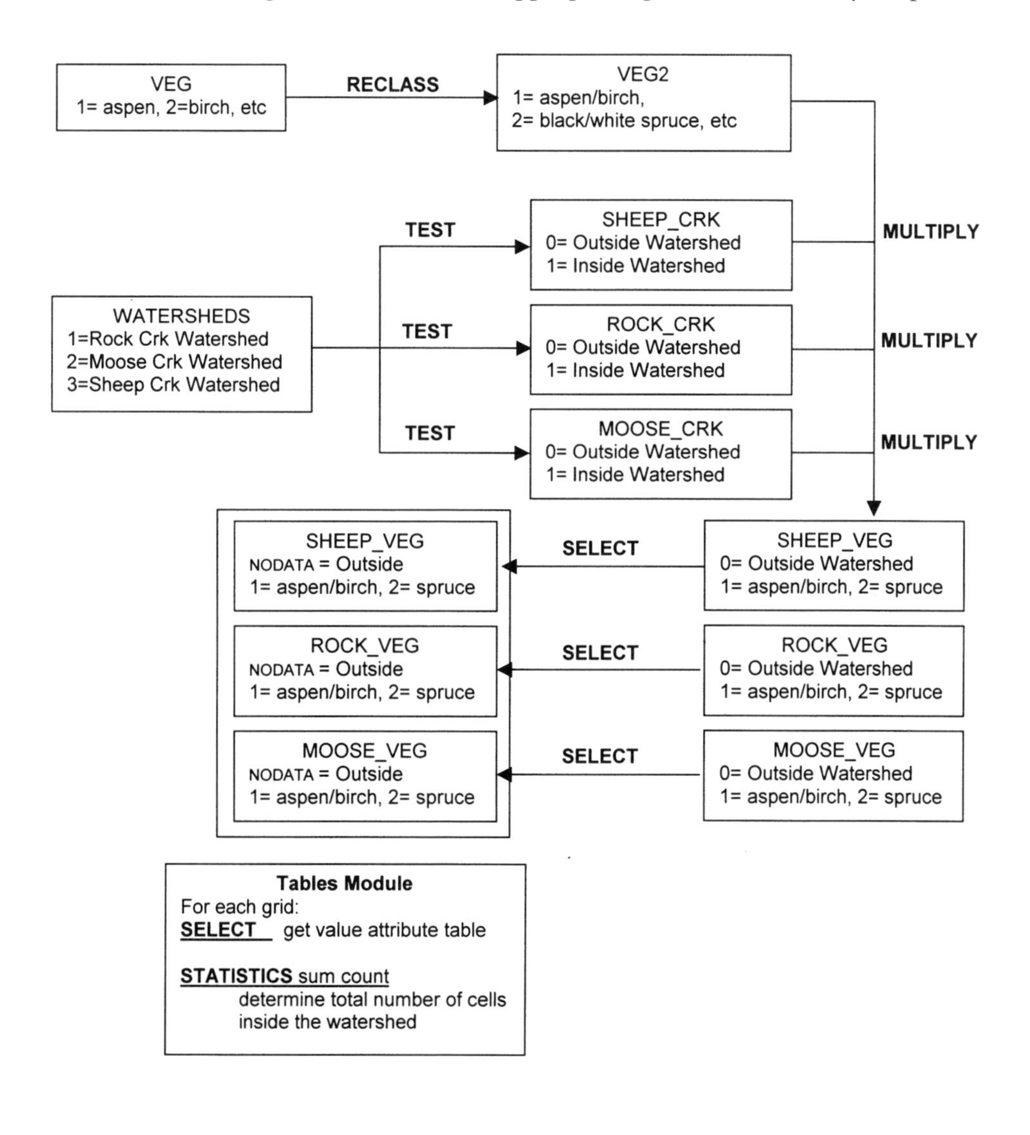

7) You have a 1/0 grid of all lightning strike locations in interior Alaska during last summer. You also have a grid of elevation values in meters for the same area. Each grid cell is 1000 by 1000 meters. You want to determine the following information:

Elevation Class	Total Number of Strikes	Total Area (ha)
Less than 500 m		
500 – 1,000 m		
> 1000 m		

Fill in the following flowchart with appropriate grid operations to solve your GIS problem:

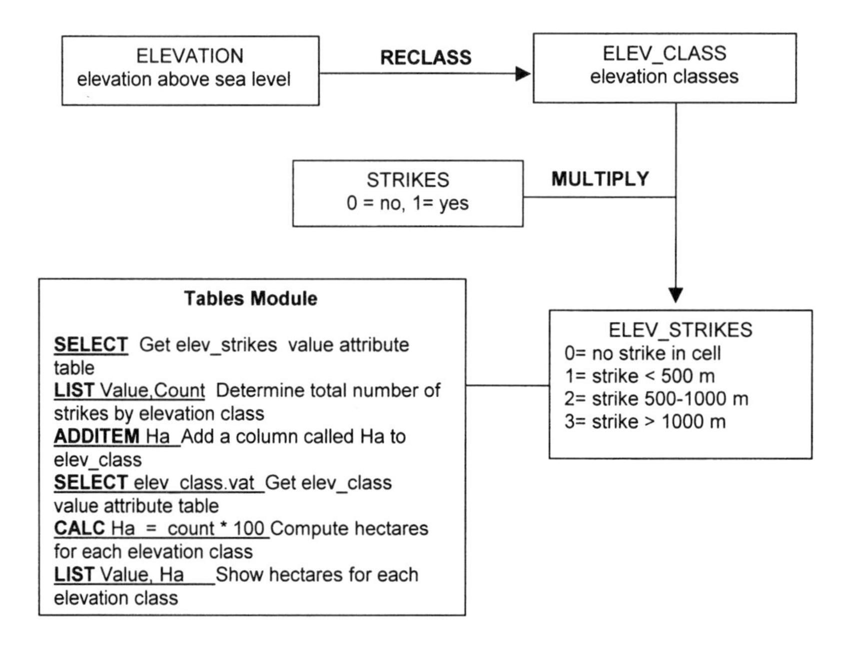

8) You have one grid (*veg*) that contains values of 71,72,73,74,75 for open water and values of 63,64,65,67 for wetland. The grids have a cell size of 25 meters. Draw a flow chart showing how you would produce a grid of wetlands cells that are on the shorelines of open water.

Fill in the following flowchart with appropriate grid operations to solve your GIS problem:

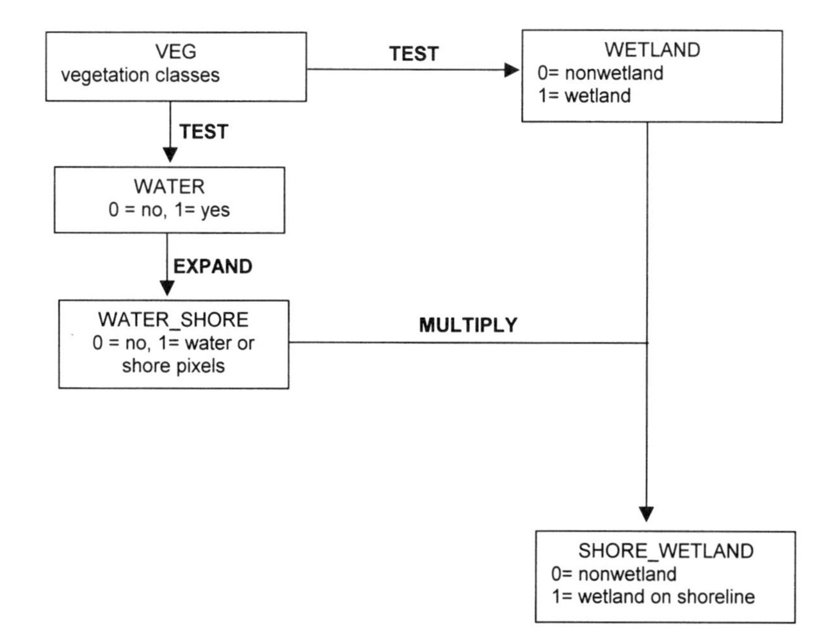

9) A forester is planning potential harvest units. She wants to find large (>100 ha) stands of closed aspen, birch, or aspen/birch forest (VEGCLASS 16-19) The SIZECLASS of the stands should be sawtimber or pole size (3 or 2) and the stands should be at least 100 meters from open water (VEGCLASS 80-89) and at least 50 meters from marsh areas (VEGCLASS 77-79). All grids have a cell size of 25 by 25 meters.

Fill in the following flowchart with the appropriate Grid tools to solve your GIS problem:

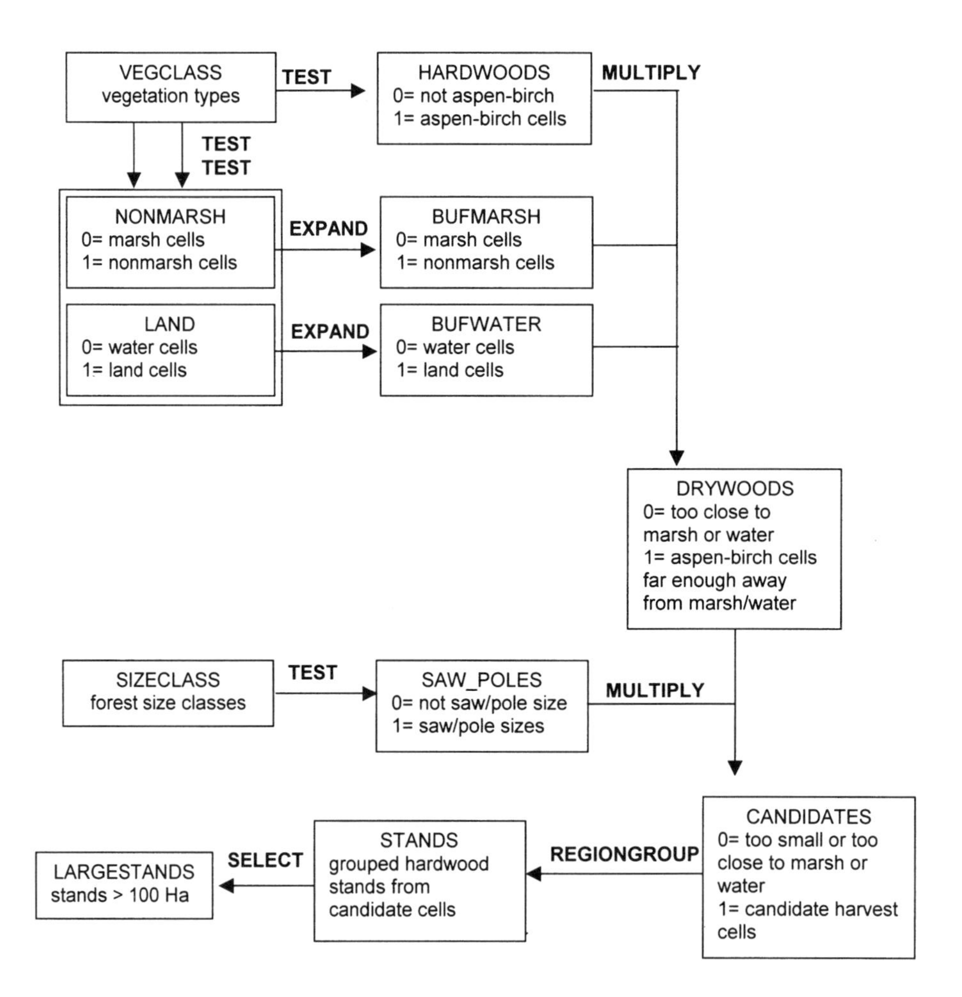

10) You have grid of vegetation types (VEG), soil drainage class (DRAINAGE) and a grid of elevation values (ELEV). You are checking for possible errors in the VEG grid. Aspen typically grows on warm sites such as coarse, well-drained soils or warm, southerly-facing slopes. You want to find any cells that have an aspen vegetation class occurring on either poorly drained soils or cold slopes (slopes with an aspect from north to east, northwest to west and a gradient greater than 10 percent).

Fill in the following flowchart with the appropriate Grid tools to solve your GIS problem:

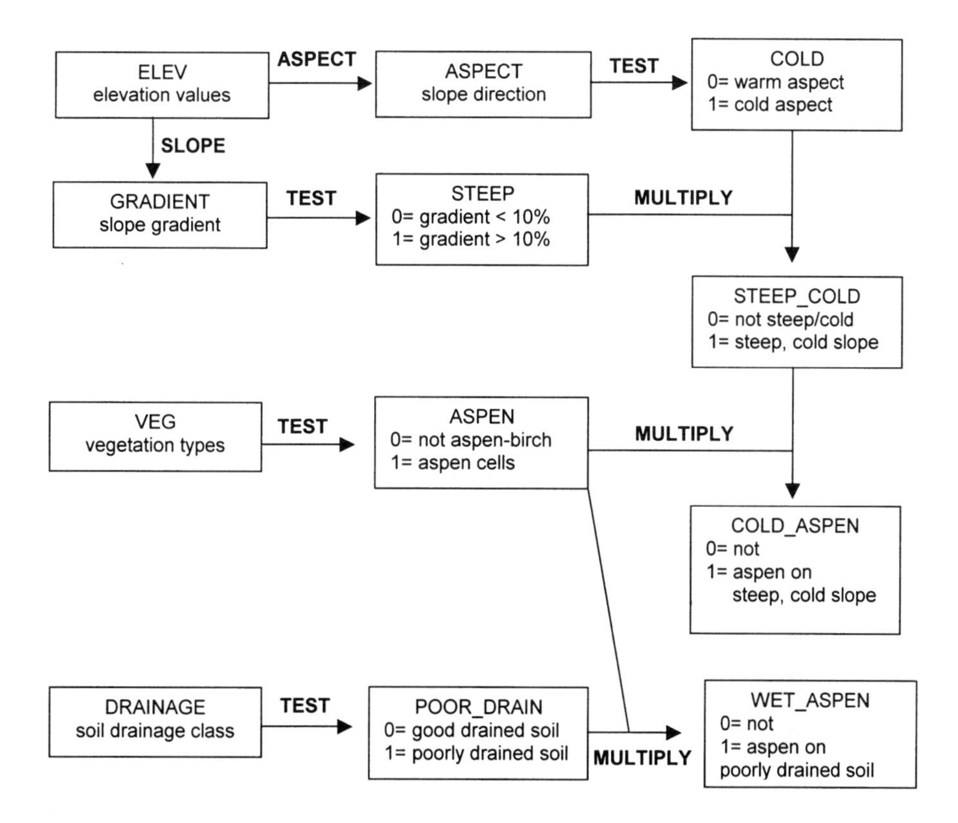

IMAGE ANALYSIS EXERCISE SOLUTIONS

1) Match the images with their companion linear contrast stretch functions:

Image#1

Image#2

Image#3

Image1 since it is the lightest, and all pixels above the mean are displayed at a video intensity of 255

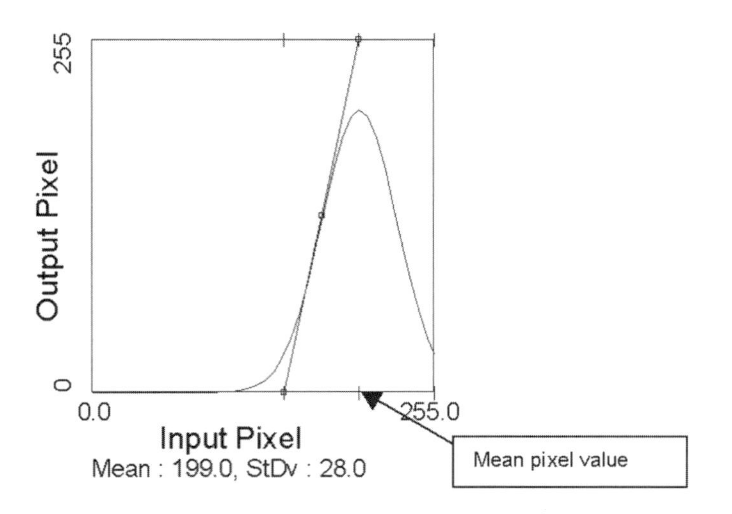

Image2 since it is the darkest, and all pixels below the mean are displayed at a video intensity of zero

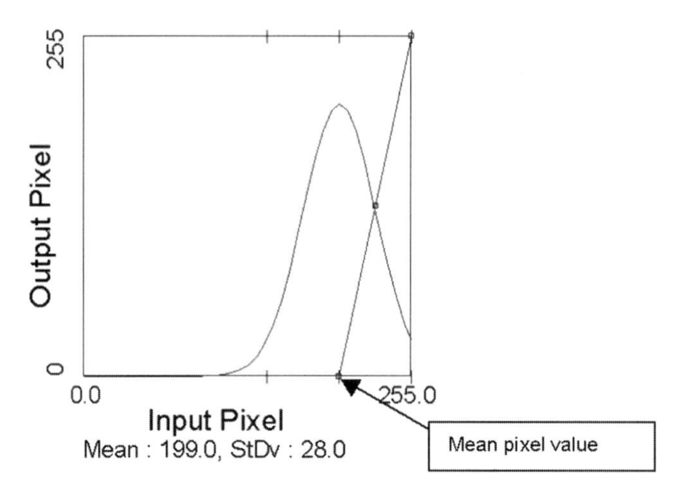

Image3 since it has the best contrast and the stretch is for most of the image pixels

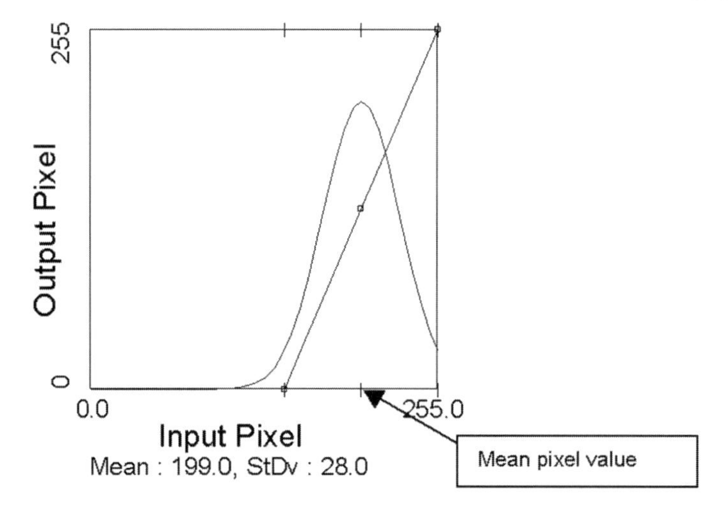

2) Circle the appropriate resampling method (either nearest neighbor or cubic convolution) for the following two rectified images:

Cubic Convolution Resampled Image . . .
smoothest due to weighted averaging of pixels

Nearest Neighbor Resampled Image since linear features have staircase shape.

3) You are trying to co-register a 30-meter Landsat Thematic Mapper image with a geo-rectified digital raster graphics topographic map. Using five ground control pixels from your 30-meter TM image, you find the following:

Map Coordinates		Image Coordinates	
UTM X	UTM Y	Column	Row
518,000	5,190,940	1	3
518,030	5,190,880	2	5
518,060	5,190,970	3	2
518,090	5,190,910	4	4
518,120	5,191,000	5	1

Using these data, develop a linear transformation model for image rectification. Fill in the proper transformation coefficients.

The slope of a line is the rise/run. For the X model, the slope would be rise/run or change in UTM X relative to a change in image columns:

slope = (518,120—518,000) / (5 – 1) = 120 / 40 = 30

Once you determine the slope of the line, you can use algebra to determine the intercept:

518,000 = ____________ + 30(1)
518,000 – 30 = __________
517,970 = intercept

UTM_X = <u>517,970</u> + <u>30</u>(Image Column) + 0(Image Row)

As a check, you can plug in any column value to see if you get the correct X Map Coordinate:

517,970 + 30(2) + 0(Image Row) = 518,030
517,970 + 30(3) + 0(Image Row) = 518,060
517,970 + 30(4) + 0(Image Row) = 518,090
517,970 + 30(5) + 0(Image Row) = 518,120

For the Y model, the slope would be rise/run or change in UTM Y relative to a change in image rows:

slope = (5,190,880—5,191,000) / (5 – 1) = -120 / 40 = -30

Once you determine the slope of the line, you can use algebra to determine the intercept:

5,191,000 = ____________—30(1)
5,191,000 + 30 = __________
5,191,030 = intercept

UTM_Y = <u>5,191,030</u>—<u>30</u>(Image Row) + 0(Image Column)

As a check, you can plug in any column value to see if you get the correct X Map Coordinate:

5,191,030— 30(2) + 0(Image Column) = 5,190,970
5,191,030— 30(3) + 0(Image Column) = 5,190,940
5,191,030— 30(4) + 0(Image Column) = 5,190,910
5,191,030— 30(5) + 0(Image Column) = 5,190,880

4) You have the following table of ground control links for developing a linear rectification model. The output image grid cell size is 25 meters. What is the RMS error of the rectification model in image pixels?

GCP#	Total Model Error
1	26.81
2	11.21
3	26.03
4	66.43
5	45.31
6	37.25
7	58.60
8	47.80

RMS error = Square Root [(Sum of Total Errors Squared)/ Number of Links]
= SQRT [(26.81^2 + 11.21^2 + 26.03^2 + 66.43^2 + 45.31^2 + 37.25^2 + 58.60^2 + 47.80^2) / 8]
= SQRT [15094.305 / 8] = 43.44 = 1.7 pixels

5) A consulting group supplies you with rectified satellite imagery. The contract specifies the rectification RMS error to be less than 1 pixel. The rectified imagery has a pixel size of 25 meters. Circle any of the following statements if they are correct:

- The positional accuracy of each pixel in the rectified image is +/- 1 pixel.
- Every pixel in the rectified image is within 50 meters of its true map location.
- The positional accuracy of the rectified image is 25 meters.

None of these statements are correct. The positional accuracy of an individual pixel is unknown and can be either less than or greater than 25 meters depending upon the pixel you choose. The model RMS error is an estimate of the expected error but the actual distribution of errors is unknown and the RMS error is an estimate based on relatively few ground control locations. The ground control locations may have been selected from flat, roaded areas of the image. If this was the case, then the positional error for pixels from mountainous areas of the image may be much greater.

You are delineating training fields using a parallelepiped seed-pixel approach. You pick the shaded pixel as your seed pixel and the parallelepiped is based on the eight neighboring pixels. Circle all pixels that would be included in your training field.

0 0	64 161	6 6	6 6	6 6	50 110	60 120	6 6	6 6	6 6
0 0	5 6	54 163	5 6	5 6	5 6	5 6	5 6	5 6	5 6
4 5	0 0	4 5	52 161	4 5	4 5	4 5	4 5	134 124	4 5
6 6	0 0	6 6	60 170	54 153	48 144	6 6	6 6	6 6	0 0
5 6	0 0	5 6	50 155	50 150	51 152	5 6	5 6	0 0	5 6
6 6	0 0	6 6	52 157	49 144	40 130	42 137	0 0	6 6	6 6
0 0	5 6	30 100	5 6	5 6	5 6	0 0	43 138	41 136	50 110
0 0	24 95	40 105	6 6	4 5	4 5	0 0	0 0	0 0	45 106
26 106	46 106	6 6	5 6	6 6	6 6	40 105	0 0	0 0	0 0
55 116	5 6	5 6	5 6	5 6	5 6	42 107	44 108	45 106	50 110

Parallelepiped: 40,130 to 60,170

0 0	64 161	6 6	6 6	6 6	50 110	60 120	6 6	6 6	6 6
0 0	5 6	54 163	5 6	5 6	5 6	5 6	5 6	5 6	5 6
4 5	0 0	4 5	52 161	4 5	4 5	4 5	4 5	134 124	4 5
6 6	0 0	6 6	60 170	54 153	48 144	6 6	6 6	6 6	0 0
5 6	0 0	5 6	50 155	50 150	51 152	5 6	5 6	0 0	5 6
6 6	0 0	6 6	52 157	49 144	40 130	42 137	0 0	6 6	6 6
0 0	5 6	30 100	5 6	5 6	5 6	0 0	43 138	41 136	50 110
0 0	24 95	40 105	6 6	4 5	4 5	0 0	0 0	0 0	45 106
26 106	46 106	6 6	5 6	6 6	6 6	40 105	0 0	0 0	0 0
55 116	5 6	5 6	5 6	5 6	5 6	42 107	44 108	45 106	50 110

7) The following likelihood contours were developed using these training classes:

Training Class	Band 1 Mean	Band 2 Mean
Aspen	40	75
Spruce	20	65
Water	5	45

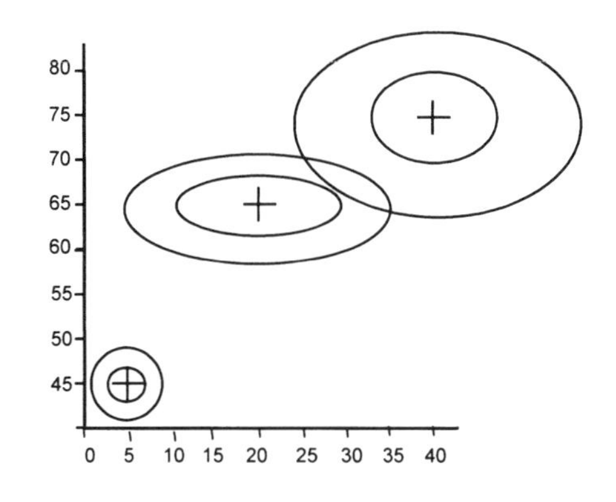

Using the maximum likelihood rule, classify the following image, by plotting the values:

Original Image:

6 44	5 43	25 67	29 60
6 43	4 43	25 65	30 61
10 65	14 68	40 77	43 78
12 66	15 69	39 76	44 78

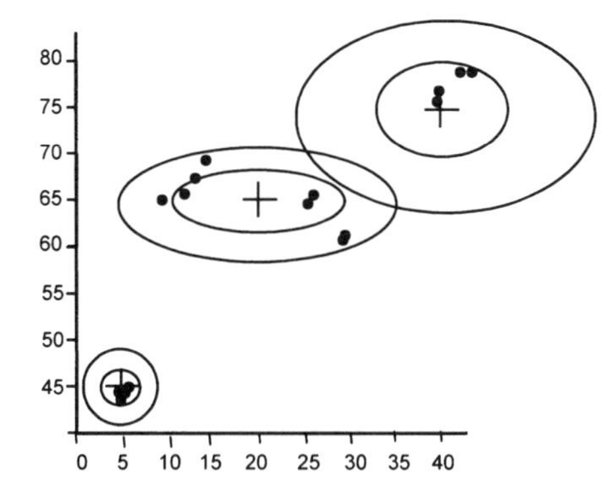

Classified Image:
1 = Water
2 = Spruce
3 = Aspen

1	**1**	**2**	**2**
1	**1**	**2**	**2**
2	**2**	**3**	**3**
2	**2**	**3**	**3**

8) A consulting company conducts a 2001 image classification to locate all sawtimber white spruce stands. The image pixel size is 25 meters.

The accuracy of the classification is assessed by several groups. Match the following accuracy assessment scenarios with the most likely overall classification accuracies.

_____50 percent overall classification accuracy . . . **Group B since they used old reference data and the land cover may have changed between 1979 and the 2001 imagery date. Also since they used polygons with a minimum mapping unit, many of the 0.0625 hectare pixels could have been correctly classified but incorrectly assessed as misclassified because of the large polygons used in the classification accuracy assessment. Both the temporal difference and the minimum mapping unit area difference between the reference data and the classified image would lead to a conservative estimate of the classification accuracy.**

_____90 percent overall classification accuracy . . . **Group A since they sampled exclusively areas that would be easy to classify . . . homogeneous areas near training fields and ignored the heterogeneous areas that would be difficult to classify.**

Group A: A guided approach was conducted in conjunction with the fieldwork for training data collection, thereby reducing the costs of data collection. Reference data were collected adjacent to training field areas. To minimize nonclassification errors (such as slight positional shifts) only homogeneous polygons of at least 5 hectares were used as reference data. The center pixel of these polygons was selected for each reference data sample location. The photos were taken the same month and year as the satellite image was acquired.

Group B: Using high altitude color infrared photographs (taken by NASA in 1979), we stereoscopically interpreted cover types to polygons with a minimum mapping unit of 2 hectares. The cover types were transferred to the USGS 1:63,360 quadrangle series using a digitizing tablet and stored in a vector GIS. For accuracy assessment these GIS polygons were rasterized to correspond to the classified image. Classification accuracy assessment was conducted on a pixel by pixel basis between the reference data and the classified image.

9) You have an error matrix as follows. What is the overall classification accuracy?

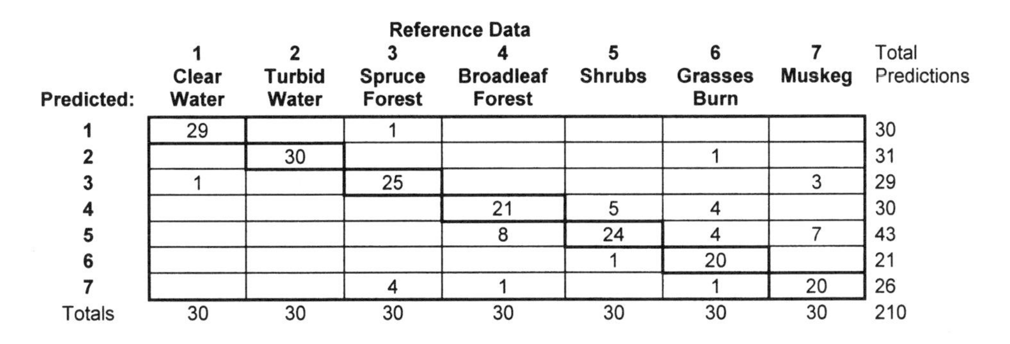

	Reference Data							
Predicted:	**1 Clear Water**	**2 Turbid Water**	**3 Spruce Forest**	**4 Broadleaf Forest**	**5 Shrubs**	**6 Grasses Burn**	**7 Muskeg**	Total Predictions
1	29		1					30
2		30				1		31
3	1		25				3	29
4				21	5	4		30
5				8	24	4	7	43
6					1	20		21
7			4	1		1	20	26
Totals	30	30	30	30	30	30	30	210

The overall classification accuracy is the sum of all correct predictions divided by the total number of predictions expressed as a percentage:

= [(29 + 30 + 25 + 21 + 24 + 20 + 20) / 210] * 100
= [169 / 210] * 100 = 80 percent

10) You have an error matrix as follows. What is the Producer's and User's accuracy for the muskeg class?

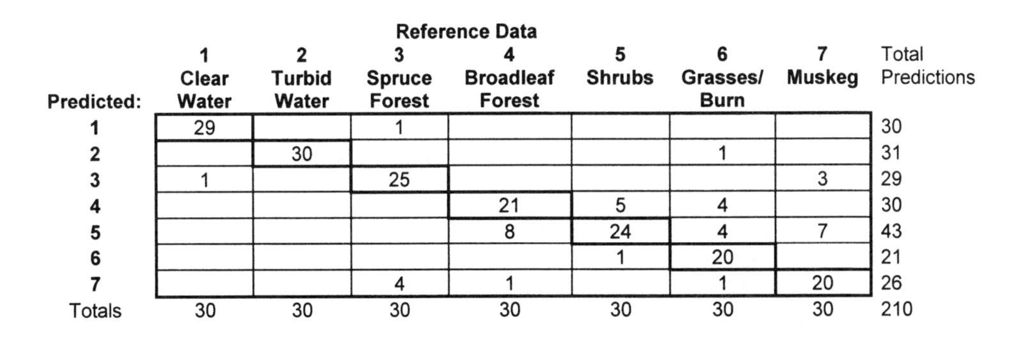

	Reference Data							
Predicted:	**1 Clear Water**	**2 Turbid Water**	**3 Spruce Forest**	**4 Broadleaf Forest**	**5 Shrubs**	**6 Grasses/ Burn**	**7 Muskeg**	Total Predictions
1	29		1					30
2		30				1		31
3	1		25				3	29
4				21	5	4		30
5				8	24	4	7	43
6					1	20		21
7			4	1		1	20	26
Totals	30	30	30	30	30	30	30	210

Producer's accuracy is computed by looking at the 26 predictions produced for a muskeg and determining the percentage of correct predictions. Consumer's accuracy is computed by looking at the reference data of 30 true muskeg pixels and determining the percentage of correct predictions for these samples.

Producer's Accuracy = (20/26) * 100 = 77 percent
Consumer's Accuracy = (20 / 30) * 100 = 67 percent

VECTOR EXERCISE SOLUTIONS

Real Estate Application. You are interested in purchasing property where you can build a retreat cabin as a vacation home. You have the following polygon themes: parcels, elev_zone, vegetation. You also have a line theme of roads. You want to find land that is privately owned so that you can make an offer to purchase. You want to purchase an area of at least 1 hectare, that totally is above 500 feet elevation (to avoid flooding problems). You want decent road access : at least partially within 100 meters of a road. And you want the area to contain some aspen vegetation (which typically grows on warm sites). Fill in the following flowchart with the appropriate vector GIS tools to solve your problem:

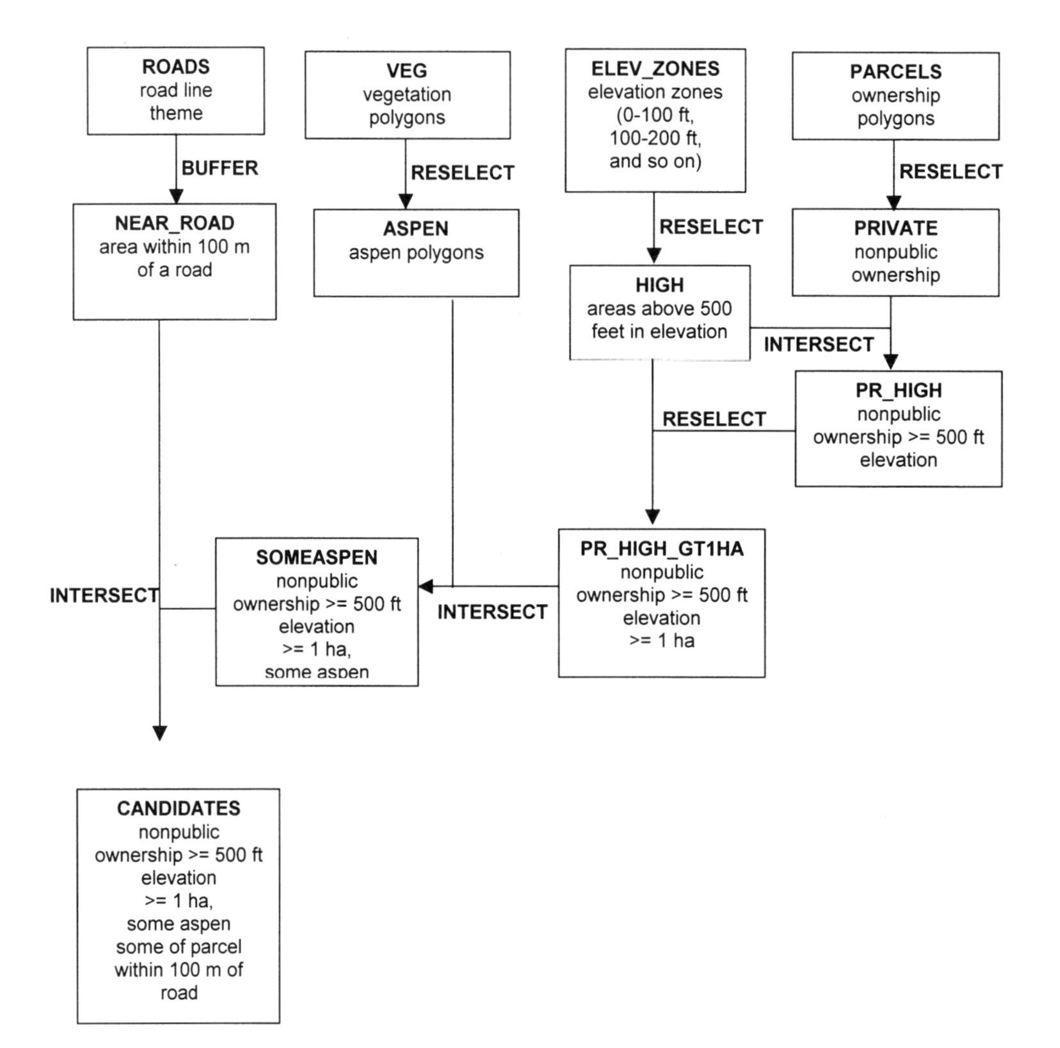

Moose Habitat Analysis. You are hired as the GIS guru in a moose habitat study. You have the following polygon themes: rivers, vegetation. You also have a point theme of radio-collared moose locations from the winters of 1995-2001. You are to test the hypothesis that mature bull moose spend most of their time in large riparian *Salix alexensis* (willow) stands whenever there is deep snow. The winters of 1995, 1997, and 2000 were deep snow winters while the winters of 1996,1998,1999, and 2001 were light snow winters. The moose biologist wants the following GIS analyses:

1) For mature bull moose, compute the mean distance to riparian *Salix alexensis* (riparian is defined as within 100 meters of a river) for light versus deep snow years.
2) Compute percentage of time that mature bull moose locations that were within 100 meters of riparian *Salix alexensis* for light versus deep snow years.
3) For the *Salix alexensis* polygons that had mature bull moose locations in them, compute the mean *Salix alexensis* polygon area in hectares for light versus deep snow years.

Fill in the following flowchart with the appropriate vector GIS tools to solve your problem.

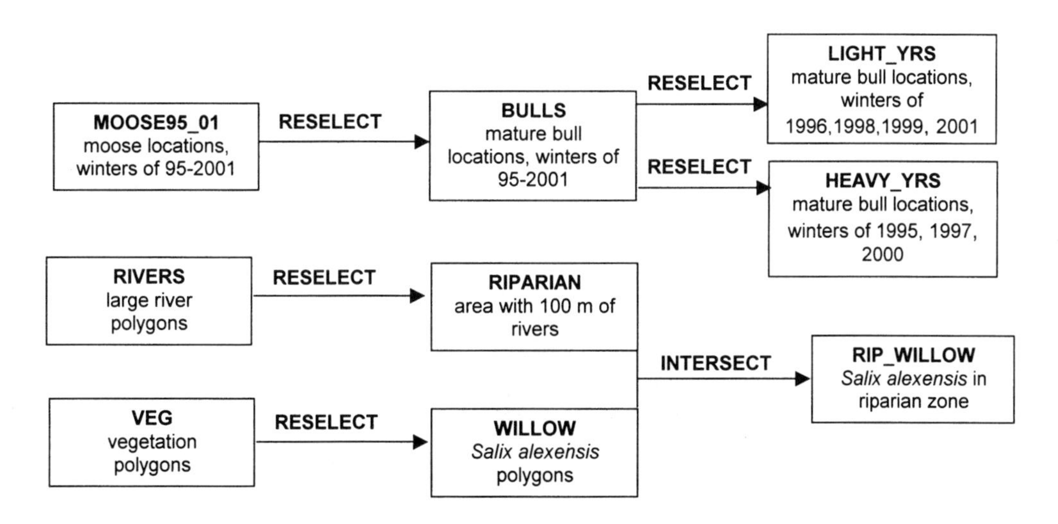

For mature bull moose, compute the mean distance to riparian *Salix alexensis* (riparian is defined as within 100 meters of a river) for light versus deep snow years.

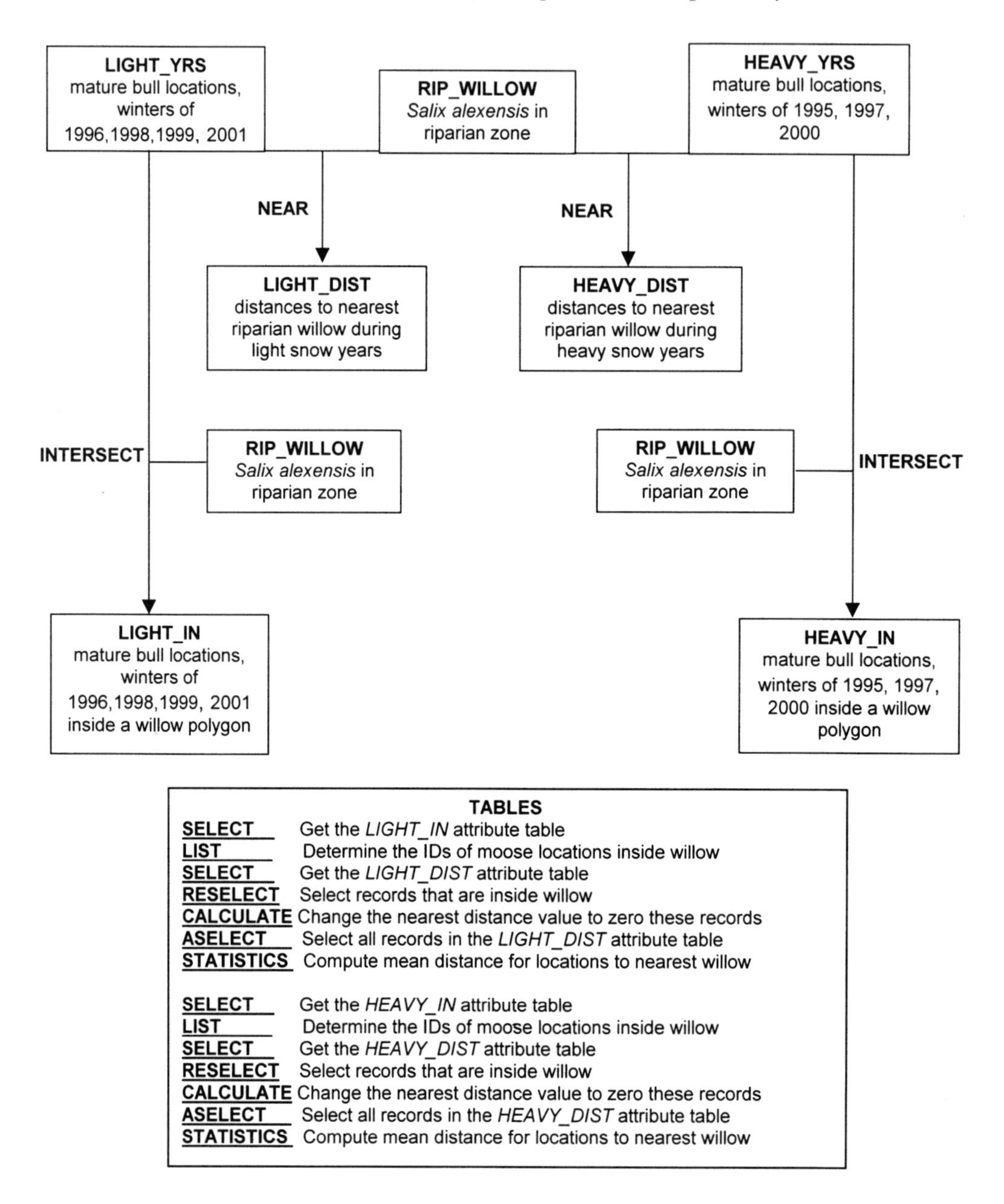

Compute percentage of mature bull moose locations that were within 100 meters of riparian *Salix alexensis* for light versus deep snow years.

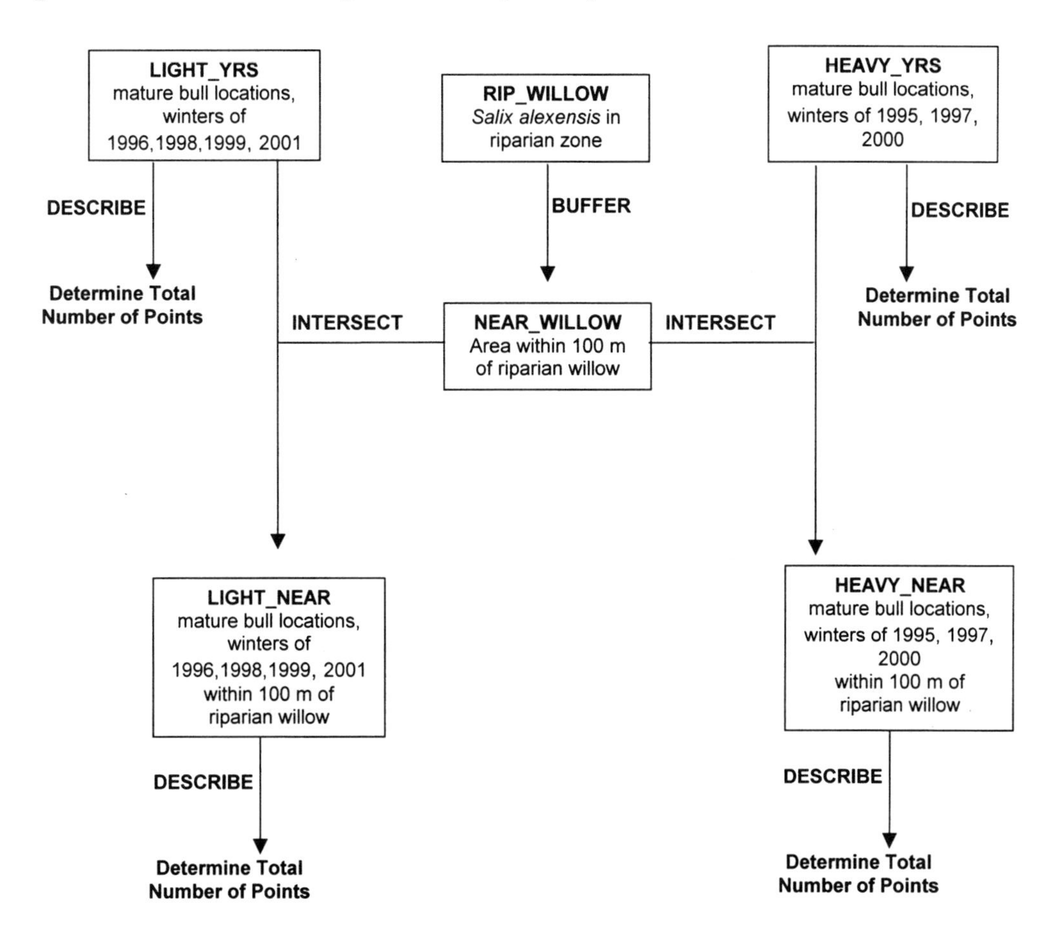

For the *Salix alexensis* polygons that had mature bull moose locations in them, compute the mean *Salix alexensis* polygon area in hectares for light versus deep snow years.

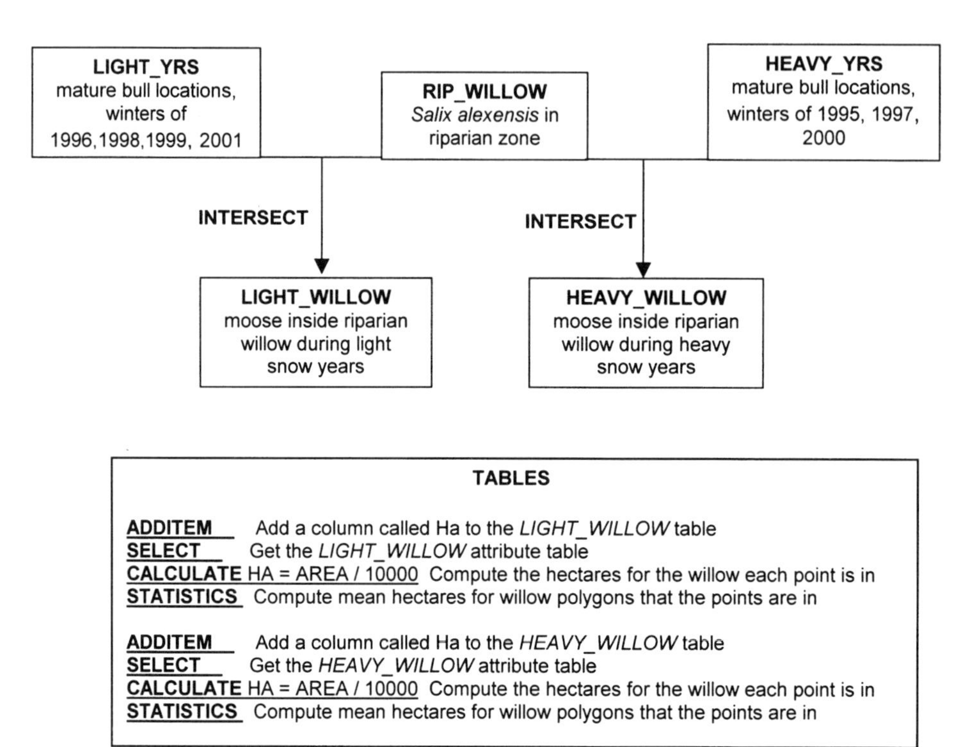

Fire Hydrant Inspection Application. You have a table of county-owned fire hydrants and their mileage recorded by driving along all county roads. The table contains information such as date of last inspection, age of hydrant, maximum flow rate, and so on. You also have a county-wide roads line theme and a parcels polygon theme. You also have a table of parcel owners; the table contains parcel-IDs, owner names, addresses, and phone numbers. All hydrants in the county are inspected on a five-year basis. Your job is to create a text file of names and addresses of parcel owners with property within 100 meters of a fire hydrant that is scheduled to be inspected in 2001.

Fill in the following flowchart with the appropriate vector GIS tools to solve your problem.

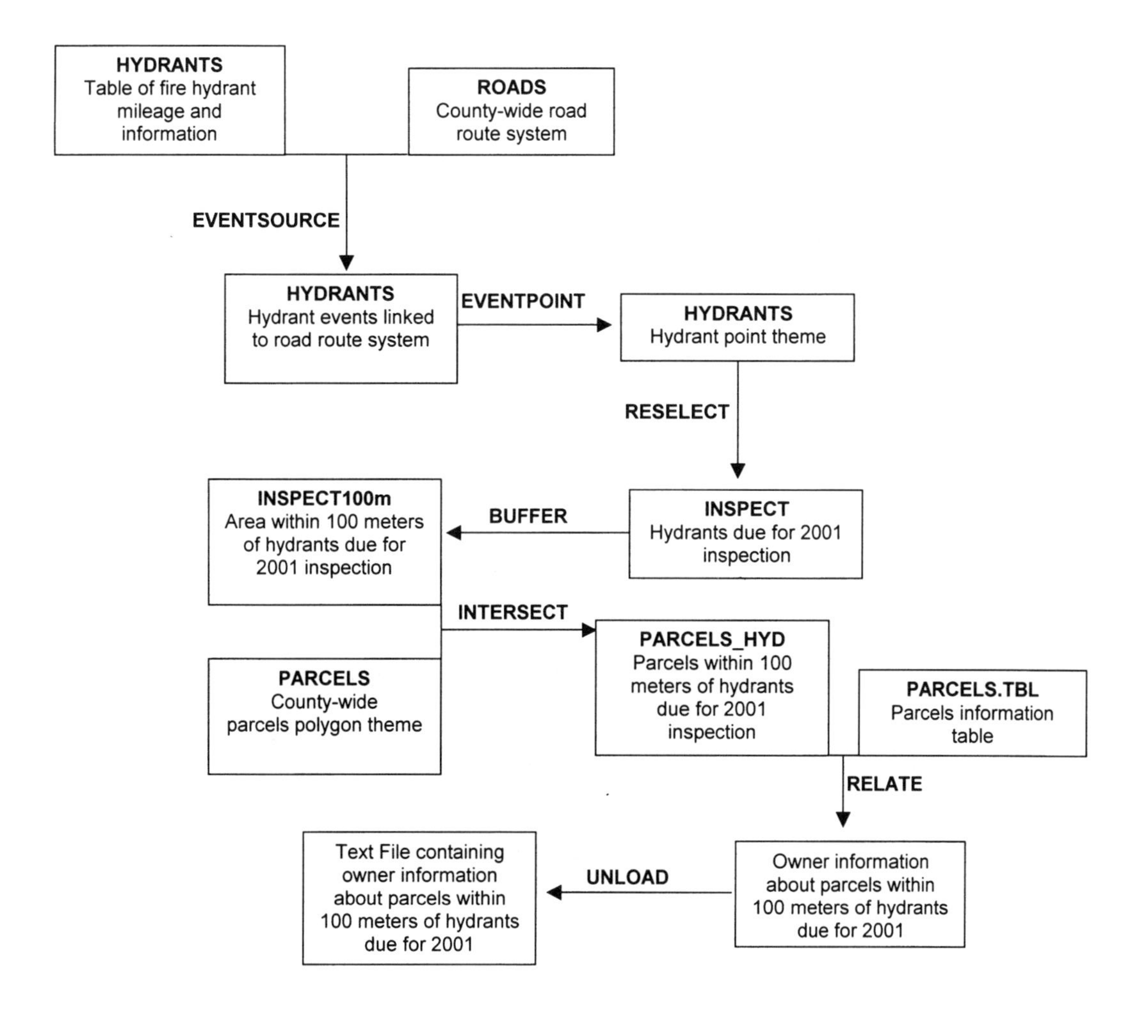

Forest Fire Application. You have a line theme of flight lines over the center of lakes, a polygon theme of lakes, and a point theme of the centers of five currently burning wildfires. In order for tankers to reload water, they need a lake that is at least 3 km long. Develop a map showing all lakes available for tanker reloading and which fire they are the closest to.

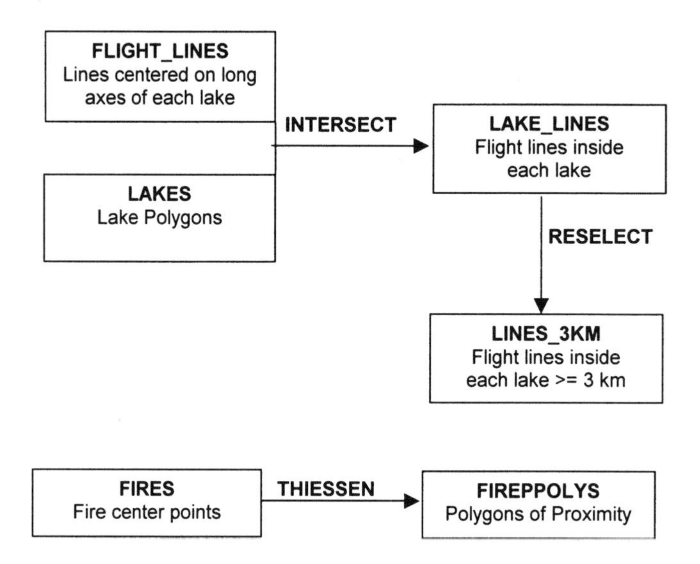

Emergency Phone Application. You are hired as a GIS Guru for the facilities management department of a large university. You have a polygon theme of buildings, a line theme of sidewalks, and a point theme of outside emergency phones. Your job is to show the area of the campus that is most lacking in emergency phone service. Fill in the following flowchart with the appropriate GIS tools to solve your problem.

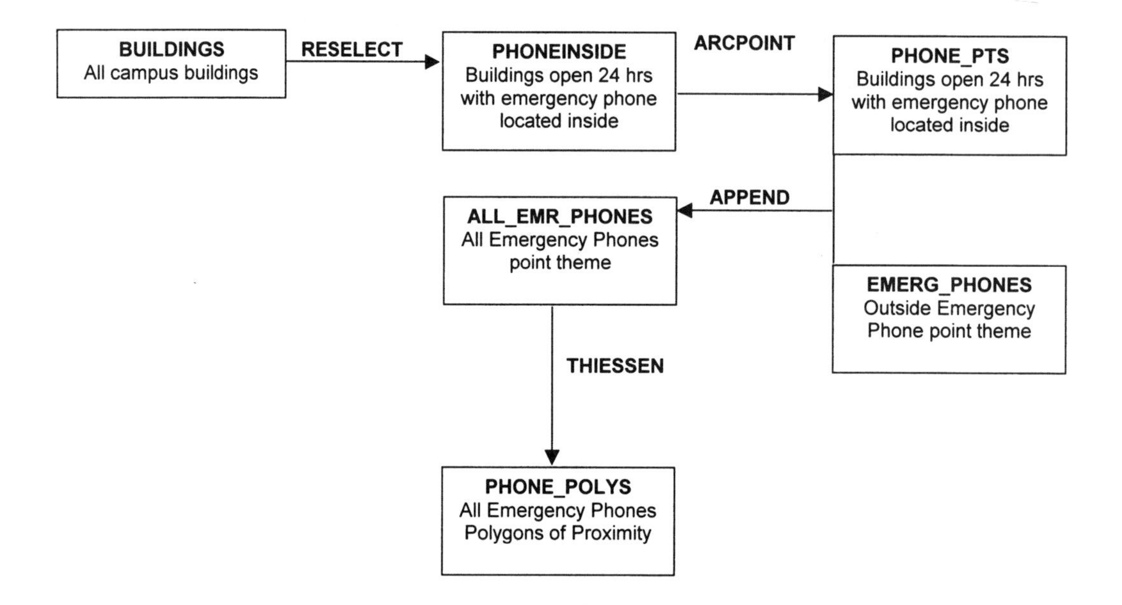

CHAPTER 11 GRID EXERCISE SOLUTIONS

Real Estate Application. You have grids of elevation, roads, and ownership. You want to find land owned by the Fairbanks Northstar Borough that is above 500 feet elevation (out of the ice fog), on a easterly, southerly, or westerly slope, and within 100 meters of a road. Your elevation values are in meters.

Fill in the following flowchart with the appropriate Grid tools:

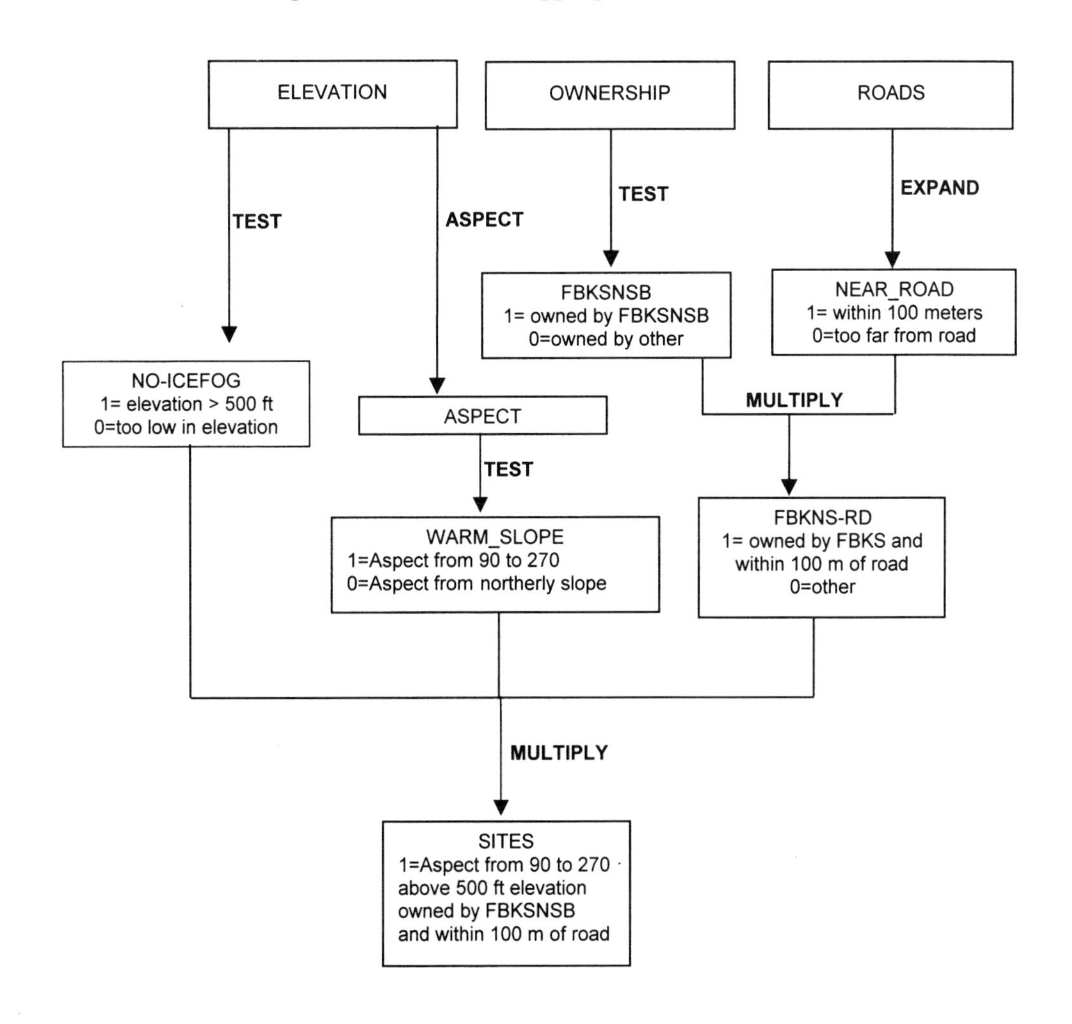

Watershed Application. You have a grid of vegetation types, and a grid of watersheds. You want a table showing the area of each watershed and the area of black spruce in each watershed. Each grid cell is 25 by 25 meters.

Fill in the following flowchart with the appropriate grid operations to solve your GIS problem:

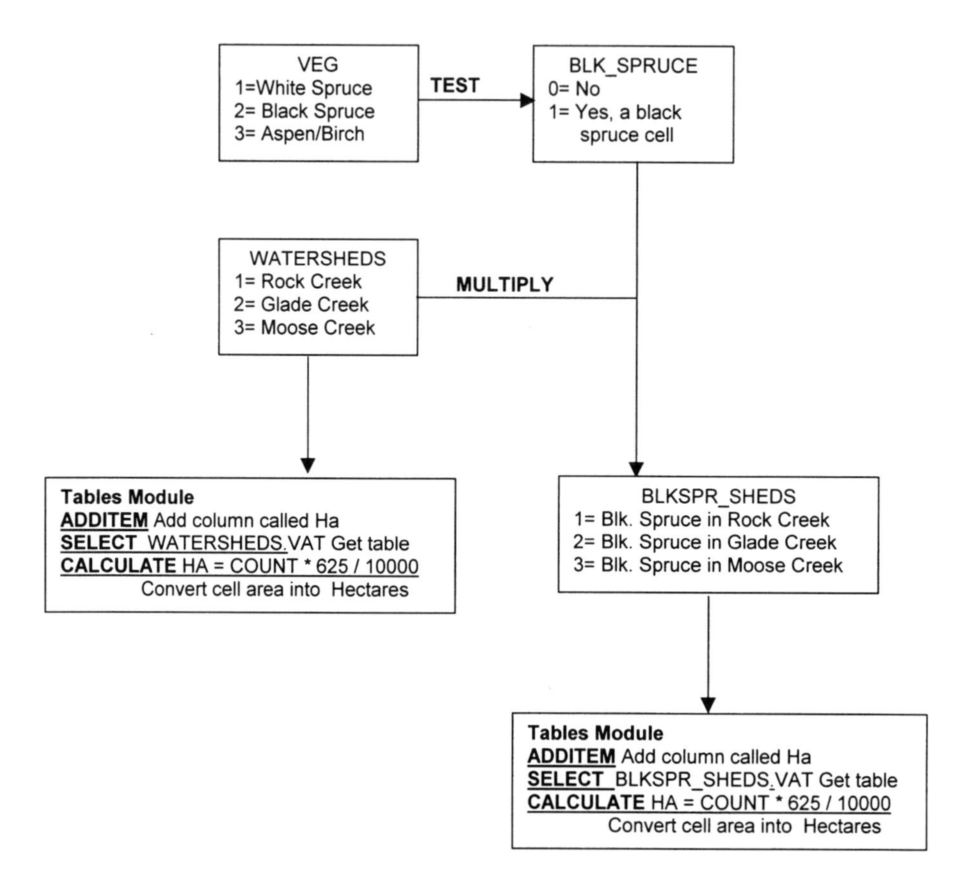

Waterfowl Habitat Analysis. You have grids of lakes. One grid is from early summer and the other grid is from late summer. You want to create a grid of the best lakes for waterfowl habitat . . . lakes that had at least 1 hectare of exposed mudflats in late summer due to drawdown of the lake water level. The cell size of each grid is 10 meters. Fill in the following flowchart to solve your GIS problem.

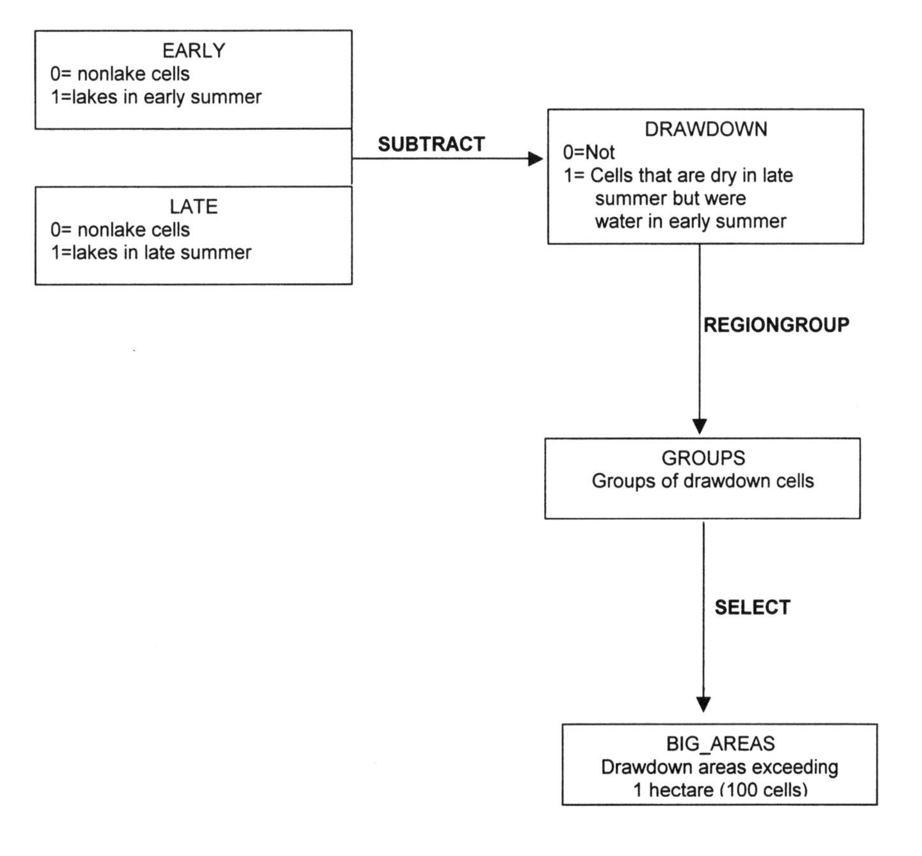

Groundwater Application. You have a grid of well locations, gas stations locations, soils, and parcel ownership that contains the address of the owner of each parcel.

Outline what tools you would use to develop a text file of all well owners that have wells on sandy textured soils within a kilometer of a gas station.

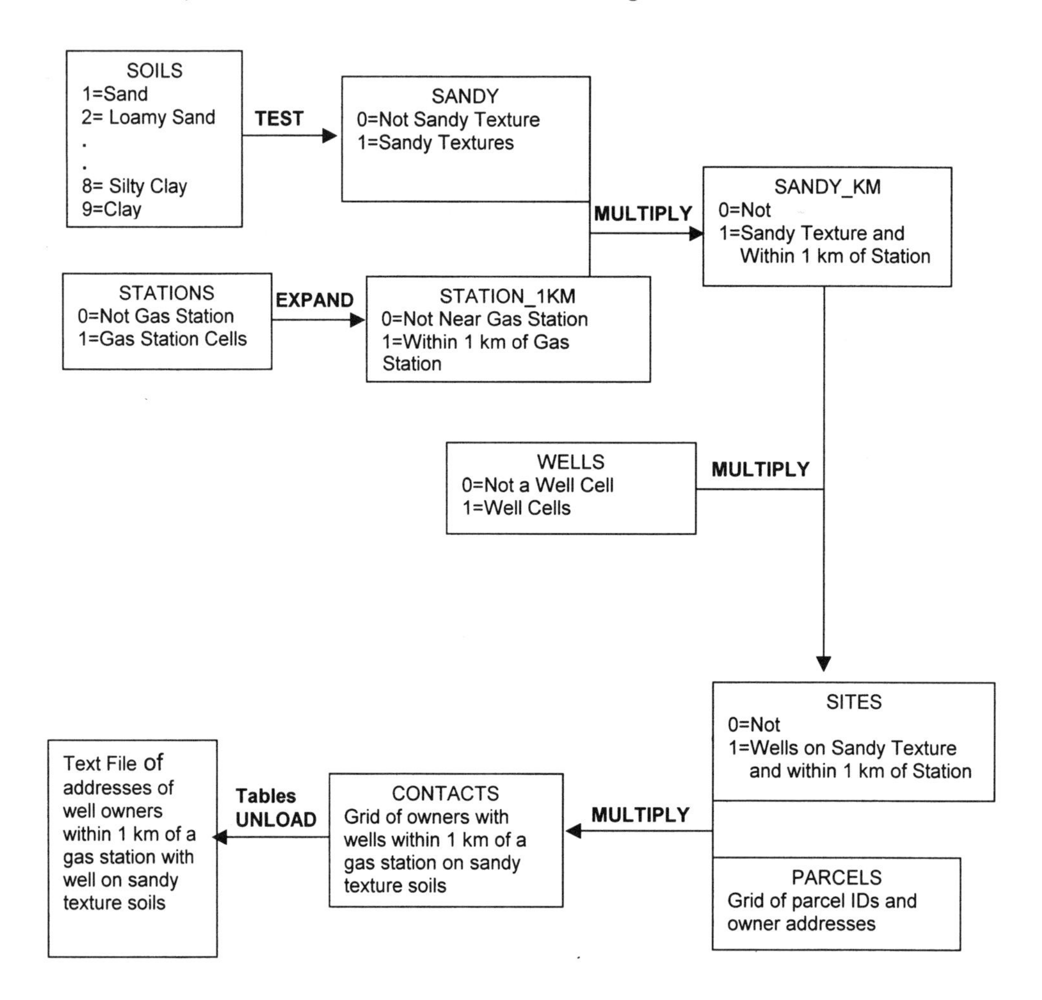

Moose Habitat Analysis. You have grids of moose locations, random locations, and vegetation types. You want to estimate the mean distance of moose and random locations to the nearest willow cell. All grid cells are 10 meters in size.

Fill in the following flowchart with the appropriate grid operations to solve your GIS problem:

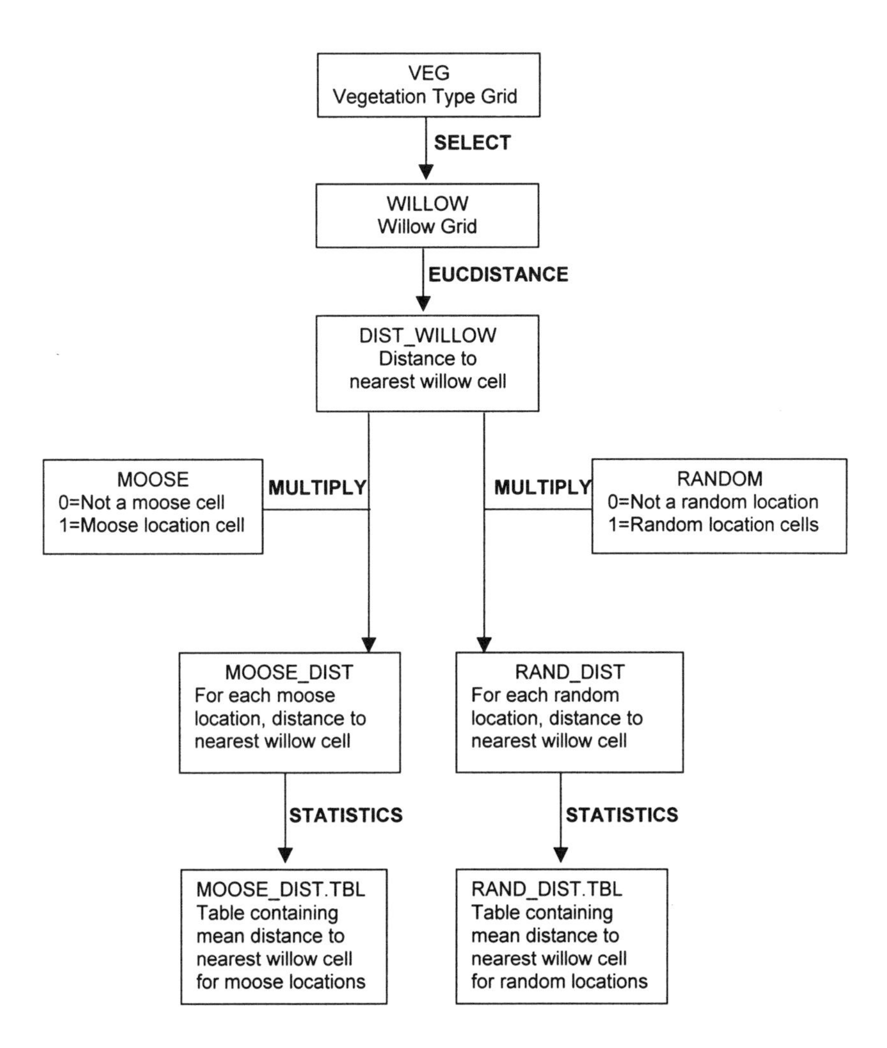

Index